APOLLO'S EYE

PUBLISHING FOR THE WORLD
125 Years

THE JOHNS HOPKINS UNIVERSITY PRESS

Published in Cooperation with

the Center for American Places,

Santa Fe, New Mexico, and

Harrisonburg, Virginia

APOLLO'S EYE

A Cartographic Genealogy of

the Earth in the Western Imagination

DENIS COSGROVE

THE JOHNS HOPKINS UNIVERSITY PRESS   Baltimore and London

© 2001 The Johns Hopkins University Press
All rights reserved. Published 2001
Printed in the United States of America on acid-free paper

Johns Hopkins Paperbacks edition, 2003
9 8 7 6 5 4 3 2 1

The Johns Hopkins University Press
2715 North Charles Street
Baltimore, Maryland 21218-4363
www.press.jhu.edu

Library of Congress Cataloging-in-Publication Data
Cosgrove, Denis E.
    Apollo's eye : a cartographic genealogy of the earth in the
western imagination / Denis Cosgrove.
        p.   cm.
Includes bibliographical references and index.
    ISBN 0-8018-6491-7 (hardcover : acid-free paper)
    1. Geographical perception. 2. Globes. I. Title.
    G71.5 .C68 2001
    910´.9—dc21
    00-009623

    ISBN 0-8018-7444-0 (pbk.)

A catalog record for this book is available from the British Library.

Frontispiece: Realizing the global vision from space: *Earthrise*, 1968.
Photo courtesy of NASA.

For Carmen

Tous affaires humains, avecq leur fraisle vie
Pendent à un filet bien tendre & bien menu,
Que Dieu peut de sa main trancher comme un festu:
Soit donc ton ame en temps de piëté garnie.

CONTENTS

PREFACE

Globalization is a driving idea of our times. Powered by technological innovation, by capital's restless search for investment opportunities, by geopolitical ambition, by ideological or religious fervor, even by touristic desire and adventure, globalization is a hydra of modernity. Whether pictured as a networked sphere of accelerating circulation or as an abused and overexploited body, it is from images of the spherical earth that ideas of globalization draw their expressive and political force. The fascination that global images exercise over the millennial imagination is apparent from even the most casual glance at newspapers and magazines, television, or advertising.

Geography lays particular claim to the globe. Its intellectual task is, by definition, to describe the globe's surface. This book originated in a request for a geographical reading of a photographic exhibition containing satellite images of the globe. What initially seemed a simple task became rapidly overwhelming as I began to glimpse the historical depth and cultural complexity of the earth's deceptively simple form. My fundamental question is simple: what have been the historical implications for the West of conceiving and representing the earth as a unitary, regular body of spherical form? Earthbound humans are unable to embrace more than a tiny part of the planetary surface. But in their imagination they can grasp the whole of the earth, as a surface or a solid body, to locate it within infinities of space and to communicate and share images of it. Only in the late twentieth century did any humans physically witness the full Earth turning in space, and the impact of that achievement echoes through the language and imagery of globalization. But the slightest reflection reveals that for all its radical newness, actually witnessing the globe culminates a long genealogy of imagining and reflecting upon the possibility of doing so. The meanings of the photographed earth were anticipated long before the photographs themselves were taken.

Contemporary globalization is Western in its origins and integrated into processes of modernization through which the very idea of "the West" has been differentiated on a single global surface. "Globalization" disrupts such stable, naturalizing geographical models as "West and rest," core and periphery, "First," "Second," and "Third" Worlds. Globalized space is characterized by circulation, exchange, confluence, moving unevenly across networks of greater or lesser density and efficiency, with diverse and hybrid consequences. And while historically this has long been true of many interactions, local, territorialized experience has meant that in the imagination, cultural worlds and identities have been organized around assumed sociospatial centrality.

Historically, *cosmography* has been the name given to the integrated study of earthly and celestial spheres. As a science, cosmography attracted intense interest among early modern Europeans as they grappled with the sudden expansion of their knowledge space and reflected a desire to grasp unity in the diversity of creation and to place humans within it. Variously powered by spiritual desire, lust for universal knowledge, dreams of global enlightenment, and geopolitical imperatives of empire, cosmography is an enduring seduction that may even flicker unacknowledged behind contemporary discourses of globalism and globalization, figured in the image of a spherical earth floating in space. Cosmography's genealogy has many strands, every one of them the subject of a large and specialized literature. Its own enduring problem, the sheer volume of knowledge and skills it generates and demands, also challenges attempts to write its story. There are specialist literatures in the history of science, philosophy, cartography, literature and art, exploration, geography, and material culture. None is independent of broader economic and political contexts, all are profoundly inflected by the continuous ethnographic interaction between the "West" and its "Others," both imagined and actual. At the simplest technical level of representation, the forms of the earth and extraterrestrial space themselves present complex historical problems. Disk or sphere, modeled, pictured or mathematically projected, the globe is known through its representations. And representations have agency in shaping understanding and further action in the world itself.

In the face of such cosmic problems, I have elected to trace images of the globe and the whole earth as they have constructed and communicated the distinctive Western mentality that lies behind the universalist claims of contemporary globalism. As I have said, both "West" and "Western" are themselves historically made and altered constructs, shaping and differentiating an already signified globe. The paradox of a universality that is necessarily proclaimed for a positioned location yields one main theme of this book: that of the Apollonian eye, the viewpoint above the earth, proclaiming disinterested and rationally objective consideration across its surface. The limitations of both rationalist and empiricist claims to objective knowledge are widely acknowledged, but their intellectual and material effects have been critical in shaping Western ideas of the globe. Closely linked to the Apollonian vision and its universal claims is the shifting discourse of the self and human distinction. Conceptions of human unity, closely connected to the image of a whole earth, have shaped much of the communication and commerce between peoples.

As sun god, Apollo was directly associated with the form and regular movement of the heavenly bodies, which long offered a model or metaphor of harmony, apprehended both visually and audibly. Music, mathematically expressed, has consistently been associated with planetary form, and movement, with planets' unmediated connections to the individual soul through ecstatic states of transcendence. The idea and experience of the cosmic dream, or *somnium,* is a recurring feature in the comprehension and representation of the globe, forming a third strand within the study.

The Apollonian gaze, which pulls diverse life on earth into a vision of unity, is individualized, a divine and mastering view from a single perspective. That view is at once empowering and visionary, implying ascent from the terrestrial sphere into the zones of planets and stars. The theme of ascent connects the earth to cosmographic spheres, so that rising above the earth in flight is an enduring element of global thought and imagination. Belief in the ascent of the soul—that the destiny of human life is transcendence to a heaven above the earth's surface—connects to the metaphysics of harmony embraced by the *somnium.* Alternatively, the Apollonian gaze seizes divine authority for itself, radiating power across the global surface from a sacred center, locating and projecting human authority imperially toward the ends of the earth. In the narrative of Christ as God-man, refracted through the heritage of Greece and Rome, these two strands have been braided together into a universalizing teleology of Western Christianity. The imperial imperative has been figured through the image of the globe and centered on its surface at axes of temporal and spiritual power. For two millennia, Rome, city of the caesars and the popes, has figured in the Western imagination as the paradigmatic global city.

These themes coalesce in images, meanings, and moral readings of the globe in Western imagination and thought. In writing of them I have tried to remain true to the inseparability of imagination, moral reflection, and practical action apparent in representations of the globe. The nine chapters that follow construct a genealogy of global images and meanings from ancient Greece and Rome to the twentieth century. They do not imply a linear or progressive historical narrative, although such a narrative is always implicit, and often explicit, in the ideas of the globe and globalization themselves. I focus most closely on the early modern period, pivoting on the sixteenth century, when Europeans first circumnavigated the earthly sphere, relocated it within their understanding of the cosmos, and revolutionized its representation in models and maps. The chapter titles refer to significant associations between the globe and Western experience: imperial and poetic, classical, Christian, oceanic, visionary, emblematic, enlightened, modern, and virtual. Each chapter focuses on a limited number of specific images of the globe or whole earth in order to structure discussion of the cosmographic idea. My intention is to reveal the deep roots of contemporary global thinking and to acknowledge something of the richly complex cosmographic tradition in which today's geographical imagination is rooted. That representations of the globe have exercised an especially powerful grasp on the Western imagination, especially in the past millennium, is unsurprising given the active globalization of Western culture.

ACKNOWLEDGMENTS

In the course of writing this book I have gathered more debts than I can hope to acknowledge. An initial period of study funded by the Nuffield Foundation at the Harry Ransom Humanities Research Center at the University of Texas at Austin provided me with the opportunity to consult a selection of humanist texts and images of the globe. Subsequently I have drawn on the resources and skills of the staff at the Warburg Institute in London, especially the seminar on the History of Cartography organized by Catherine Delano Smith at the British Library Map Room and at the Bibliothèque Nationale in Paris. Various parts of the book have been improved by their presentation at seminars in university departments and conferences in Britain and the United States.

Specific ideas and help have been given generously by friends and colleagues, among them Steve Daniels, Bernard Debarbieux, Anne Godlewska, Mike Heffernan, Rik Jazeel, Patty Kellner, Luciana de Lima Martins, Susanna Morton-Braund, Allessandro Scafi, Peter Taylor, and Arturo Tosi. A special thanks is owed to Christian Jacob, who provided an inexhaustible supply of ideas and a copy of Le Large's commentary on Coronelli's globes. Jerry Brotton and Felix Driver read large parts of the manuscript, and their comments filled blank spaces on my map of knowledge while saving me from a few incautious claims. Orly Derzie was wonderful in finding and securing permission to reproduce images. All the errors and misinterpretations represent my personal failings. Staff members and graduate students of the Royal Holloway geography department's Social and Cultural Group provided both a stimulating intellectual environment and, with the department and the college, a sabbatical period to complete my writing. But the first and most consistent stimulus to the work has been my wife, Carmen. Not only was it her family's house and boat at Houston that accommodated work on the NASA images, but she has been the longest supporter of "the globe book." In dedicating it to her, I hope she will feel that it was worth the wait.

APOLLO'S EYE

ONE *Imperial and Poetic Globe*

Qui peut comprendre? Ici le cercle parle,
Lui sans début et sans fin, sans ailes
Et séparant du monde son espace
Pour un jour pur.

    Ici l'on ne sait pas
Ce qu'est mensonge. Un concept inconnu
Clair en ses lieux dans le cercle parfait.

Qui le traça? L'oiseau, le géomètre?
Un homme court après lui-même. Athlète
Sans un relais, sans autre but que d'être,
De se rejoindre à travers ses contraires.

Il est prunelle. Il est soleil et lune.
Son territoire est tracé par les astres.
Qu'un dieu l'efface et naît le mot néant.

Nous célébrons, nous vénérons le cercle.

—Robert Sabatier, from *Icare*

In Greek and Roman mythology Phoebus Apollo drives the sun's golden chariot above the terrestrial sphere, tracing the diurnal arc. His arrows, unleashed dispassionately from the firmament, bring unexpected calamity to mortals. In Jacopo de'Barbari's engraving, Apollo bestrides the globe, while Diana, lunar goddess of the night, slides eastward to sylvan shade (Fig. 1.1). In Christian iconography Apollo became the risen Christ, redeeming humans and reuniting heaven with earth through bodily resurrection and ascension. Renaissance princes competed rhetorically for Apollo's synoptic grasp of a circumnavigated globe, while dashing air aces and astronauts have figured as modern Apollos, eyewitnesses to an orb previously visible only in

1

1.1. *Apollo and Diana,* engraving by Jacopo de'Barbari, 1500. The sun god Apollo bestrides the celestial globe as the lunar goddess Diana declines with the dawn. © Copyright The British Museum.

dream and imagination. Separated but not disconnected from the earth, Apollo embodies a desire for wholeness and a will to power, a dream of transcendence and an appeal to radiance.

The Apollonian eye is synoptic and omniscient, intellectually detached. Below its gaze the earth is surface or film. Abraham Ortelius's *Theatrum orbis terrarum,* of 1570, the first systematic atlas claiming to encompass the whole earth, conjures Apollonian associations to preface a novel cultural project. A prefatory verse rhapsody seats its maker in Apollo's chariot, while a caption to the 1579 edition's opening planisphere recalls Seneca's words: "Is this that pinpoint which is divided by sword and fire among so many nations? How ridiculous are the boundaries set by mortals" (Fig. 1.2).

Today, the globe continues to sustain richly varied and powerful imaginative associations. Globalization—economic, geopolitical, technological, and cultural—is widely regarded as a distinguishing feature of life at the second millennium, actualizing the Apollonian view across a networked, virtual surface. Resistance from the solid ground of earth, characteristically located at the spatial and social limits of Apollo's conventional purview, proclaims the limitations of its male-gendered Eurocentrism,[1] a globalism hopelessly bound to exercising and legitimating authority over subordinate social and natural worlds. The criticism is well founded, both historically and morally. But the issue is by no means simple. The Apollonian perspective prompts ethical questions about individual and social life on the globe's surface that have disturbed as often as they have reassured a comfortable Western patriarchy. It also prompts a poetics of global space, an attachment beyond the material and visible surface. A cultural history of imagining, seeing, and representing the globe—Apollo's eye—stitches elements of a historically deep geographical imagination to practices of globalization that have helped define the West through continuous reworkings of an expanding archive of global images, narratives, and myths.[2]

The globe is a figure of enormous imaginative power; until 1968 "seeing" the spherical earth meant imagining or picturing it, an activity often inseparable from visionary experience. To achieve the global view is to loose the bonds of the earth, to escape the shackles of time, and to dissolve the contingencies of daily life for a universal moment of reverie and harmony. Reverie is the closest English translation of the Latin *somnium,* the sense of imaginative dreaming long associated with rising over the earth.[3] Apollo's company was the Muses, his lyre as significant as his bow. Apollonian music was created by the mathematical harmony of revolving cosmic spheres. In competition with earthly music, Apollo's was always victorious, its harmony exceeding the audible. The German word *Stimmung* captures this "tuning" of a vital earth to a resonant, universal harmony.[4] It complements the lucent geometry of solar light. The figure of Apollo thus prompts the conception of a unified world, a sphere of perfect beauty and immeasurable vitality, bathed in a beatific gaze. Plato describes such a perspective in *Phaedo:*

1.2. *Typus orbis terrarum*, Abraham Ortelius's world map of 1570 and its Stoic epigram. Harry Ransom Humanities Research Center, The University of Texas at Austin.

First of all the true earth, if one views it from above, it is said to look
like those twelve piece leather balls, variegated, a patchwork of
colours of which our colours here are, as it were, samples that painters
use. The whole earth is of such colours, indeed of colours far brighter
still and purer than those: one portion is purple, marvellous for its
beauty, another is golden, and all that is white is whiter than chalk or
snow; and all the earth is composed of other colours likewise, indeed
of colours more numerous and beautiful than any we have seen. Even
its very hollows, full as they are of water and air, give an appearance
of colour, gleaming among the variety of other colours, so that its
general appearance is one of continuous multi-coloured surface.[5]

The idea of seeing the globe seems also to induce desires of ordering and
controlling the object of vision. At the opening of his earthly ministry the
Christ-Apollo was removed to a desert vantage point to be offered dominion over the terrestrial globe. Emperors, kings, states, and corporations have
yielded to similar temptations, picturing globes and global panoramas to
proclaim territorial authority. Harsh realities of rule have been softened into
apparent harmony by the peaceful coherence of the synoptic vision.

Realizing the Apollonian vision has required the representational assis-
tance of spherical geometry, graphic and literary artistry. The history of such
representations is complex, connected as closely to lust for material possession, power, and authority as to metaphysical speculation, religious aspiration, or poetic sentiment.

Naming the Globe

That the earth does indeed have the geometrical form of a globe and that
it revolves on an axis in determined spatiotemporal relationship with other
spherical bodies and thus sustains the conditions for life are facts so familiar
as to veil the full scope of their imaginative significance. Some knowledge
of terrestrial and celestial rotation and the astronomical determination of
time and season are the intellectual heritage of every culture, the starting
point for instruction in natural philosophy, cosmology, or geography.[6] Their
observation yields theoretical structures of fixed compass points, lines, and
coordinates, abstract networks with considerable symbolic power. The principal vehicles for cosmographic understanding and speculation are calendrical images and models, in Greco-Roman tradition the armillary sphere,
celestial and terrestrial globes, and the planisphere or world map. Like all

images and representations, these objects are not innocent copies of a fixed external reality; they are fabricated and used in specific contexts for various and diverse purposes, and they readily acquire value as desirable possessions and iconic objects.[7] They play active roles in constituting knowledge and meaning for the phenomena they depict. Spherical geometry denotes unity and perfection, even divinity, so that representing the roundness of the earth can appropriate to the representational object meanings attributed to the globe's actual form. The Copernican revolution was secured through the circulation of cosmographic images that challenged ways of imagining and experiencing not only planetary arrangement and movement but the entire cosmic arrangement in which human existence was created and performed.[8] Twentieth-century photographic images of the earth have stimulated equally profound changes in perceptions of society, self, and the world. Both sets of images demarcate key moments in the evolution of the "globalized" earth.

Geometric, pictorial, textual, and numerical techniques have worked together in complex ways to represent the globe and its spatial and temporal relations.[9] Cosmography, geography, and cartography are intimately connected, although today they are distinct practices with distinct histories. While I draw upon these histories, my intention here is not to replicate or add significantly to them.[10] My focus is the terraqueous globe and a limited set of ideas and images through which a globalizing Western culture has "placed" itself in relation to universal space and time, both on the globe's surface and within the greater cosmos.[11] Contemporary English words such as *globe* and *earth* or *empire* and *humanity,* which structure my discussion, mask complex, shifting and often contradictory experiences and cultural practices whose meanings have been made and remade in specific contexts of place and time. Attention to geographical and historical context, however, cannot ignore long genealogies of meanings and practices whereby cultures rework identity through memory, learning, and imitation. Such a heritage of texts and images relating to the globe underpins social and environmental globalisms today.

Images of nature and society are always folded into each other and reach also into individual self-reflection, so that histories of global representation touch the depths of individual unconsciousness. Thus the physicality of the globe is captured by the word *earth,* which frames it as nature *tout court,* while *world* denotes a more social universality to which empire and humanity more readily attach. A connecting theme in globalism, originating in Stoic philosophy, is that human existence and agency, when set in Apollonian perspective against the vastness of global space, are petty and insignificant, subsumed by the greatness of its "nature." While this is a foundational idea of contemporary, universalizing environmentalism, it also has roots in a notion of *sublimity,* that sense of awe, even terror, in the face of cosmic magnitude and regularity described in John Milton's poetry and pictured in the works of John Martin, the nineteenth-century painter of biblical scenes. The globe's sublimity prompts reflection on life's origins and destiny, so that eschatology and teleology are shaping themes in globalism.

Three English words commonly describe this planet: *earth, world,* and *globe.* Used interchangeably, each has distinct resonance. Earth is organic; the word denotes rootedness, nurture, and dwelling for living things. It also implies attachment and habitation: earth is the ground from which life springs, is lived, and returns at death. Earth was the central, stable element that, with water, air, and fire, composed the terrestrial sphere of premodern thought. Earth is also soil, especially productive soil plowed for cultivation. Both Germanic and Latin languages gender the earth female, attaching to her the capitalized *Mother.* The Renaissance iconographer Cesare Ripa represented *Terra* as a woman seated on a globe to indicate the immobile earthly sphere. Surrounded by animals, in her right hand she holds a second globe, and she is clothed in various herbs and wears a headdress made of plants and a cornucopia.[12] In contemporary parlance, *earth* is *environmental* rather than spatial, and when applied to the planet, it attaches, through its agrarian connotations, to the sense of locale, actually counterposed to the global.

In contrast to *earth, world* has more of a social and spatial meaning. *The world* implies cognition and agency. Consciousness alone can constitute the world: humans go "into the world," they may become "worldly"; they create life-worlds or worlds of ideas, worlds of meaning. World is a semiotic creation; in the New Testament it proceeds from the Word and encompasses all creation. We are born into the world, may engage it or retreat from it, and die perhaps into another world. Worlds may coincide or collide, and we may imagine past or better worlds. In Latin (although not in Germanic) languages *world* is male gendered. Ripa's *Mundus* is a man standing in a powerful position with a long robe of diverse colors, a golden globe on his head to reflect the strength of the world.[13] *World* implies mobility and communication across the global surface, together with the power and authority this brings. But there are moral ambiguities: In early modern culture the emblematic Lady World was a finely attired female standing upon, or closely associated with, a globe. She represented the moral dangers of too close an attachment to this world.[14]

Neither *earth* nor *world* denotes the spatiality of *globe. Globe* associates the planet with the abstract form of spherical geometry, emphasizing volume and surface over material constitution or territorial organization. Unlike the earth and the world, the globe is distanciated as a concept and image rather than directly touched or experienced. As a globe, the planet is geometrically constructed, its contingency reduced to a surface pattern of lines and shapes. Thus the globe is visual and graphic rather than experiential or textual. As a spherical object, the globe of Earth can be associated with other spheres, such as the crystalline spheres that revolved in the Ptolemaic planetary system or crafted by the fortuneteller. The form of the globe finds anthropomorphic expression in the human eye or the female breast (see Fig. 7.1), generating a poetics of form that connects the microcosm of a gendered human body to the macrocosm of the planetary globe. The term *globalism* itself draws upon the abstraction of *globe* to generate associations quite distinct from those of *earth* or *world*. Its rich symbolic potential makes *globe* the most apt of the three terms for a study of images and symbolic meanings.

The closing decades of the last millennium witnessed further shifts in the relationship between these three planetary descriptors. Seeing (and photographing) the planet from space was one stimulus for reworking physical and conceptual spatialities and the poetics of the globe. Given the constitutive role of representations in framing as well as communicating social meaning, the problems of representing the globe seem to increase in proportion to our technical capacity to do so. Today the globe appears at once mechanical and organic, localized and deterritorialized. James Lovelock's self-sustaining *Gaia* thesis of the 1970s has mutated through 1990s cyborg—part mechanical, part organic—planet to the continuously reforming virtual spheres of the global Internet.[15] Such evolution in global imagining, however, is novel more in its rapidity than in its occurrence and often unconsciously borrows or restates poetic insights of considerable antiquity and longevity. One of the questions driving this study is the extent to which contemporary global imaginaries are indeed historically novel.

Cosmography and Global Geography

Planetary relations among the celestial orbs is conventionally the concern of cosmology and cosmography. Today, celestial globes are individual planets in and perhaps beyond the solar system. Until the seventeenth century most Europeans attributed closed global form to the universe itself. At its center were subcelestial and mutable spheres of earth, water, air, and fire materially enclosed within immutable celestial and crystalline spheres (Fig. 1.3). The

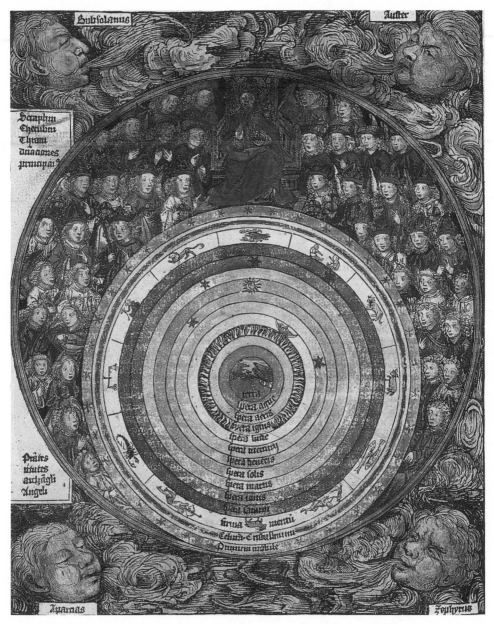

1.3. The Christianized cosmos, from Hartmann Schedel's *Liber chronicarum,* of 1493. Harry Ransom Humanities Research Center, The University of Texas at Austin.

revolutionary movement of these concentric spheres expressed a poetics of creation to which metaphysics attributed universal harmony—that divine music, inaudible to the human ear, whose study paralleled that of geometry and number.[16] Cosmos was long represented by the armillary sphere, a mathematical realization described by Ptolemy, possibly already manufactured by Greek globe makers *(sphairopoiia),* of the five astronomically observed circles of equator, tropics, arctic, and antarctic, the polar axis and the line of the ecliptic containing the zodiacal band. From the late fifteenth century, European globe makers and illustrators implicitly ruptured the cosmic unity of material and celestial worlds by pairing distinct terrestrial and celestial globes. Thenceforth the full scope of cosmography would be represented by three global representations, with the armillary whose axes and circles are common to both earth and heavens as the mediating symbol of unity between celestial and terrestrial orbs, the icon of cosmography.

The enduring philosophical influence of Pythagorean thought and Platonism has sustained belief in the incorruptible perfection of mathematical relations and conceptual forms. As the regular form whose circumference is equidistant from its center and whose spatial relations are governed by the mathematically ineffable π, the globe, or sphere, is the most geometrically perfect three-dimensional body. It seems inescapable that the sphere would be a symbol of totality and universality, iconographically significant across human cultures. In the West it has personified faith, justice, hope, and fortitude; it has denoted chance and fortune; it has signified both liberal arts and the mechanical sciences. The globe is the unchallenged symbol of astronomy, and it lies at the feet of a personified Philosophy. Its metaphysical associations determine its use, in crystalline form, for predicting the future.[17] If the sphere's form is perfect, however, and the revolutionary motion of celestial spheres predictable, the globe's instability on solid ground suggests the opposite. In contrast to the cube or the pyramid, for example, bodies whose flat bases place them firmly on the ground, the sphere is mobile. The globes of Fortune and Fate thus signify unpredictability as well as universality, conditions of life in the mundane world.

Connections between the conceptual globe constructed by geometry and the physical and metaphysical planetary spheres generate further symbolic associations. Among the iconographic attributes of the classical gods the globe belongs to Apollo. It would become an attribute of the Christian Trinity. The Father may be represented within a sphere; more commonly, Christ is shown standing upon or holding a globe. Secular rulers have appropriated the symbolic potency of the globe to signal the universality of their pretensions. It was adopted in Rome about 75 B.C.E. to signify empire; with the addition of the cross, its authority was borrowed by the Frankish and German Holy Roman emperors and later projected across the spaces of oceanic discovery by the Habsburg kings. In the lavishly symbolic court of Louis XIV of France the iconic integration of globe, Apollo, and Sun formed the core of imperial rhetoric (see Fig. 6.8). More recently, at a Brussels exhibition in 1935 a gilded sphere six meters in diameter demonstrated that the sun never set on the British Commonwealth and Empire.[18]

The geometric globe is smoothly undifferentiated; coloring its surface can amplify its meaning. In the Western iconographic tradition white signifies purity and perfection, while gold marks power, rule, and commerce. Further symbolic elaboration derives from inscription: the armillary's circle and bands signify cosmic harmony; the radiating networks of airline-route globes, the geographical reach of commercial influence. The parallels and meridians forming the graticule on terrestrial globes and world maps also denote significance beyond their practical cartographic, calendrical, and navigational uses. Not only have they shaped an evolving mathematical visualization of the globe but they have territorialized the planet. The 1493 papal meridian dividing Spanish from Portuguese imperial pretensions in Atlantic oceanic space set in train a globalizing geopolitical discourse that reconstructed the meanings of East and West and accounts today for the linguistic geography of Latin America.[19] The poles, topographically invisible, have become actual places, locations of heroic personal self-sacrifice in the cause of national self-aggrandizement. The lines of tropics and equator and of the climatic belts drawn between them have denoted distinctions between natural and human life-worlds since antiquity. The cardinal directions, points on the planisphere—north, south, east, and west—attain powerful teleological and anthropological resonance. Constructed in large measure by these conceptual points and lines on the globe, the geographical imagination in the West has expended untold physical energy and passion in rendering them actual across terrestrial space. In the process, meanings and identities of whole peoples and environments have been transformed.

Beyond the inscription of pure geometry on the terrestrial globe, the Apollonian gaze inscribes its surface with elemental division of earth and water. This elemental distinction is particularly apparent to peoples of Mediterranean and Atlantic Europe, who inhabited and navigated a fragmented space of peninsula and islands, as well as to other archipelagic cultures. But the pattern of earth and seas is not the only possible global geography. From space, color and tonal variations and the mobile patterns of clouds dominate the globe's surface patterns. *Geography* has conventionally been distinguished from *cosmo*graphy and *choro*graphy because of its attention to such patterns as opposed to astronomical and local spatial representations, respectively.[20] The primary geographical order of continents and oceans was not finally determined until the twentieth century—a globalizing intellectual imposition of the European geographical imagination.[21] The extent and disposition of lands and seas and their mutual enclosure long remained in question. The globe's geographical naming is necessarily arbitrary and conventional rather than logical or empirical, its apparent order and stability in atlas or world map deceptive: the nominal globe is a space of contestation rather than of concord. Even the location and numbering of meridians has a cultural politics, imposed in 1884 by an imperial Britain and challenged in recent years by postcolonial proposals to relocate the prime meridian in the Pacific Ocean.[22] Europe remains nominally a continent despite the invisibility of its eastern land boundaries, while India and Greenland, both similar in size to Europe, are granted subcontinent and island status, respectively.

Representations of such geographical divisions are also fields of symbolic appropriation. In Western iconography Europe has been crowned with the tower of urban civilization and military power, garbed in the purple cloak of empire and adorned with imperial symbols of scepter and orb. Her accompanying bull refers to the rape of Europa, carried from Athens to Crete by the disguised Jupiter, a myth recalled and elaborated as historical foundation for the cultural construction of a European humanist project.[23] The colors of the continents and identification of their highest land and deepest sea points, their geographical center, their antipodal locations, and their longest rivers are among the many knowledge constructions inscribed on globes and world maps. Together with actual locations of natural phenomena, sacred sites, and historical events, these frame and construct poetic geographies, shaping meaning across the global surface. The history of the river Nile and its African sources, extending from the ancient world to the age of steam railway and electric telegraph, is only one example.

The search for regularity or formal symmetry in global distributions registers an enduring cosmographic urge to parallel the metaphysical with the geographical global order. Medieval representations of a tricontinental terrestrial disk, the long-imagined *terra australis* of early modern globes, and the insistent conviction that navigable northern sea passages should balance southern circumpolar navigations indicate that the imaginative hold of global symmetry is not historically confined. Rationalist schemas of global symmetry quickly seized on the changing pattern of continents and oceans revealed by European navigators: Ortelius sought to fit the coast of Africa to that of South America, while in his *Novum organum* Francis Bacon was among the first to note their similarities as isthmian continents narrowing from the equator to southern capes, "a circumstance not to be attributed to mere accident."[24] The geologist Charles Lyell theorized antipodal asymmetry in lands and seas, while in the 1990s the Russian mathematician Alexei Shulga claimed that the eight-sided cube helps disclose congruence and symmetry in continental masses and deep ocean trenches, which recur every ninety degrees along the equator.[25] Claims of elemental order in the globe's surface patterns are more than amusing myths and fantasies readily banished by observational science. They indicate a metaphysical urge to harness geographical diversity—and often thereby cultural diversity as well—to a dream of order. They play a significant role in shaping human identities and actions.

Of equal significance to physical location within the evolution of Western global imaginings has been the boundary between known and unknown space. On a flat map the known can be extended to the very edges of representational space, leaving implicit the question what lies beyond the frame; on the globe the "ends of the earth" cannot be ignored. It is not accidental that terrestrial globes became common in Europe in the same decades as the first circumnavigations.[26] Practically, on the global surface ends are also beginnings; psychologically, boundlessness implies the chaos that attends the dissolution of form. The ancient world left no unambiguous evidence of globe making; knowledge of the spherical earth alone necessitated a boundary to its known spaces. The ends of the earth lay beyond the Greek *oikoumene,* that region determined theoretically by climatic zones and the pattern of Mediterranean lands and seas (see Fig. 2.3).[27] From antiquity the bounds of the earth have been drawn to distinguish humanity from the rest of nature and to register imperial claims over both nature and peoples.

The meanings of global representations are elaborated in associated texts and images, within or beyond their spatial framing. Ortelius's world map is titled *Typus orbis terrarum.* To place such words within an elaborate cartouche in which they emerge from surrounding clouds, accompanied by personified wind heads at the cardinal points, establishes a relationship between the image and the rhetoric of imperial Rome, just as captioning the 1972 NASA globe photograph "Spaceship Earth" makes a connection to ideas of mission and isolation current in American culture of the mid-twentieth century. Within terraqueous space itself meaning can be similarly elaborated. On Vincenzo Coronelli's great seventeenth-century globes, for example, detailed texts offer complex historical, ethnographic, and allegorical commentaries coherent only in the specific context of the globes' production (see Fig. 6.9).[28] Their significance lies as much in their presence and location as in their narratives, the former implying that nothing of greater significance than the word occurs at that point. Pictorial images perform a similar function. It was long common to elaborate oceanic spaces on globes and planispheres not only with linear representation of waves but also with fishes, sea mammals, and monsters, many of them allegorical. On Giulio Sanuto's 1570s globe, for example, a sea monster migrates from his engraving of Titian's *Perseus and Andromeda* to the southern oceans, a mythological moment transferred to geographical space.[29] Complex symbolic commentaries on globes and planispheres serve as residues of knowledge and belief to be remembered and reappropriated on later images, continuously informing the geographical imagination.

Globe, Globalism, Globalization

The images in these chapters are all in different ways *global*.[30] *Global* refers to an expanding range of environmental, economic, political, and cultural processes. Much of its force derives from the arresting concept of the earth as a single space made up of interconnected life systems and a surface over which modern technological, communications, and financial systems increasingly overcome the frictions of distance and time to achieve coordinated simultaneity. Even negative evaluations of global processes as uneven and locally disruptive rely upon an implicit commitment to globalism in the sense of equality of rights and opportunities over global space. The conviction that disruption of the natural world by deforestation, atmospheric pollution, or desertification threatens a "global" balance in nature has intensified as much through instantaneous global dissemination of information as through such activities themselves. The discursive impacts of global processes seem to be magnified in precisely those regions long placed imaginatively on the margins of a Eurocentric *ecumene*.[31] Thus the contemporary resonance of *tropical rainforests, polar icecaps, deserts,* and *deep oceans* owes much to the history of Western global visualization and imaginings.

Globalization gives to contemporary processes of accelerating globalism a socioeconomic and political dynamism, even teleology. Economically, the integration of financial, commercial, and trading markets responds to the demands of hypermobile capital, disconnected from attachment to a specific location. The process has deep roots in Western imperial history, pioneered in modern form by "global" trading bodies such as the Dutch East India

Company or the Hudson's Bay Company. Socially, it is increasingly meaningful to speak of a global division of labor, global production, global marketing and consumption of goods and information, and to contrast these
with life processes that remain localized in time and space. Political phenomena such as "world" wars ("hot" or "cold"), international religious, humanitarian and nongovernmental agencies, superpowers with global diplomatic and military reach, and even the international agreement to remove
Antarctica from competing territorial claims render the term *globalization*
geopolitically as well as economically meaningful. Culturally, the global scale
seems increasingly appropriate for describing and explaining such phenomena as the Internet, cinema, video and popular music, news programming,
fast food, dietary preferences, clothing and personal fashion choices, tour-
ism, and even claims for scientific and artistic integrity, human rights, or ethical behavior.[32]

A significant insight for which contemporary globalization is at least partially responsible is the *de-centering* of global geographies, in recognition that
visualizing the globe from a central region whence knowledge and influence
flow outward towards margins is inadequate to capture the mutual, if uneven,
shaping that occurs at both ends of every connection. Spatial metaphors such
as network and rhyzome have thus replaced core and periphery, which depended upon an imperial and Eurocentric vision. It remains true, however,
that expanding populations, increased demand for new consumption goods,
competition for territory and status among powerful elites, and a doctrinal
intolerance shaped a strongly Eurocentric globalism over a half-millennium,
from about 1450.[33] And this simultaneous expansion at the margins of peninsular Europe and restructuring of alternative "centers" drew upon and reshaped a memory of global imperialism in European antiquity.[34] Images and
ideas of the globe and the earth as an integrated totality, developed over the
course of European expansion, have both shaped and been shaped by the
actuality of a globalized world. Globalization, of course, remains always partial and contingent, always in tension with more localized perspectives and
experiences.[35] Thus, as the landscapes of imperial cities reveal, Rome most
dramatically, centers always have to be constructed, continuously remade in
the imagination and actually to express the global spatialities of caesars,
popes, and imperial states.

Global models and cartographic and pictorial images have been patent
and potent tools in these globalization processes. To imagine the earth as a
globe is essentially a visual act, as the Apollonian gaze implies. Such a gaze
is implicitly imperial, encompassing a geometric surface to be explored and

mapped, inscribed with content, knowledge, and authority. Yet the Apollonian image also recalls the earthly globe enclosed within other spheres, a home or dwelling, thus implicitly local and rooted. Discussing this paradox, the anthropologist Tim Ingold has criticized the visualist assumptions of current global discourse for subordinating and demeaning local knowledge, regarding such discourse as a triumph of technology over cosmology.[36] The victory of ocular vision over other forms of knowledge parallels the history of modern colonialism, and the processes are not unconnected.[37] But the dialectic of detachment and engagement is as phenomenological as it is historical. Seeing the globe and sensing the earth have both shaped and been shaped by the Western imperial and colonial project of making the modern world. Eurocentrism itself indicates that the imaginative experience of the local has shaped the meanings of the global.

Global Empire, Colony, and Humanity

Both historical and geographical knowledge of global space in the contemporary West represent the outcome of a continuous reworking of sources. Until very recently in Western cultural history Hebrew, Latin, and Greek texts offered the most consistent references, although these texts themselves were hybrid products. To present them as simple tributaries entering a broadening river of "Western culture" masks complex contexts in which such sources as Babylonian astronomy, the Hebrew Scriptures, Greek testaments, Euclidean geometry, or Socratic moral philosophy have been deployed.[38] Ignoring the passage of ancient Greek texts through Islamic readings en route to the twelfth-century Latin West was itself part of the nineteenth-century construction of a European Renaissance.[39] I do not mean to imply that context and the constructive role of memory erase all continuities. However altered in their passage across two millennia, two connected discourses—those of empire and humanity—have recurrently figured in Western knowledge and imagination of the globe. The acceptance of Christianity, a branch of Hebrew monotheism, as the official faith of the Roman Empire at its height meant that Rome's geopolitical claim—"imperium orbis, cui imperio omnes gentes reges nationes . . . consenserunt"[40]—became also that of the empire's official faith. Christian proselytism into the gentile world and the textual expression of its principal tenets in the language of Hellenistic humanism meant that Christianity emphasized human unity rather than differentiation. Finally, Christianity adopted from Rome an organizational structure comprising papacy and episcopate, wherein authority flowed imperially from the city to the world, *urbe ad orbem*.[41] Via the institution of a monarchical papacy and subsequent seizure of its moral legitimacy by individual nation states, a discourse of empire and humanity underpinned European imaginative geographies of the globe. The global expansion of European power, spearheaded by merchants' improving access to material goods, stimulated and was structured by continuous reworking of the literary and artistic culture of ancient empire, including its secular humanism. The missionaries of Victorian imperialism and even contemporary humanitarianism have continued to use global images to signify the universalism of their projects.

Empire

The drive for territorial supremacy need not be global, but it is imperial insofar as it depends upon the process of "othering" those over whom supremacy is exercised. The conceptual association of civilization with the city, indicated by their common Latin root, *civis,* makes explicit the idea that extension of territorial authority beyond the confines of a city is over "other" people, either another civilization or *un*civilized beings. Historically, empires have emerged wherever autonomous urbanization occurred—in China, India, South and Central America, for example. Imperial spatiality of center and frontier figures a landscape of self and home by othering people and places.[42] Unless imperial expansion is checked by a sufficiently powerful Other, the territorial question of empire is that of determining its geographical bounds. Imperial cultures tend to figure their opponent within a Manichean discourse of a life struggle against an "evil empire," as in the case of Persia and Greece, Rome and Carthage, Britain and Russia, the United States and the Soviet Union. Imperial bounds may be set in stone, for example, the Great Wall of China or the boulders of the *limes germanicus* set across the Danube lands.[43] The actual location of imperial limits shifts with changing geographical knowledge and cartographic representation. Until the fifteenth century the imagined bounds of Christian empire remained the Aristotelian *klimata,* delimiting the northerly and southerly bounds of the temperate zone at sixty-six degrees and twenty-four degrees from the equator, respectively. In the west, the Pillars of Hercules, with their proclamation *non plus ultra,* separated the Mediterranean from the "Ocean Sea," which bounded the three continents. To the east, frontiers were set by social Others, peoples whose humanity was provisional. The East was at once an opportunity and a threat, to which various Western "Orientalisms" represent a recurrent response. The cardinal directions south, west, and north have thus played distinct roles in Europe's encompassing spatialities and have helped frame the boundaries of citizenship and difference within and beyond imperial space.[44]

Ancient Greece both universalized the idea of empire and introduced into it distinct *colonial* space. Initially, Greece's was a sea-based empire; its epic myths relate the struggles of city-states—Minoan, Spartan, Athenian, Trojan—to establish Aegean hegemony, and its victory over Persia was achieved by naval superiority. Alexander's march toward "the ends of the earth" from 334 to 323 B.C.E. constructed a true Hellenic empire; an empire that not only covered a larger geographical area than any of its predecessors but also drew upon and extended a Babylonian and Athenian science to relate Greek territorial expansion to a cartographic concept of the globe.[45] That Alexander's empire could be mapped onto a picture of a habitable earth whose extent was preordained by the sphericity of the globe and the latitudinal order of the climates opened the conceptual possibility of imperial control extending to territorial limits imposed by *nature,* thus prompting the idea of a universal empire incorporating a "universal humanity," which significantly complicated any simple dualism of self and other. Resolving this contradiction between human difference constituted through geographical diversity and a universal humanity constituted through the spatialities of empire resonates across the history of global thinking.

Universal empire was quite specifically articulated in Rome under Augustus at the time of Christ. In his *Res gestae,* posted posthumously near his mausoleum, the emperor's achievements in framing the territorial shape and limits of Roman rule were recounted. Divine will "had assigned to Rome the destiny of conquering, of dominating, but also of pacifying and organizing the whole world."[46] Two centuries earlier Polybius had introduced the idea that Rome might exceed the reach of the four empires it had succeeded—Assyria, Media, Persia, and Macedonia. Pompey claimed to have stretched the frontiers of Rome's dominion to the very limits of the earth, and by Augustus' time the concept of *urbs et orbis terrarum,* "city and earthly globe," had become commonplace in Roman imperial rhetoric. It would "foster the nostalgia of ecumenism until Byzantium or Charlemagne,"[47] and be regularly resurrected in the Latin West in a vision of empire beyond the Pillars of Hercules, a global space upon which "the sun never sets."

As an imperial symbol, the globe appears on Roman coins from about 75 B.C.E. Unmarked, the sphere may represent equally the celestial or terrestrial globe, permitting a symbolic link between the divine order symbolized by the former and the territorial order imperially imposed upon the latter. Thus Augustus's imperial declaration was placed within an architectural complex at the heart of his imperial city that may also have incorporated a world map and certainly included calendrical monuments establishing a coincidence between the emperor's nativity and the order of cosmic time. Thus the terrestrially located emperor is imaginatively transformed into the celestial Apollo, omnipotent because omniscient across global space.

If the emperor and empire are cosmographically located and legitimated by reference to the predictability and order of the heavens, the terrestrial orb denotes direct territorial authority. Territorially, empire is in key respects a cartographic enterprise. The practical roles of cartography and geography in establishing imperial rule are well recorded. Alexander's imperial expedition to the East was promoted in part as scientific exploration and Alexander was accompanied by scholars. Subsequent imperial ventures—Bougainville's and Cook's Pacific navigations, Napoleon's Egyptian survey, Livingstone's journey into Africa, and Peary's advance to the North Pole—were similarly legitimated. The eponymously imperial city of Alexandria became home to a great school of geographical and cartographic learning, its library the principal source for Strabo's synthesis of Augustan geographical science.[48] The museums and libraries of Europe's colonial capitals have played the same role in gathering and archiving the intellectual empire. The Place Royale in Brussels, for example, is a complex of spaces and buildings dating from the brief period of Belgian colonialism when a tiny European buffer state, created to meet the strategic needs of post-Napoleonic Europe, ruled a vast region of tropical Africa. In design, function, and iconography the square coordinates global time and space and gathers universal knowledge at the symbolic center of Belgian rule. The architectural design and the contents of the surrounding museums and galleries simultaneously connect and differentiate the European and African material culture, art, and science they house. At the center of the square Geoffroy de Bouillon, first Crusader king of Jerusalem and a local Brabant hero, raises an imperial sword over the city to recall Belgium's role in Christendom's imperial conflict with its Islamic other. The church of St. James in Kondenburg, on the *place,* was the coronation site for Belgium's kings and emperors; its principal painting is *The Assumption of St. Catherine,* by G. de Crayer (1584–1669), a baroque work in which the Christ child holds a prominent terrestrial globe.

Seen from the center, imperial geography is primarily concerned with territorial defense and the ordering and coordination of internal movement. Local subjectivities within imperial space are a secondary concern. The cartographic measure of the administrator and the trader is distant and controlling, radiating from the center along axes that pay little heed to local specificity. All roads lead to Rome, the world's seaways converge on the port of London, global air routes radiate from New York. In reality, of course, imperial space is always bounded, dreams of universal empire by necessity confronting the specifics of geographical difference and the constraints of distance; but speaking abstractly, we may describe a fourfold spatial structure for the imaginative inheritance of Western empire. At the center is the imperial city, claiming authority over a territory theoretically extensible across the *oikoumene* and thus bounded only by the limits imposed by nature. To the imperial city are attributed the qualities of an *axis mundi,* a point where terrestrial space connects with celestial time.[49] The city's claim to universal centrality is performed and expressed rhetorically, through architecture, landscape, ritual, collection, and display.[50] Narratively, the empire sets its own limits (for Alexander it was the River Hyphasis; for Rome before Caesar, the Rubicon). Spaces beyond the authority of the imperial center are either active frontiers within the ecumene or wilderness beyond it, uninhabitable by fully human beings. The combined geography of urban, imperial, frontier, and wild spaces constitutes the globe over which the emperor claims, if not exercises, authority. The terrestrial globe lies within a celestial sphere whose regular movement translates as the paradigm of imperial order on earth. Enthroned in the imperial city, globe in hand, the emperor personifies (indeed often divinely embodies) the spatial and temporal order that commands this spatial hierarchy. The imperial perspective is thus appropriately Apollonian: fixed in time yet mobile, globally central and divinely celestial.

Colony

Distinct from Alexander's imperial project but related to it was the Greek Mediterranean colonization that began in the eighth century B.C.E. and quickened from the fifth. This was driven as much by desire for land as by desire for military conquest and commerce. In Sicily and southern Italy, in the Black Sea and Egypt, settlements were established whereby Greek civilization was spatially transferred to a new land. Colonists were *apoikia,* people from home dwelling in a different physical environment. The Latin *colonia* is a direct translation of this Greek word, maintaining its close association with *dwelling.* Greek colonies such as Syracuse in Sicily or Neapolis on the coast of Italy were modeled on the polis: autonomous and self-governing. Citizenship in the polis depended upon ownership and cultivation of land; in later Roman usage *colonus* came to mean a tiller of land.

The sense of dwelling that adheres to *colonia* is apparent in Virgil's account of Aeneas's epic journey from the destruction of Troy to the founding of Rome. This process is narrated less in terms of spatial conquest than of forced emigration, exile, and yearning for a lost home. It includes Aeneas's key rejection of Dido at Carthage (Rome's "other" and false "home" to the Trojan heroes) and the cosmogonic act of planting a community in a new, promised land.[51] The Tudor English usage of *plantation* for the settlement of Ireland and Virginia conveys a similar sense of rooting culture in a newly native soil.

Virgil wrote his epic at the key transitional moment in Roman imperial history, after Augustus's declaration of empire but before significant colonial occupation of subject territories. Imperial Rome would establish *coloniae,* mainly for former soldiers as payment for their services, an astute recognition that dwellers' attachment to the soil made them its fiercest defenders against autochthonous claims, a tactic used throughout imperial history up to twentieth-century soldier settlements in Western Australia and western Canada. While empire and colony are closely related, they involve distinct spatialities and environmental relationships.[52]

The empire constructs a *world:* global, urban-centered, hierarchical, and visually distanciated. The local spaces of the plantation or agrarian colony are those of *earth*. There may indeed be a ruling urban center, spatial and social hierarchies, clearly marked boundaries against the wild or uncultivated, and an imposed cartographic geometry inscribed upon the land of a colonial plantation, but these tend to be imperial impositions, often resented and commonly ignored or overthrown in the colonists' drive for autonomy. Repeatedly in British colonial America, for example, Virginians, Pennsylvanians, or Georgians ignored or abandoned schemes for urban concentration conceived in London in favor of a looser, more autonomous pattern of farms, and even Thomas Jefferson's continental settlement scheme, which sought to regulate the republic's colonial urge, met with limited success. There is evidence of a similar colonial response later in the colonial experience of territories such as Australia, New Zealand, and East Africa. The essence of the agrarian colony is attachment to land rather than territorial order. Preexisting occupancy brings the *planted* colonist into more intense and brutal conflict with other *natives* than is faced by the imperialist, a conflict commonly figured as a struggle between culture and nature, humans and animals, in which colonists represent the former and autochthons the latter. Not uncommonly, colonists find themselves in conflict not only with prior inhabitants but with the imperial center itself, where a universalist rhetoric of protection is more readily articulated, a rhetoric that is perceived by the colonists as being more sympathetic to "savages" than to themselves.[53] This conflict introduces a second, related idea connecting the globe and the imperial concept, that of *humanity*.

Humanity

A declaration of empire entails a claim to rule over subject peoples, a claim moralizing the imperial gaze cast dispassionately over a family of nations; thus Victoria's rhetoric of imperial mother. Empire gestures toward a global construction of humanity. Ancient Greek discourse, in addition to founding the language of empire and colony, is in large measure responsible for Western concepts of humanity and humanism, abstractions predicated on fixed qualities differentiating human from nonhuman life forms. Historically, humanism has been closely related to global imagining and representation, in part through the connection first established in Greek thought between human habitability and the globe's climatic zones. Although climatically defined, the concept *oikoumene* "in its most essential meaning can be defined as a region made coherent by the intercommunication of its inhabitants, such that, unlike the radius of this region, no tribe or race is completely cut off from the peoples beyond it."[54] Life beyond the *oikoumene* must by definition be nonhuman and Other.

The current belief that differences in language, customs, belief, or skin color are less significant than shared *human* characteristics, with consequential implications for moral status and rights, has a long and complex history. Categorical boundaries within and between the human and the nonhuman are historically and culturally fluid; indeed there are arguments for their almost total erasure.[55] Pressures to extend "human" rights to other life forms, advances in genetic manipulation, replacement surgery, and in vitro fertilization all demonstrate the continued mutability of such boundaries. But over most of European history attempts to delimit "humanity" have been determined largely by geographical boundaries.

Take, for example, the world map in Hartmann Schedel's 1493 *Nuremberg Chronicle* (Fig. 1.4), located immediately after the illustrations of Creation, thus representing the postdiluvian earth delivered by God to Ham, Shem, and Japhet, sons of Noah. Schedel's world map illustrates the classical *oikoumene* at the very moment when its geography was to be altered irrevocably. Indeed the text actually records the discovery the previous year of islands in the western Ocean. At the corners of global space are Noah's sons, to whose progeny Christian thought traditionally ascribed the populations of the three

1.4. The Ptolemaic ecumene with the monstrous races at its textual margins, from Hartmann Schedel's *Liber chronicarum*, of 1493. Harry Ransom Humanities Research Center, The University of Texas at Austin.

continents, while at the edges of mapped space are arrayed bizarre creatures. Despite constituting different nations—some yet to be redeemed—the population of the *oikoumene* constituted *humanitas*. All were sons of Adam and thus capable of salvation. In the terrestrial spaces beyond the frame of the *oikoumene* are other, hybrid creatures having human and nonhuman characteristics: tails, rabbits' ears, dogs' heads. These "monstrous races" dated back at least to Pliny's *Natural History* and had long appeared on *mappae mundi* and in medieval encyclopedias.[56] Their physical forms denote the moral tensions between the universalism inherent in the Judeo-Christian idea of global creation and a single human family, on the one hand, and the apprehensions of a European localism whose limited geographical knowledge denied the ascription of humanity to the inhabitants of "other" places. The practical consequences of working out these tensions during the European encounter with non-Western peoples over the half-millennium since this illustration have been at once tragically destructive and mutually transformative, even liberating. They continue to produce ethical dilemmas today. I suggest that they are inherent in a Western imperial conception of a globe geographically greater than direct experience of it, in the urge to legitimate territorial dominion over a global surface, and in colonial projects for migration and settlement.

Judeo-Christian inheritance is again critical. Cosmographically located outside his creation, the Hebrew God's epiphany was not geographically ordained. Certainly sites of actual epiphany became sacred, and nowhere more so than Jerusalem. But this God was not geographically or environmentally confined, as were those on Olympus, for example, who took responsibility for defined and limited aspects of the material world and were manifest in predetermined locations. When the Judaic God was sacrificed as Christ-man, the redemptive act itself was geographically extensible. Resurrection signifies disconnection from a specific location and is registered on behalf of all human creatures. The mobility of consecrated bread signifies a radically different spiritual spatiality from that of autochthonous faiths. And inherent in Christianity is a missionary imperative: to spread the supposed benefits of redemption to all peoples. The recurring question raised by this imperative concerns the limits of humanity, that is, of souls capable of salvation.

Western conceptions of what it means to be human represent a sustained dialogue with Greek thought, especially its emphasis on reason and logic as defining qualities of free men—Apollonian gifts, often themselves symbolized by the paradigmatic compass measuring the global sphere. If the limiting racial, gender, and other assumptions of that philosophy are today blind-ingly apparent, it is arguable that such recognition is itself an extension of the philosophy and its doctrine of rights. Cultivation of the mind and its capacity for abstraction, classification, categorization, logical induction and deduction, theoretical analysis and synthesis, indeed all those things that made up *scientia,* were defining qualities of humanity for Greek thinkers, intro-duced into Christianity through Hellenism's influence in Rome's empire.

In antiquity, full human status and rights derived from citizenship; they were quite literally grounded. In the Greek polis citizenship derived initially from ownership of cultivated land, and ownership of immobile property— "real" estate—remained for millennia the foundation of political franchise. In part, property released *men* from manual labor, freeing them to cultivate the mind, whose reason signified full humanity. A hierarchical order that mapped space, society, the idealized body, and its faculties to a scale of hu-manity and opposed human "culture" to nature has been continuously re-worked in Western thought and practice.[57] The city was regarded as the spa-tial expression and locus of a fully developed humanity. Gender assumptions long restricted this status to mature adult males characterized by *scientia,* rational thought connected with the head.[58] The fully human status of women and children, consistently deemed closer to nature, was ambiguous. In the symbolic geography of the body the domestic was related to the heart. Geo-graphically, uncultivated nature or wilderness and its animal inhabitants lay beyond the limits of the human world. Its inhabitants were deemed inca-pable of intellectual reason and heartless in their savagery, living according to the dictates of their loins. Those who lived off the land, especially if they were not settled cultivators, and those whose activity was restricted to un-skilled manual tasks constituted the lowest social class in such a hierarchy. For them, the boundary between human and animal nature has been blurred, whether by Athenian philosophers, Renaissance humanists, or Victorian so-cial thinkers.

The very openness of the ascriptions in such schemas of course invites contestation of ascribed status. In the Enlightenment, for example, the con-cept of "natural man" and his capacity for reason became bound directly to debates over the rights of man,[59] whereas nineteenth-century racial science constructed an evolutionary "ladder" rising from childlike, feminized, nat-ural "dark" "races" to adult, masculine, cultured "white" "races." It goes with-out saying that the categorizers consistently rank themselves at the top of such hierarchies. And the fiercest challenge to such confining schemas of "humanity" has come from globalizing and leveling tendencies within the

same philosophical tradition. Formally, slaves in the ancient world were ex-
cluded from full human status because they owned no property, not even
their own bodies. For Christianity, this contradicted the universal embrace
of redemption, which ascribed primacy to an immortal soul rather than to
the physical body. Slavery was long tolerated for those who had not accepted
Christ's message, for example, Slavs, on the eastern frontiers of Christendom,
but it was a morally dubious activity. Similarly, beyond the confines of Rome's
empire barbarians (whose beards betrayed their animal affinities) also raised
difficulties for Christianity.[60] The contradictions between slavery, salvation,
and humanity remained live moral issues well into the modern period as the
Christian Church struggled with the spiritual status of peoples located beyond the limits of the ancient *oikoumene.*

The issue became one of critical moment with the European discovery
of worlds unknown to biblical or early church authorities. Theoretically,
Europeans' treatment of native peoples would depend on its outcome. Paul
III's 1537 encyclical *Sublimis Deus* proclaimed all such peoples potentially
capable of redemption. Practically, colonial *plantation* has been a consistent
response to the imagined and actual threats posed by wilderness to "humanity." Conversion of native peoples and the introduction of sedentary agriculture followed a single path: "Wild animals flee the clearing where the
colonist and priest set up their cabins; the devil retreats along with the
woods."[61] Enlightenment writers such as Jean Jacques Rousseau and Thomas
Paine secularized the discourse of a common humanity as one of "rights."
Nineteenth-century global missionary, antislavery, and humanitarian societies and present-day nongovernmental organizations (NGOs) have sustained this ambiguous response to the cultural Other.

Poetics of the Globe

If Western thought has defined human distinctiveness by the capacity for
abstract, rational thought, the purest cultural expression of such thought is
mathematics, expressed in numbers, proportions, and geometrical forms.
Geometry gives spatial expression to mathematical relations; it originates in
measurement of actual terrestrial and celestial spaces.[62] For nearly two millennia the Aristotelian distinction between mutable, elemental, and un-
changing celestial spheres dominated cosmography, while from Platonic philosophy derived the idea that if perfection cannot be found in corruptible,
finite nature, it must exist in a realm of pure intelligence and is most closely

approximated at the scale of the cosmos and in the (implicitly male) human body itself. Speculation about relations between global macrocosm and human microcosm has commonly been pursued by mapping images of each onto the other, rendering visible fundamental mathematical homologies.[63] The eye itself is a globe, an often noted feature of such discourse, while vision has been the privileged sense in Western science, certainly since the twelfth century.[64] The Apollonian link between global vision, graphic representation, and the abstract intellectualism that characterizes humanity is forged through geometry.

Morphological relationships between the geometrically defined body and the cosmos lead readily to the analogical belief in processual connections between macrocosm and microcosm, invisible to the material eye but made intellectually apprehensible by their mathematical harmony. Music, at once sensual and intellectual, was long regarded as the most direct access to cosmic harmony. The German word *Stimmung,* with its sense of "tuning," directly captures resonance between the human soul *(anima)* and that of the world *(anima mundi).*[65] Apollonian music, unlike that of Pan, which sings of localism and attachment, is universal and transcendental.[66] The long tradition of the *somnium* relates the dream of spiritual ascent and interstellar flight to the effects of music. In the *somnium* the human mind can achieve the Apollonian perspective over the earth denied to the physical eye.

Together with a Stoic recognition of human insignificance in the vastness of creation, the implications of cosmic transcendence include the synoptic vision of the earthly globe and the preternatural, possibly magical capacity to know and intervene in the harmonies between celestial and elemental worlds. Western globalism is closely bound theoretically and graphically to this hermetic, Neoplatonic tradition. Such a complete and harmonious vision implies rising above the mundane; thus it is a mark of the exceptional being, the call to heroic destiny of the paradigmatic human. Plato's *Phaedo* is among the earliest texts containing such a vision; Plato's merging of philosopher, king, and priest suggests that the connection of intellectual transcendence and temporal power is a sacred act that will actualize universal harmony. The most influential narrative of the *somnium* has been Cicero's "Dream of Scipio," in which the conqueror of Carthage, hero of Rome's imperial expansion into Africa, dreams of ascent to the realm of the stellar gods and recognizing the limits to even so great an empire as his own. New Testament narratives of Christ's desert temptation, transfiguration, and ascendance have similar Apollonian overtones. In the opening years of Euro-

pean global imperialism the Portuguese poet Luis Camões's and the Eng-
lishmen John Donne and John Milton adopted the conceit of the *somnium*
to reflect upon global expansion. The following is from Donne:

> Could I behold those hands which span the Poles,
> And turn all spheres at once, pierced with those holes?
> Could I behold that endlesse height which is
> Zenith to us, and to'our Antipodes
> Humbled below us?[67]

In Rome, center of the Jesuit missionary empire, Athanasius Kircher, lulled
by sacred music, also dreamed a celestial journey (see Fig. 6.5), while in the
twentieth century the poetics of both aerial and space flight have similarly
woven together imperial and Apollonian imaginaries.

Geometric metaphors, symbols, and images have been the characteristic
representational modes of *anima mundi*. The poetics of the globe have consistently inflected the formal mathematics of measure and survey with hermetic meaning, while recognition of sacred geometry has often been regarded
as itself evidence of enlightenment and vision.[68] It is therefore unsurprising
that global models and maps have represented the metaphysics as much as
the physics of earth and heavens.[69] Manuscripts and printings of Macrobius's
second-century *Commentary on the Dream of Scipio,* for centuries a key text
on the interpretation of dreams, were regularly illustrated by global maps,
while founders of modern cartography such as Ortelius, Mercator, and
Hondius marked their globes and planispheres with metaphysical references.[70]

A distanced, rational gaze is thus placed in dialogue with cosmic harmony and transcendental vision; imaginings of empire and its geographical
limits are ranged against the practicalities of localized experience and measure; belief in a common humanity faces the challenge of difference and
alterity; the poetics of the whole cosmos connect through the earthly globe
to the individual human body. The following chapters trace these connected
themes through expressive moments of Western global imagining. They
suggest something of the complex genealogy which lies behind contemporary global meanings and the images upon which, often unconsciously, these
continue to draw.

TWO *Classical Globe*

All boundaries have shifted, and cities have set their walls in a new
land; the all-travelled world lets nothing remain in its previous station;
the Indian drinks from Araxes' cold waters, the Persians drink from
the Elbe and the Rhine.[1]

Designer T-shirts worn by New Guinea hill people, Hindu temples in Scottish suburbs—such spatial juxtaposition of cultural markers is often taken as a marker of a uniquely modern globalization. But these words of Seneca, written at the height of Roman imperial power, indicate that the sense of diversity and exchange as upset or distortion of a preexisting, more "natural," order characterized by geographical localism is not new. The stasis of a contemplated globe contradicts the contingency and mobility of global patterns generated by human interaction, providing fertile material for geographical imagination. In antiquity, poets and philosophers speculated on the implications of the form and cosmic location of the earthly globe, while a literature concerned with heroic travel to the edges of space "seems to have held a unique fascination for Greek and Roman readers."[2] Seneca's Rome was heir to a long and complex tradition of global speculation and calculation from which claims to empire *ad termini orbis terrarum* (to the ends of the earthly globe)—the foundational claim of later Eurocentrisms—drew meaning.

Poetic narrative, measured observation, and rational speculation of celestial motion and terrestrial pattern variously shaped the sense of global order in ancient Greece. Poetry, unlike mathematics or prose, registers divine inspiration rather than human authorship as the earliest foundation for Greek knowledge.[3] Poetic knowledge is less closely bound to ocular vision than the Greek *theoria,* or rational knowledge. The narrative quality of myth gives temporal structure to the natural order, marking beginnings, becomings, and ends, while observational description and speculation construct a more fixed spatial and geographical frame of being.[4] But space and time are representationally interdependent, and from antiquity we can trace the connection between global space-time and ideas of human destiny. The sphere combines these discourses into a single image of order and, unsurprisingly, from the earliest times offered divine legitimacy to human will.[5] Thus the Farnese Atlas, an almost unique surviving example of a globe from late antiquity, sculpts a human figure supporting the cosmos, delineated by the armillary circles and forty-three constellations.[6] Whether Atlas or Heracles, the figure stands at the junction of heaven and earth, at once divine and human. Atlas guarded the western edge of the Mediterranean world; Heracles, armed with club and lion skin, embodied god, man, and beast, his heroic destiny marked by his passage beyond the western, sunset end of space and time to the garden of the Hesperides on the slopes of the Atlas Mountains. From Heracles' almost reckless stretching of boundaries and his temporary shouldering of Atlas's cosmic burden originates a complex genealogy stretching through Alexander of Macedonia, who claimed Heraclean descent to structure his own myth of global empire, and Augustan Rome to Portugal and Spain in the sixteenth century and beyond.[7]

Creation Myths and Spatial Order

Regular motion of the heavenly bodies and the circularity of the cosmos, if not the sphericity of Earth, were studied by the ancient Sumerian, Persian, Egyptian, Babylonian, and Greek cultures. Poetic cosmogonies offer structurally similar accounts of the origin and evolution of natural order.[8] The earth is a parturition of preexisting chaos through the agency of divine force.[9] In Hesiod, for example, a complex series of couplings produces the principal elements of the visible universe:

> Chaos was born first and after her came Gaia
> the broad breasted, the firm seat of all
> the immortals. . . .
> Chaos gave birth to Erebos and black Night;
> then Erebos mated with Night and made her pregnant
> and she in turn gave birth to Ether and day.
> Gaia now first gave birth to starry Ouranos,
> her match in size, to encompass all of her,
> and be the firm seat of all the blessed gods.[10]

A living earth breeds giants and heroes—including Atlas himself, who "supports the broad sky / on his head and unwearying arms, / at the earth's limits"[11]—and ultimately mortals, whose social evolution Hesiod outlines in the companion song, *Works and Days*. Golden, silver, and bronze Ages are characterized by successive and progressively violent and destructive beings, culminating in a race of men/gods, including the Homeric heroes of the Trojan War. Some of these remain

> . . . settled at earth's ends,
> apart from men . . .
> They lived there with hearts unburdened by cares
> in the islands of the blessed, near stormy Okeanos.[12]

The past is thus present at the furthest edges of space: heroic destinies press to terrestrial peripheries; heroes are immortalized in the heavenly constellations, while earlier mortals are buried in the depths of the earth.

Homer gives a landscape expression to the imagined cosmic order in book 18 of the *Iliad,* in his image of Achilles' shield (Fig. 2.1). The work of the divine artificer, Hephaestus, its embossed center is decorated thus:

> There shone the image of the master mind.
> There earth, there heav'n, there Ocean he design'd;
> Th'unweary'd sun, the moon compleatly round;
> The starry lights that heav'n's high convex crown'd;
> The *Pleiads, Hyads,* with the northern team;
> And great *Orion*'s more refulgent beam.

A 1749 illustration of Achilles' shield from Alexander Pope's translation of the *Iliad* pictures the Homeric geographical narrative in a calendar of landscapes. Pastoral scenes of dancing, hunting, and shepherding are succeeded by landscapes of the Mediterranean agrarian year showing plowing, reaping, and gathering of the vintage. Two cities are described, at peace and at war, completing the universal social evolution, while at the edges of continental space

> . . . pour'd the ocean round
> In living silver seem'd the waves to roll
> And beat the bucklet's verge and bound the whole.[13]

Living around 900 B.C.E., Hesiod and Homer mark the origins of the Greco-Roman poetic tradition. Close to a millennium later, under the rule of an imperial Augustus, the Roman poet Ovid (43 B.C.E.–17 C.E.) used a structurally similar cosmogony to narrate global order.

> Before there was any earth or sea, before the canopy of heaven
> stretched overhead, Nature presented the same aspect the world over,
> that to which Man have given the name of Chaos. . . . There was no
> sun, in those days, to provide the world with light, no crescent moon
> ever filling out her horns: the earth was not poised in the enveloping
> air, balanced there by its own weight, nor did the sea stretch out its
> arms along the margins of the shores. Although the elements of land
> and air and sea were there, the earth had no firmness, the water no
> fluidity, there was no brightness in the sky. Nothing had any lasting
> shape.[14]

Ovid's warring elements await a god, "a natural force of a higher kind," who separates and sorts them, binding them to distinct form within "a harmonious union":

> His first care was to shape the earth into a great ball, so that it might
> be the same in all directions. After that, he commanded the seas to
> spread out this way and that, to swell into waves under the influence
> of the rushing winds, and to pour themselves around the earth's
> shores. . . .
>
>   As the sky is divided into two zones on the right hand, and two on
> the left, with a fifth in between, hotter than any of the rest, so the
> world which the sky encloses was marked off in the same way, thanks
> to the providence of the god: he imposed the same number of zones
> on earth as there are in the heavens. The central zone is so hot as to
> be uninhabitable, while two others are covered in deep snow: but
> between these extremes he set two zones to which he gave a temper-
> ate climate, compounded of heat and cold.[15]

Across this global surface blow winds, each designating a cardinal point, and above "appropriate inhabitants, stars and divine forms occupied the heavens [while] the waters afforded a home to gleaming fishes, earth harboured wild beasts, and the yielding air welcomed the birds."[16] A harmonious sphere thus awaits the appearance of humans, creatures of earthly elements but seeded by gods, their divine right to mastery "over all the rest" signified by their upright stature and intelligence. Like Hesiod, Ovid proceeds to a narrative of universal ages that connect social and environmental evolution: a golden age of everlasting spring, a self-generating nature, and social harmony; a silver age of seasonal cycles, plow cultivation, and cave dwelling; a bronze age of winter and warfare. In the present "iron" age, nav-

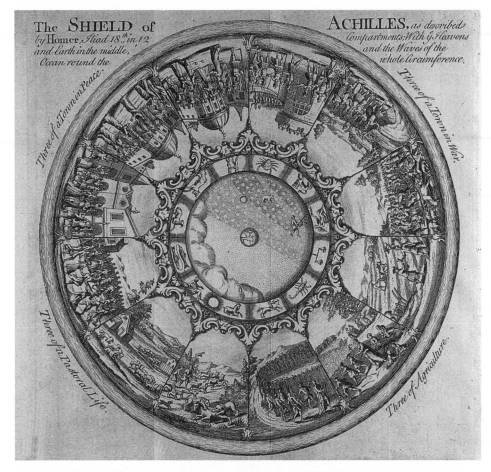

The SHIELD of *by* Homer, *Iliad* 18.*th* *in* 12 *and* Earth *in the* middle, Ocean round the

ACHILLES, *as described* Compartments; With *if* Heavens and the Waves of the whole Circumference.

Three of a Town in Peace

Three of a Town in War.

Three of a Pastoral Life.

Three of Agriculture.

2.1. Homer's shield of Achilles, engraving from *Gentleman's Magazine* 19 (1749): 392. Courtesy Royal Holloway, University of London.

igation appears and "the land, which had previously been common to all . . . was now divided up far and wide by boundaries, set by cautious surveyors," while humans explore the very bowels of the earth in search of gold and jewels. Reflecting with a critical eye processes shaping the Roman imperial world, Ovid connects the spatial promiscuity of commerce and navigation to the cartographic inscription of colonization and territorialization. The passage opens Ovid's *Metamorphoses,* a key cultural source for Western humanism from later medieval times to the present.[17]

Ovid's more orthodox contemporary, Virgil, harnessed a similar narrative to celebrate Augustus's proclaimed global imperial order. From the pastoral

Eclogues, through the cultivated world of the *Georgics,* to Aeneas's founding of Rome, Virgil's poetry naturalizes Rome's ideological claim to direct descent of urban civilization from Greece and Troy. The apotheosis of the cosmogonic narrative and succeeding universal history is Augustan empire and a universal Pax Romana. Virgil's description of Aeneas's shield, mirroring Homer's earlier cartographic account, reworks the Greek's map of the cosmos to celebrate Roman triumph over the *orbis terrarum.* The final passage recounts Augustus's entry into Rome after his defeat of Antony and capture of the eastern part of the ecumene at the naval battle of Actium:

> He himself was seated at the white marble threshold of gleaming white Apollo, inspecting the gifts brought before him by the peoples of the earth and hanging them high on the posts of the doors of the temple, while the defeated nations walked in long procession in all their different costumes and in all their different armour, speaking all the tongues of the earth. Here Mulciber, the God of Fire, had moulded the Nomads and the Africans with their streaming robes; here, too, the Lelegeians and Carians of Asia and the Gelonians from Scythia with their arrows. The Euphrates was now moving with chastened current, and here were the Gaulish Morini from the ends of the earth, the two-horned Rhine, the undefeated Dahae from beyond the Caspian and the river Araxes chafing at his bridge.[18]

Virgil's Apollonian vision tracks a temporal and spatial teleology from chaos to order, from the earth's periphery to Rome at its center.

The ages of "universal" history in antiquity correspond to a mythical map in which the domed structure of an ordered cosmos is focused on the microcosm of domestic and community space *(domus).*[19] As Apollo gazes across the earth, so the poetic eye looks outward from domestic space over the cultivated realms of Dionysus and Demeter, deities at once of reproduction and social order, equally capable of tempting the citizen away "into the mountain wilds, beyond the pale of civilisation."[20] Further from the city, beyond the cultivated realm, a pastoral zone yields to uncontrolled natural forces; it is home to shepherds and goatherds and divinities who merge with their animals: Pan and the satyrs. Further still the woods and caves of the hunting zone are home not only to wild beasts but also to Diana and her retinue of nymphs, desirable but unattainable symbols of an alternative order to the patriarchy of household and city. At the ends of the earth are places where the past's natural and social perfection and the chaos of the world's origins might both be projected.

Conceptual Description of the Globe

Achilles' shield might be taken as the founding figure of a Western carto-graphic imagination, cosmic and terrestrial space articulated by the liminal zone of Apollo's circuit. The view that the material world realizes pure, un-changing numerical forms and relations is associated with sixth-century Pythagoreans, a mathematical poetics of space. The concept of a spherical earth is Pythagorean. For Plato, the unbreachable diurnal, mensual, and an-nual regularities of celestial geometry signified a *kosmos,* the harmonious cre-ation of a rational mind. While this word conveyed the aesthetic sense of "ornament," it was also a characteristically "scientific" feature of both Pythag-orean and Platonic thought: "On this point Heraclitus, Anaximander, and all the *physiologoi* would stand united—a handful of intellectuals against the world. Everyone else, Greek and barbarian alike, would take it for granted that any regularity you care to mention could fail, and for a reason that ruled out *a priori* a natural explanation of the failure: because it was caused by supernatural intervention."[21]

Plato's strictures on the mendacity of representation indicate his episte-mological distance from Hesiod or Homer, although his own cosmology is powerfully poetic.[22] In *Timaeus* he offers a metaphysical narrative of the cos-mos as the conscious creation of a craftsmanlike intelligence. In both *Phaedo* and the final book of the *Republic* Plato describes a spherical universe, unique and temporal, attributes derived from moral propositions concerning its cre-ator. The cosmos is animated by a soul; it moves in the circular motion "most appropriate to reason and intelligence," its motion progressively less perfect as it is communicated from the pure circle of fixed stars to the planetary eclipses and finally the static earth itself.[23] This rational cosmos is animated by divine love, vital force of a living cosmos that, properly accessed and di-rected, might be the foundation of universal social harmony.

Actual measurement of the terrestrial globe in antiquity is reasonably well recorded. Anaximander of Miletus (610–546 B.C.E.) is often attributed with the first recorded attempts at global mapping, employing a calendrical gno-mon to fix a line of longitude between the poles and determine latitudes by the midday solar shadow. Anaximander's mid-latitude line passed from the Pillars of Hercules to the Taurus Mountains, intersecting at Delphos with a prime meridian determined by the mythological meeting of two eagles released by Zeus, one from either pole.[24] In Hellenistic culture Delphos was the navel of the world, the earth's *omphalos,* where a vertical *axis mundi* joined the heavens and the underworld. In a dramatic physical landscape, where a cleft on the rocky slopes of Mount Parnassos gives views across the Gulf of Corinth, earth and sky, land and sea, come together physically and mythologically. Arguably the most sacred cultural site in Greece, where the sky god defeated the chthonic Python and the Oracle delivered her cryptic forecasts, Delphos was the location of a temple to the Pythian Apollo. Delphos's mythical association with the Hyperboreans, peoples from the polar ends of the earth, beyond the source of the north wind, served to link this global center to the periphery of antiquity's imaginative world map.[25] Anaximander's native Miletus lay at the frontiers of the Greek empire, in his time a key departure point for colonial expeditions. His map was an imaginative exercise in bringing the center and the limits of the world into a coherent, controlled unity—"a global speculative project."[26] Whether Anaximander understood the earth to be a cylinder or a sphere remains unclear. Pythagorean writing assumes a spherical cosmos, and Archimedes' intellectual circle in third-century Sicily produced mechanical spheres *(sphairopoiia)* as part of their mathematical studies. The key geographical question, however, concerned the distribution of life across terrestrial space. Herodotus (489–425 B.C.E.) and Democritus (460–370 B.C.E.) debated the concept and shape of an *oikoumene,* a delimited area habitable by humans, rooted in the idea of home *(oikos).* Posited relationships between *oikoumene,* terrestrial globe, and spherical cosmos thenceforth structure the evolution of a Western geographical imagination.

Equally significant in founding Western globalism was Aristotle's fourth-century conceptual synthesis of celestial and terrestrial spheres. His texts *Physics, Heavens, Generation and Corruption,* and *Meteorology* constituted the foundations of Western cosmography from antiquity through Hellenistic, Islamic, and Byzantine scholarship and the Latin West from the twelfth century. Aristotle impresses a celestial spatiality onto the terrestrial sphere, thus generating a conceptual global geography, a template into which the more contingent ecumene is (sometimes violently) pressed. The fundamental nomenclature of astronomical geography—*axis* and *poles,* equatorial, tropical, arctic and antarctic *circles, colures,* and *climates*—is ultimately Aristotelian.[27] An unchanging celestial sphere characterized by perfect circularity in form and revolutionary movement and constituted by a fifth element, ether, and a corruptible, contingent sublunar sphere characterized by rectilinear movement and a static center share a common geometry that allows graphic representation and connection. An axis running from the celestial boreal, or arctic, pole, named for the Great Bear, passes through the center of the earthly sphere to an antarctic celestial pole, forever invisible to the Mediterranean.

Five celestial circles are projected onto the terrestrial sphere: the Apollonian solar circle describes an equator of diurnal and nocturnal equivalence, while annual solar movement produces the tropical circles of Cancer and Capricorn, Arctic and Antarctic. Angling across the celestial equator and contained within the tropics is the twelve-degree zodiacal band, whose central line is the ecliptic. These circles generate five celestial *zones,* poetically described by Virgil:

> Therefore the heaven is strictly portioned out
> And told by twelve stars to obey the sun.
> Five zones possess it; one reflects the sun's
> Perpetual splendour and perpetual heat;
> To right and left two keep the utmost flanks,
> Steel-blue, regions of ice and murky rain;
> Twixt these twain and the first, two were vouchsafed
> By God's grace to poor mortals and a path
> Was cleft between them through the midst, that here
> The stellar host might slant its rolling march.[28]

As Virgil indicates, the regions of the earthly sphere located under these celestial zones are differentially habitable. The shorter distance and longer duration of insolation in the intertropical zone renders it uninhabitable because of the heat, while the cold of the two arctic zones makes them similarly devoid of human life. Each zone is further characterized by specific meteorological phenomena, above all by winds, which in Aristotelian theory either descend from the heavens or rise from the earth and gain their characteristic directions from rotation around earth.[29] Further latitudinal subdivision produces a variable number of *klimata* banding terrestrial space. Claudius Ptolemy, writing in the first century C.E., named some twenty-eight such belts after prominent geographical features such as mountains, rivers, or cities. Each had the width described by a half-hour lengthening of daylight on the longest day. Twenty-one lay between Thule, at the boundaries of the Arctic, and the equator; fifteen in the temperate zone; and seven "anti-climates" were named south of the equatorial line. Despite such elaboration, the simplicity of the basic fivefold schema long gave it primacy in global representation.

Deductively, the global zone habitable by humans was a mid-latitude belt surrounding the earth, further distinguished by its characteristic meteorological phenomena.[30] Above all, the wind blowing from each cardinal direction has specific characteristics of temperature and humidity. Wind characteristics were known empirically, but Aristotle theorized them according to his conceptual understanding of the four corruptible elements.[31] Thus, winds from the south are connected with those from the east as hot and dry, those from the west with those from the north as cool and wet, producing an implicit meteorological model of the global surface (Fig. 2.2). Hippocrates' humoral theories related these conditions to human physical and emotional life, opening possibilities for a global anthropology in which groups living in different regions were assumed to display characteristics and conduct associated with their natural environment, an idea offering considerable justificatory scope for Western ethnographic prejudices.[32] Named winds, originating at the cardinal points of the globe would become graphically significant in the context of the complex eschatological iconography of later global mapping.

Aristotle seems to have shared with Homer the image of a terrestrial land area completely delimited by Ocean. Within this space, the habitable *oikoumene* occupied a median space whose latitudinal dimension was three-fifths that of its longitudinal extent. From textual references to actual locations, Jean-François Staszak has constructed the world map as Aristotle might have pictured it (Fig. 2.3).[33] A hemispheric "Middle Sea" at the midpoint of the temperate zone occupies also the heart of a single landmass comprising Europe, Asia, and Libya (Africa). Aristotle's resolutely culture-centric schema locates peninsular Greece at the geographical center of the *oikoumene*.

The Globe and a Common Humanity

Aristotle adopts the Apollonian perspective to "map" conceptual patterns onto a terrestrial globe, the rational mind replacing a divinity or hero in the cosmic ascent to disclose order in the contingencies of empirical geographical knowledge. His philosophical schema provides a suitable frame for Aristotle's equally conceptual understanding of humanity in the *Politics*. This is significant because the philosophical influence of Athens long dominated ideas of a cosmopolitan humanity as distinct from localized senses of *domus* and community attached to the polis.

Aristotelian social theories are outlined in the *Politics,* a key text in antiquity's evolving distinction of humanity as a distinct class of creatures from both gods and animals. From Homer's time, Greek literature reveals strong cultural distinctions: Trojans, for example, could be regarded as human, while "other," often half-mythical peoples—"blameless" Ethiopians; "happy" Hyperboreans; *kunokephaloi,* or dog-headed inhabitants of remote India; or

2.2. Tower of the Winds, Athens. Photograph by Carmen Cosgrove.

the one-eyed Cyclops—were lumped together as at best quasi-human. True humans were differentiated from gods by their mortality and from animals by their speech, upright gait, and consumption of bread (signifying settled agriculture).[34] The Greek *logos* signified both the spoken word and reason, inseparable and defining characteristics of humanity. A term such as *leleges,* applied to nonhumans, is onomatopoeic, characterizing those whose speech is unintelligible. Like most such "othering" terms, it served dialectically, at once excluding those who were Other and constructing the cultural identity of those who used it.[35] Aristotle's thinking legitimated a long-evolving cultural differentiation not only between Greeks and non-Greeks but within Greek society between men and women, adults and children, *aristoi* and other classes.

Classical notions of humanity changed over time; indeed, the "striking growth of geographical and anthropological knowledge . . . had an impact on the Greek mind comparable to the effects of exploration on the Elizabethans."[36] Ethnographic descriptions by such "objective" reporters as Herodotus sought to challenge fancy by revealing the true customs of non-Greeks, although his descriptions too often reinforced the special status attained by his compatriots. It was the Persian Wars, the rise of Athens, Alexander's conquest, and the increasing presence of imported slaves and traders in the Greek city, making Athens a *cosmo*polis, that underpinned Aris-

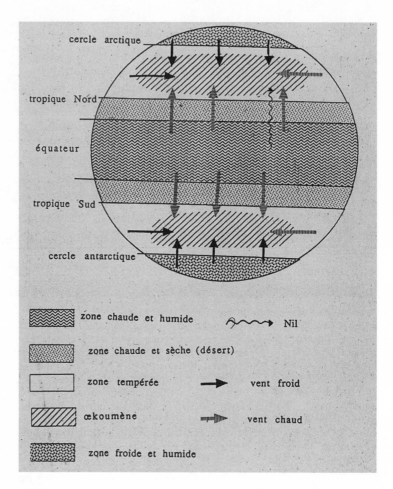

cercle arctique

tropique Nord

équateur

tropique Sud

cercle antarctique

zone chaude et humide	Nil
zone chaude et sèche (désert)	
zone tempérée	vent froid
œkoumène	vent chaud
zone froide et humide	

2.3. Contemporary reconstruction of Aristotle's meteorological model of the ecumene, from Jean-François Staszak, *La géographie d'avant la géographie: Le climat chez Aristote et Hippocrate* (Paris: L'Harmattan, 1995).

totle's categorization of humanity. In the *Politics* the philosopher establishes community (the polis) as the natural state of human existence. Only beasts or gods may choose to live outside such a state. Within it, the canonical human is an adult, able-bodied, male citizen. Although they possess *logos,* females, slaves, and "lesser" males all fall short of *telos,* the rational intentionality supposedly embodied by male citizens. Although they are *anthropos* and thus differentiated from the semihuman races at the earth's ends, barbarians are fitted by nature for slavery. However, "clanless, lawless, heartless" tribes—the Cyclops, for example—are incapable of reason, live outside society, and do not merit the designation *anthropos.* Leaving aside the contradictions and unacknowledged self-interest of this schema, so blindingly obvious to a modern observer, we should recognize Aristotle's alignment of

physical embodiment with fully human *telos.* Physical deformity and otherness are mutually reinforcing categories, mapped onto both geographical
space and living bodies. Yet *logos,* the defining quality of human *telos,* need
entail no visible, physical attributes. Thus, if reason can be disengaged from
language and physiology, an inclusive notion of humanity implicit in this
mode of thinking may be released. It was the figuring of *logos* as a principle
of rationality implanted in the human soul rather than visible in the body
that characterized Stoic philosophy and was associated especially with the
teachings of Heraclitus.

In Milan's Brera Gallery a Renaissance domestic interior has been reconstructed, one of whose panels represents two figures, one smiling, the other
shedding a tear, gesturing toward a terrestrial globe (Fig. 2.4). A work by the
artist Donato Bramante, it was completed toward the end of his stay in Milan,
in 1477–99. The righthand figure is the Stoic philosopher Heraclitus, the left
the Epicurian Democritus. Epicurians rejected Aristotle's emphasis on the
collectivity of the polis in favor of individualism, a vision of humanity consistent with Epicurus's more contingent cosmology, built upon the chance
collision of individual atoms rather than proceeding from a divine *logos.* Epi-
curianism opened a moral space for the *cosmo*politan individual such as Aris-
tippus, who rejected the polis for a mobile humanity at home in the world
rather than rooted in the locale. Faced with the globe, Democritus states:
"Since we share humanity, we should sorrow over the misfortunes of human
beings, not laugh at them."[37] Democritus's world map emphasizes the open-
ended parallelogram of the *oikoumene.* The Stoic Heraclitus's laughter, by
contrast, reflecting amusement at the folly of individual pride when the
global space available to humans is such a tiny part of the surface, emphasizes collective interest. Bramante's image, painted in the early decades of
Renaissance globalism, gestures toward the moralizing power of the globe in
an era of geographical expansion, its emblematic status in a moral debate that
threads through Western thinking. The story of Heraclitus and Democritus
denotes an openness in Greek thought to the moral implications of mapping humanity onto global space, prompted in part by the Hellenistic experience of empire and colony.

Classical Empire: Globe, *Oikoumene,* and the Ends of the Earth

The area shown on Bramante's globe is the *oikoumene* as mapped by Claudius
Ptolemy (90–168 C.E.). The *oikoumene* represents the combination of spherical geometry and empirically derived coordinates to picture the parts of the

2.4. *Heraclitus and Democritus,* fresco by Donato Bramante, Brera Picture Gallery, Milan. Stoic philosophers view the globe of human affairs, here mapped according to Ptolemy's *Geography* but extended by early European navigation.

globe known in late antiquity.[38] Between Aristotle's conceptual geography and Ptolemy's image lie two phases of imperial expansion that generated an explicit rhetoric of universalism, three centuries of connecting the image of the globe to theories of humanity. Ptolemy himself worked in Alexandria, imperial city and cosmopolis of Hellenistic learning. Ptolemy's own science synthesized and extended a lineage of speculative and empirical cosmography, from Aristotle through Euclid's second-century B.C.E. geometry, Eratosthenes' circumferential calculation of a spherical earth and measurement of the *oikoumene* along the parallel of Athens, Archimedes, Crates of Milo, Hipparchus, and Posidonius, originator of fixed armillary circles. Acknowledgment of this intellectual pedigree contributed to Ptolemy's authority and the persistence of his ideas during later periods of geographical discovery. The role of Greek and Hellenistic mapping was primarily conceptual; its images did not serve navigation or administration so much as abstract intellectual discourse. But its Apollonianism cannot be disconnected from imperial imperatives in ancient Greece and Rome.

Gaining the measure of the globe's dimensions and theorizing the relations between humanity and nature at the edges of the *oikoumene* gives imaginative significance to that implicitly imperial border zone, the "ends of the earth." Aristotle's most celebrated pupil was Alexander, whose brutal, alcohol-driven conquest was rapidly idealized within the imperial culture of antiquity as the quest of the youthful, heroic, and semidivine figure driven toward the edges of space and time, who returns with the experience and wisdom necessary for rule.[39] Most of our knowledge of Alexander derives from later, largely Roman accounts, thus reflecting the gloss of Rome's own imperial ideology. Alexander's narrative is consciously modeled on the journeying Heracles; his eastward march across India parallels Heracles' western passage, "an eschatological journey which parallels the young hero's quest for mortality and divine enlightenment."[40] Its heroic mode echoes Achilles: it is claimed that in Asia Alexander slept with a copy of Homer's *Iliad* annotated by Aristotle.[41] The altars and columns marking oriental space from Asia Minor to the Indus River valley identify Alexander's eastward expansion of Mediterranean space. Like Caesar's and Pompey's conquests in Europe and Africa, they were recorded on global maps well into the modern period. Alexander's youth, physical beauty, daring, and learning constructed an individual *telos,* a personal destiny figured as a spatiotemporal movement toward the ends of the earth. This is a culturally recurrent narrative of social spatiality in which authority is secured through the myth of heroic travel to the horizon line of a curving earth, outward from the center of space and backward in time, from the civilization of the polis to chaos at the outermost limits.[42] The simultaneous confrontation with human and natural limits defines the poetics of global empire.

The Greek term *apeiron* signified originating, elemental chaos, the opposite of cosmos. The same term was applied to the stream of Ocean that circumferenced the world island: *epi peirasi gaies,* or "at the bounds of the earth." Ocean signified both origin and end of time and space; in Ocean the sun both rose and set. Ocean was boundless, its very consistency a sluggish compound of undifferentiated elements. This eschatological, chaotic quality accounts for Ocean's role in defining the heroism of an imperial *telos.* Since Aristotle's climatic theory delimited the northern and southern boundaries of *oikoumene* and Ocean marked its western edge, the *telos* of Greek heroism was directed eastward. Alexander's defeat of Darius's Persian Empire at Issus (see Fig. 5.6), precisely the easternmost point on the Mediterranean Sea, opened a trajectory of empire toward Ocean at the eastern extremity of the *oikoumene.* Alexander's own journey was terminated in 326 B.C.E. at the river Hyphasis in Punjab, despite the appeal to his troops to press onward to

Ocean.[43] Columns were erected to parallel those of Heracles, signifying the coincidence of empire, *oikoumene,* and Alexandrine destiny.

Herodotus, recorder of the global knowledge produced by Alexander's expedition, replaces the linear *peirata* with *eremoi* to denote the boundary of the world island, implying a more diffuse, open frontier zone, an "empty space" or uninhabited waste potentially available for further human expansion. This represents a critical shift toward a more imperialist vision. *Oikoumene* "in its essential meaning can be defined as a region made coherent by the intercommunication of its inhabitants, such that, within the radius of the region, no tribe or race is completely cut off from the peoples beyond it."[44] Herodotus's usage, *eschatiai tes oikoumenes,* implies a contiguity of habitable space and the bounds of the earth. Ideologically reversing actual power relations, Herodotus regards the semimythical peoples occupying liminal, edging spaces as threatening to a civilized center (which their very otherness constructs) by virtue of their pressing to occupy the *oikoumene*'s climatic optimum. Such peoples are at once barbarous and valorous, the harshness of their environment shielding them from the physical and moral degeneration produced by the comfortable environment of the imperial center.[45]

The cosmopolitan ideal of a common humanity within the *oikoumene* is also associated with Alexander. In stark contrast to his quasi-genocidal tactics of conquest,[46] his legend has him ignoring Aristotle's advice to care for Greeks as friends and kinsmen, while adopting "the same attitude to barbarians as to animals and plants."[47] Alexander's story is frequently associated with gestures toward human unity within the rule of a universal empire, such as encouraging union between his own men and widows of defeated enemies. Hellenistic lore imagined "other worlds" that Alexander might have brought within a single orbit. "All earthly affairs are subject to one principle and one government, that all men are one people. If the divinity that sent Alexander's spirit had not speedily recalled it, one law would hold sway over all men and they would have turned their faces to a single form of justice as if towards a common light. But as it is, that part of the world which did not behold Alexander remained in darkness."[48] A discourse of universal empire through which a "Western" culture has consistently constructed itself is mapped onto a global image of light penetrating obscurity to the edges of the earth. Colonial expansion across the Mediterranean and the Black Sea inscribed Greek reason onto the settlement landscape, mapping the Aristotelian polis across the *oikoumene.* Scant attention was paid in this process to Alexandrian notions of universalism; rather, colonialism intensified Aristotelian distinctions between Greek culture and its Others. The closed social space of the planted polis contrasts sharply with the belief that but for his untimely death, Alexander would have united the world under a single polity.

Alexandrian romance was in large measure a literary product dating from Roman expansionism following Augustus's victory over Pompey at the battle of Actium (30 B.C.E.). The rhetoric of benevolent empire attributed to the Macedonian comes from Plutarch's writing four centuries after Alexander's death. An appearance of continuity masks the reconstruction of the past as a pedigree for contemporary events. Rome's defeat of Carthage and the Hellenistic empire produced a convergence of Greek and Roman cultures. By Plutarch's time Eratosthenes' remarkably accurate calculation of the globe's dimensions had paralleled his mapping of empire within an *oikoumene* stretching from Thule to the Horn of Africa and from the Pillars of Hercules to those of Alexander. The imposition of Roman hegemony over this space allowed Polybius (220–124 B.C.E.) to compose a *Universal History,* a supposedly empirical account replacing Homeric myth that was in fact a scholarly conceit of a single historical experience framed by a geography of globalized space. Polybius's work originated a tradition that would yield the medieval encyclopedia, the Renaissance cosmography, and the modern atlas.

The coincidence of Rome's imperial orbit with the *oikoumene* and its descent from the Assyrian, Median, Persian, and Macedonian empires underpinned Augustus's claim to *imperium ad termini orbis terrarum.* For all the diversity of peoples within Hellenist space, Eratosthenes' global calculations indicated that the known world was but one of four possible *oikoumenoi.* Crates' studies suggested three more: two southern and one western *(antoikoi, antipodes, peroikoi).* Practically as well as theoretically, imperial and global space failed to coincide: the *limes imperii* represented a shifting zone of threat and potential. Julius Caesar, Pompey, and Antony pushed them toward the edges of the *oikoumene.* Against the German tribes they were physically marked by stones, and across northern Britain, by walls. Fulfilling the role of liminal peoples, the *Germani* were figured in Roman discourse as both barbarian savages and perfect communitarians living in a sylvan golden age.[49] Less familiar regions offered greater imaginative possibilities, for example, Ultima Thule and the mist-shrouded Atlantic islands beyond Britain, the upper Nile, and the *antipodes,* whose location determined the complete othering of inversion of the known. Practical recognition of the globe's scale combined with the rhetorical claim to empire to the edges of the earth could only be rec-

onciled in the Apollonian conceit and realized in the oneiric space of dreams.
Whole-earth dreams flourished in the literature of late antiquity, a product
not of the edges of empire but of its center, the imperial city of Rome.

The Imperial City

"Gentibus est aliis tellus data limite certo: Romanae spatium est urbis et orbis
idem," Ovid's claim that in Rome city and world coincide, is a defining statement of imperial spatiality; the rooted culture of place yields to "civilization,"
the sphere centered through poetic rhetoric.[50] Having secured in Pompey's
defeat the Mediterranean as *mare nostrum*, or "our own sea,"[51] and established
administrative unity across Roman imperial space, Augustus formally proclaimed his empire in 27 B.C.E. He set about reconstructing Rome as a
microcosm of the earth, physically and ideologically, by initiating a process
of urban projects—forums, mausoleum, and monuments—that would be
elaborated by successive emperors. While the globe appeared on insignia and
coins as an imperial emblem, Rome was being reshaped as the embodiment
of the Augustan *telos,* a material map of the world theater.[52]

The House of Augustus appropriately abutted the Temple of Apollo on
the Palatine, and temples on the Quirinal, the Palatine, and the Capitoline
were designed to connect Augustus's rule to Rome's initial foundation. Columns *(betylos)* honoring Apollo Agyieus, protector of roads and cities, were
erected at street intersections, and laurel—an Apollonian plant—planted
throughout the city served as a metonym for Augustus himself.[53] A colossal architectural complex north of the *Campus Martius,* intended to commemorate the emperor, explicitly appropriated global space. Echoing the
calendrical spaces at Alexandria, a set of monuments surrounded the Ara
Pacis, erected by Augustus to denote universal peace. A solar calendar inscribed in brass on the ground, its gnomon an obelisk brought from Egypt
after the battle of Actium, aligned the solar shadow to this altar on the
evening of Caesar's birthdate, 22 September, thus proclaiming him son of
Apollo and embodying him as *axis mundi.* On the Mausoleum itself Augustus inscribed his personal proclamation: "Rerum gestarum divi Augusti,
quibus orbem terrarum impero populi Romano subiecit."[54]

The *Res gestae* was a sober, factual record of Augustus's subjection to
Roman imperial rule of the lands, seas, and peoples of the known earth, "a
genuine geographic survey," possibly a commentary on the *mappa mundi* that
contemporary accounts suggest may have decorated a portico close to the
mausoleum, with its iconography of Roman heroes and elogia to the Latin

homeland. The text referred to Rome itself, Italy and fourteen Italian provinces, twenty-four defeated countries and nations, four "world" rivers (the
Danube, the Rhine, the Don, and the Albis), three seas, and six cities. "The
spatial extension of [imperial] conquest to the limits of the world is shown
as being linked to the new world order that was established and guaranteed
by Augustus."[55] As in the case of Alexander, the globe's variety is unified
through the *telos* of a single individual. The term *orbis terrarum* enters Roman
rhetoric as a claim that the borders of Roman rule coincide with those of
the habitable earth itself, a claim of ecumenism that will recurrently articulate Western globalism, its iconographic expression the image of the globe.
Roman imperial usage effaces a clear distinction between a celestial and a
terrestrial globe, signifying cosmocratic authority over both spheres. On a
coin of 56 B.C.E. a globe records Pompey's victories in Africa, Spain, and
the East; in his theater Pompey is sculpted holding a globe in his left hand.
A stone statue of Julius Caesar at the Capitoleum has him stepping on a
bronze female figure of the *oikoumene* crowned by a tower, a figure to be
resurrected in early modern Europe to represent the continent's humanized
status, in contrast to the personifications of other continents, which are accompanied by plants and animals. The bronze globe held in the hand of Constantine's colossal statue from the later years of the Western Empire would
become one of the *mirabili* of pilgrim Rome, a direct iconographic passage
from antiquity to imperial papacy.[56]

   As Seneca's statement opening this chapter indicates, Rome appropriated
the world materially as well as symbolically. Like Athens and Alexandria
before it, imperial Rome was a cosmopolis. Its administrative, mercantile,
and, above all, consumption activities attracted peoples and goods promis-
cuously from across the empire. Rome consumed the globe. Just as Aristotle and Alexander had brought biological and zoological specimens back to
Athens for study at the academy, so both Augustus and Nero commissioned
exploratory expeditions. Posidonius, Strabo, and Pliny, among others, turned
the imperial city into an intellectual center of calculation for a tributary
universe, while wilderness at the edges of the earth was produced for popular display as though it were a circus animal or a gladiatorial contestant.[57]
Strabo's *Geography* outlines a global science, inventorying the human habitation, alteration, and exploitation of the earth. Strabo's subject is rule, and
he is, appropriately, one of the few authors to offer an explanation of Rome's
shift from *libera res publica* to *imperium*.[58] Discussing marble in the mineralogical section of his *Historia naturalis,* Pliny the Elder makes Rome the
eighth wonder of the world, an architectural and engineering marvel, col-

lecting and displaying the treasures of global nature and art at the center of Italy: "the mother of all lands, chosen to gather the scattered empires, to make manners gentle, to gather all the grating barbaric tongues of different nations into one common language, and to make . . . a common fatherland for all the peoples of the world."[59] Assurance of the world's unity under Roman rule is a powerful theme in Pliny's writing; domestic spaces too reassured their occupants that global nature was subject to imperial authority. The garden room at Augustus's wife Livia's villa was decorated with a frieze of plants and birds taken from the four corners of the earth in a complex allegory of order bestowed by the benign authority of Rome: "Like the *Ara Pacis* in its urban context, the Garden Room clearly celebrates the garden world of perpetual victory and perpetual peace."[60]

Roman globalism peaked under the emperor Hadrian a century after Augustus. The city was now a cosmopolis of more than a million people. The promiscuous mixture of commercial, political, and religious functions that originally occupied the Forum Romanum was replaced by purely rhetorical uses that extended into an intimidating complex of monumental architectural spaces: the imperial forums. Hadrian himself symbolized Rome's eternal nature in the Pantheon, a universal structure of drum and dome modeled on the shared geometry of cosmic, terrestrial, and corporeal space to house the gods of all the imperial territories at the heart of Rome. Light through the circular opening in the dome projects the revolving heavens into the heart of the city. Hadrian's mausoleum, ancient Rome's greatest domed structure, created a similar vertical axis descending to a *mundus* where buried flesh propitiated divinities of the underworld.[61] These two buildings, imperial umbilicus and tomb, respectively, related city and world to the imperial destiny through the body of the emperor. At coincident edges of *imperium* and *terrarum orbis,* Hadrian's great wall against the Picts echoed the *pomerium,* the plowed furrow marked by animal entrails with which Roman *coloniae* were bounded at their founding.

An implicit tension exists between the ordered unity of Roman imperial rhetoric and the cosmopolitan mixing, even confusion, of peoples and languages, practices and products, that characterized life at the imperial center. Augustus's *Res gestae* inscribe a fixed geography of *nationes,* a natural and permanent ethnographic order ceding sovereignty to universal empire. By contrast, poets such as Juvenal expressed the fear that Rome's identity was threatened by its cosmopolitanism, that foreigners were taking over the city and that native Romans were losing their home and familiarity of place, that its citizens were becoming alienated and corrupted. For Stoics such as Cicero and Seneca, fears of an increasingly cosmopolitan world might be calmed by global dreams of universal community.

Dreaming Global Humanity

The idea of human unity inherited from antiquity by a modern West "was not Greek, but Graeco-Roman." It emerged not from localized community but from the sense of difference and the commerce among peoples "spread over the various countries of the inhabited world—mankind, in fact, viewed geographically."[62] For this perspective Stoicism offered an appropriate vision, stressing the insignificance of an individual located within the vastness of a cosmos and thus the necessity for human modesty. This theme is consistently associated in Latin literature with the idea of seeing the globe. Like Platonism and Cynicism, Stoic philosophy "bears witness to the diffusion of the exercise of *katakopos,* that 'view from above' carried across the earthly globe that leads to a relativizing of human values and achievements but also to the adoption of an intellectual perspective, the spiritual gaze that discloses the beauty and order of the world beyond the shimmering of appearances and the limitations of human knowledge."[63] A Stoic "whole-earth" literature connecting universal empire, the centrality of Rome, and the subjection of the sphere to the authority of a single God-man flourished in the Augustan age. Commonly little more than imperial rhetoric, it could at times be implicitly critical of official ideology. Oneiric writing, exploiting the simple trope of flying sufficiently high to allow a panoptic view of the whole sphere, dates back to early Greek authors. Homer's description of Achilles' shield might be taken as an example; Hesiod's *Theogony,* with its continual references to Zeus's all-seeing eye, is effectively an imagined flight above the earth; and we have quoted Plato's description of the multicolored globe in *Phaedo.* The mythological Daedalus, emblem of human challenge to divine craftsmanship, realized the dream of flight and the empowering capacity to see from above the earth's lands and seas, the fate of Icarus signifies the consequences of overweening hubris, common tropes in the whole-earth literature of antiquity.

The title of Anaximander's scientific treatise *Periodos ges* connects his circumscription of the earth to a hybrid genre of whole-earth description at once scientific, poetic, and geographical.[64] In these narratives (of psychological space as much as material space) "the terrifying *apeiron* of primal chaos was banished to the outermost edge of the globe, where flowed the stream of Ocean, so as to permit a more formal ordering of its central spaces; and this outer region was decisively fenced off from the rest of the world, by both natural impediments and divine sanction."[65] Pytheas's quasi-mythical voyage beyond the Pillars of Hercules at the time of Alexander was a *periplus,* a term that came to denote both a travel account and a graphic description of the order of the earth. Above all, *periplus* signified a *visual* survey of the earth's imaginative or actual bounds through navigation or flight, thus encompassing literary writings by Cicero and Seneca as well as the Hellenistic geographical texts upon which they drew.[66] As Claude Nicolet points out, "The world vision—which varied according to the authors, the level, the style, and the context—as conveyed by all types of geographical writings is neither Greek nor Roman: it is one and the same."[67]

The archetypical *periplus* description, hugely influential in medieval and early modern thought, occurs in the final book of Cicero's *De republica* (51 B.C.E.). Scipio, conqueror of Carthage, dreams of ascending above the earth. The text is partly a cosmogony from which "humanity" emerges as the soul of creation, "given a soul out of those eternal fires which you call stars and planets, which, being round and globular bodies animated by divine intelligence, circle about in their fixed orbits with marvellous speed."[68] Scipio's vision embraces the vastness of terrestrial space, registering the globe's climatic zones and noting the lack of communication between peoples living in the four *oikoumenoi*. In the Stoic mode, he remarks the diminutive scale of Rome's actual empire: "The earth seemed to me so small that I was scornful of our empire, which covers only a single point, as it were, upon its surface."[69] Cicero thus gives cosmographic significance to Scipio's victory over Carthage, a globalizing moment in Roman imperial history. His text combines humility in the face of the earth's vastness with a sense of imperial *telos,* the goal of universal conquest pushing the bounds of empire toward the ends of the earth. Cicero's "Dream of Scipio" captures the combination of cosmic spatiotemporal unity, common humanity, willed individual destiny, and the geographical limits of the earth—*eschatos* and *telos*—that shapes the imperial incubus of the Western geographical dream. As Romm points out, "The fact that geographic revelation is now placed in the mouth of a world-bestriding general rather than a god gives it a new significance."[70]

Other authors used the whole-earth theme more skeptically to connect the ends of human destiny to the edges of space and time in a critique of imperial overreach. Seneca, also a Stoic, cautioned against the misuse of the winds for navigation and criticized a reckless ambition for global empire:

An age shall come, in later years,
When Ocean shall loose Creation's bonds,
When the great planet shall stand revealed
And Tethys shall disclose new worlds.[71]

In lines from the same poem quoted at the beginning of this chapter Seneca represents globalization as a perversion of geography. Empire has confused the "natural" and anthropological order; the order imposed by Roman authority cannot prevent an eventual return to originating chaos, ever present at the edges of space.[72] In the later empire the Icarian theme of hubris was commonly related to views of the earth, for example, in the satirical writings of Lucian (125–200 C.E.), where Charon, the boatman of the underworld, asks Hermes, the flying messenger of the gods, to give him a view of the living world. Since it would be sacrilegious for Charon to visit the heavens, he has to be content with a vantage point on the surface. The two partners pile up four mountains and sit upon the topmost one contemplating the earth: "I see a vast stretch of land, and a huge lake surrounding it, and mountains, and rivers bigger than Cocytus and Pyriphlegethon; and men, tiny little things! and I suppose their dens . . . [but] these cities and mountains look for all the world like a map. It is *man* that I am after."[73] Although Charon and Hermes fail to attain Scipio's synopsis, their height is sufficient to erase geographical and ethnographic differentiation. Only the common aspects of humanity remain, and these are cause for contempt. Humans are insignificant, foolish, and vain, and their inability to recognize their faults arises precisely from their incapacity to gain the objectivity and distance that the global perspective provides.

Themes in Classical Globalism

Global representations in antiquity provided numerous themes to be reworked within later Western imaginations. The mathematics of the sphere and the geometries of celestial space stimulate images of order and perfection projected onto the global surface as a speculative geography. Gnomonically generated lines of latitude and the measured distribution of lands and seas around the Mediterranean generate concepts of fixed climates and counterposed hemispheres. Celestial motion imprints cosmic time onto terrestrial space and denotes harmonies that are sensed rather than seen. Stoic philosophy found in the contemplation of the globe from above a powerful vehicle for articulating its reflections on the nature of life, its critique of

human affairs, and its disinterested search for moral precepts. By late antiquity the idea of an Apollonian spiritual journey had became a philosophical commonplace, figured, as in a first-century mosaic in the Annunziata
tower at Naples, by linking the human eye to the sphere of the globe.[74]

Closely related to these precepts is the vision of the human body as a
global microcosm. Although more fully developed in the Christian era, this
vision was already implicit in Platonism, especially in *Timaeus*. Augustus's
architect-engineer, Vitruvius Pollio, whose writings conveyed the architectural principles of imperial Rome to the modern world, draws a strong parallel between the greater and lesser spheres of cosmos and human body,
using their measure as the proportional module for designing individual
buildings, their decoration, and the city as a whole.[75] The Vitruvian model
of the human body, although gendered, links a canonical, centered human
form to the spherical geometry of globe and cosmos, offering a theoretical
resolution of the logical tension between the undifferentiated space of the
sphere and the centered placement of humans on a global surface, a tension
that pits universalist ideals of common humanity and heroic *telos* against
xenophobia and localism.

Global images and meanings in antiquity work across both fertile mythological narratives and rationalist calculation through number and geometry.
Ptolemy's cosmographic synthesis, a key vehicle for transporting that knowledge into the early modern world, exemplifies this. A manual for constructing mathematically correct material spatial representations, in its regional
tabulae the *Geography* locates the cities, altars, and columns that marked imperial rule across the *oikoumene*. The absence of actual globes and maps from
antiquity makes it difficult to verify claims about the actual production, circulation, and consumption of such artifacts. Certainly, in both textual and
architectural representation global concepts and narratives articulated a language of empire. The ultimate human destiny *(telos),* incorporated in the
semidivine figure of the hero or emperor, is to secure the wisdom and authority offered by actual or imaginative passage to the ends of the earth. Monumental spaces such as Augustus's imperial Rome offer the most enduring
mappings of this vision, myth made concrete in the stone geometry of imperial forums, obelisks, arches, columns, mausoleums, and domes. In the aftermath of empire Rome continued to serve paradigmatically and rhetorically
the Western connection of *urbs* and *orbis,* signifying global order imposed
over chaos and wilderness through the realization of imperial destiny. The
ruined stones of Rome could not, of course, record the anxieties of globalism and empire—its promiscuous mixing of peoples and things, the com-

merce that mobilizes the "natural" order and harmony associated with the idea of the sphere.

Such tensions might appear to resolve themselves from the distanced, Apollonian perspective. The dream of human flight sufficiently high to offer a global perspective is an enduring theme of Stoic philosophy, in which seeing attains the dual sense of sight *(noein)* as an empirical check against speculation, an assurance of truth in the description of the earth, and of vision, the capacity for poetic grasp beyond mundane or earthbound daily life, for a truer, imaginative knowledge.[76] This is the implication of whole-earth literature from Cicero, Lucan and Seneca which offers its male heroes their destiny in synoptic vision. Their *telos* combines an imperialistic urge to subdue the contingencies of the global surface with an ironic recognition of personal insignificance set against the scale of globe and cosmos.

THREE *Christian Globe*

. . . Only do thou
Smile, chaste Lucina, on the infant boy,
With whom the iron age will pass away.
The golden age in all the earth be born;
For thine Apollo reigns. Under thy rule,
Thine, Pollio, shall this glorious era spring,
And the great progress of the months begin.

. . . .

See how Creation bows her massy dome,
Oceans and continents and aëry deeps:
All nature gladdens at the coming age.[1]

Virgil's fourth *Eclogue* was long regarded among Christian scholars wishing
to redeem classical culture from the moral stigma of paganism as an anticipation of their own sacred narrative by the greatest of the Latin poets. The
poem was composed as a paean to Augustus's Apollonian divinity. Its rein-
terpretation is evidence of medieval Christianity's cultural hybridity. Virgil
proclaims a cyclical revolution in universal history, heralding a second golden
age, a renewal of natural perfection. This turning of the spheres that govern
time and space is occasioned by the birth of a God-man, an Apollo.[2] Christians came to regard such a world-historical renewal, or redemption, as the
work of Jesus Christ, in Judaic terms the second Adam, from whose birth a
global calendar would ultimately be calculated. Christianity was officially
sanctioned within the empire by Constantine's Edict of Milan in 313 C.E.,
and its dominance of Latin culture was assured as the faith of the empire by
Theodosius in 380 C.E. In 325, at the Council of Nicaea, the first "world-
wide" gathering of bishops universalized the Christian faith, fixing a combined solar and lunar date for the key festival of Easter[3] and determining a
single creed, whose text reflects Christ's transition from savior of the humble
to imperial Apollo. Western cosmography and European globalism would

henceforth be deeply colored by Christian historical narrative and claims to universal redemption.

If Christianity inherited Rome's official soul and its imperial claim to universal dominion, the geography of papal Christianity never coincided with the spatial compass of Rome's authority. Christianity always sustained strong attachment to the scale of household and community and to local church autonomy. In the course of the first Christian millennium an ecumene of faith was effectively extended north and west to the limits of antiquity's known world through such missionary expeditions as Columba's in northern Britain and Severinus's in Danubia.[4] The churches they founded sustained only tenuous links with Rome, and their practices reflected sensitivity to local cultural traditions.[5] Only slowly and hesitantly were the centralizing goals of the papacy enforced at these margins. For the Greek-speaking eastern Mediterranean, the collapse of the Western Empire contradicted Rome's claim to spiritual centrality. The domain of "Eastern" and North African patriarchal churches reflected the fault lines of the former empire and political division between Byzantium and the fragmented Latin *nationes*.[6] In North Africa and western Asia a competitively universalist Islam rapidly expanding from the sixth Christian century swamped Christian communities in Ethiopia, Yemen, the Indian subcontinent, and the steppes of Central Asia, isolating them from papal influence. Within the circumscribed spatiality of the Latin imperial imagination, "east" and "south" came thus to be imagined as distinct global spaces, to be marked as such in their representations of the earth. Located in this "other" part of the Latin globe were the two centers of intellectual life in antiquity—Athens and Alexandria—and the sacred territories of Christian faith and history—Eden, the Holy Land, and Christianity's *axis mundi,* the holy city of Jerusalem. The textual and iconographic record of the classical globe passed into the hands of Arabic and Byzantine scholars. Its reentry into Latin consciousness was slow and intermittent from the twelfth to the fifteenth century, its eventual outcome a transcultural product of Arabic, Greek, Jewish, and Latin scholarship. It was thus a transcultural globe with which Latins would eventually transgress the oceanic bounds of Christendom and the classical ecumene.[7]

Christ, Apollo, and the Imperial Orb

In picturing cosmic and terrestrial spheres Christianity drew on both Hellenistic and Judaic precedents. The New Testament was a formally sanctified collection of texts written in Greek but drawing upon Judaic monotheism for fundamental features of its cosmology. A single creative intelligence consciously present in the world was not foreign to Greek thinking. It is implicit in both the mythological Zeus and in Plato's *Timaeus*. Neoplatonic Roman cosmologies worked with the idea of a "world soul." But the idea of a celestially located, "almighty," patriarchal divinity banishing sinful human creatures from an original Edenic landscape and later selecting a specific human group for special favor and allocating to them a geographically precise "promised land" defines Jewish belief. Judaic monotheism combined with the Greco-Roman imperial vision to give Western Christianity a uniquely global character. The Roman popes consciously adopted imperial ritual and rhetoric after the fall of the Western Empire in 476, and their sacramental consecration of feudal monarchy from the time of Pépin and the Frankish kings sustained in the fragmented west of Europe an illusion of sacred imperial continuity. Thus the emperor Henry II's blue robe decorated with the constellations represented "the universe over which its wearer reigns at the center of a cosmos revolving around a divine axis"; in the king's hand the orb represented the terrestrial earth, and his eight-sided crown prefigured the celestial city at the end of time.[8] The twelfth-century popes Gregory VII and Innocent III completed the transformation of the papacy into an absolutist and universalist monarchy located at the juncture of celestial and terrestrial space. Their crusading struggle against Slavic paganism and Islamic "Antichrist" identified eastern Others against which Christendom's cultural and territorial unity could be forged, at least rhetorically.

The physical *body* of the risen Christ is central to this theological geography. Embodying the globe or world map is a recurrent feature of medieval Christian iconography. In the *Liber divinorum* (1492), for example, the Father literally *encompasses* elemental and celestial spheres, his head reaching beyond even the enclosing frame of the universe, while the Son's body corresponds to the dimensions of the earthly sphere itself. Trinitarian doctrine proclaims God as simultaneously celestial Father, physical Son, and divine Spirit or breath of life.[9] As coeval son of God and Man, Christ can represent the true microcosm, the physical realization of the form and motion of the created universe. The "body of Christ" introduced radically different spatialities to spiritual practice than were present in classical antiquity. The Nicene Creed required belief in the corporeal ascension of Christ into the heavens and a universal physical resurrection at the end of time, making celestial space as physically significant as terrestrial space, and for Christians spiritually more important. Divinity could also be physically introduced into human time and space. Christianity's central ritual of eucharistic communion involves physical consumption of the divine body in the form of bread and wine. In both Judaic and Greco-Roman religious practice sacrificial ritual had been tied to specific locations sanctified by either *genius loci* or historical event. In Christianity the sacrificial space is mobile, realized in the performance of the eucharistic ritual itself. The consecrated host itself may be transported across space, carrying sanctification with it.[10] This mobility connects to Christianity's claim to encompass the world: like the Apollo of Virgil's fourth *Eclogue,* the risen Christ declares a redemption beyond that of a single chosen people, to a global "fallen nature." Thus Christ and Apollo are consistently conflated in Christian images, such as the thirteenth-century Ebstorf map, in which the crucified body of Christ incorporates the terrestrial sphere, his head and limbs marking the four compass points.[11]

Encompassment goes beyond the physical body of Christ. Like a mason, master builder, or architect, the Christian God holds the compass that measures the fabric he will create. Platonic and Pythagorean thought offered Christians the concept of idea preceding its material realization and of number and measure as the means for materializing the idea.[12] As fleshly Son of God, Christ physically encompassing the world signifies divine materialization of the cosmos. The compass signifies Christ's will, his *telos,* to refigure through death and resurrection the original creation. Thus Christianity allocates the *telos* of both created nature and humanity to a divinely human individual and extends it across universal space and time. At the first doctrinal gathering of Christians, the Council of Jerusalem, it was determined that Christian redemption, while fulfilling Jewish prophecies for the "chosen" people, extended also to gentiles. Christianity thereby embraced a concept of universal humanity: children of Adam, to be unified through redemption regardless of parentage or geographical location. Unlike Aristotelian or Ciceronian concepts of human unity, or even the idea of Roman citizenship extended to all free inhabitants of the empire in 212 C.E., the Christian view of humanity is potentially uncoupled from essential ties to a place-bound *community.* Christian community comes explicitly through *communion,* which, although it is practiced in specific places, is not locationally restricted. From Paul's writings above all, Christianity adopted a globalizing imperative, constructing an ideal of *mission* extended to the ends of the earth. In Beatus of Liebana's eighth-century commentary on St. John's Apocalypse, each apostle is allocated a specific region to evangelize, but to Paul is given the world, unconsciously capturing Christianity's tensions between local autonomy and imperial centralism.[13] The cosmographic Christ synthesizes many of the themes that had collected around the image and idea of globe and cosmos in late antiquity since the Latin church projects cosmopolis *ad termini orbis terrarum*.

Spiritually, belief in Christ as God made man unifies the corruptible and incorruptible parts of the Aristotelian cosmos. Christ's physical encompassing of the terrestrial globe speaks to his corporeal presence on earth, while the cosmographic image places Godhead in the incorruptible realm. The risen Christ's physical ascension brings the whole world within his Apollonian embrace. Christ's wilderness temptation, narrated in Matthew 4:8–10, emphasizes the implications of this perspective. Given sight of "all the kingdoms of the world, and the glory of them" from "an exceeding high mountain," Christ is offered global dominion in return for an acknowledgment of a distinct authority over the elemental sphere. The visit from "ministering angels" that follows his refusal indicates the unity of the cosmos and the primacy of the celestial.

The Apocalyptic Globe

Angels occupy a similar position in Judeo-Christian belief to that of the giants and heroes of Greco-Roman mythic history, predating humans and still present as liminal figures in cosmic space. The angelic hierarchy could be mapped onto spheres of the supercelestial realm corresponding to planetary spheres in the subcelestial cosmos, and angels appear at the margins of late medieval world maps, sharing cartographic space with personifications of the winds at the cardinal points of terrestrial space. Antiquity had personified Mediterranean winds, setting them imaginatively at the compass points of the earthly globe, signs of cosmic order. Homer and Hesiod associated individual winds with the seasons, Aristotle and Hippocrates with environmental humors.[14] Unseasonal or uncharacteristic winds, a warm, wet northerly wind, for example, signified dangerous alteration to the pattern of nature. Wind towers such as that in the Roman agora at Athens (see Fig. 2.2) were cosmological constructions, a measure of order in time and space, for winds were deemed to originate beyond the limits of the ecumene.[15] Like Ocean itself, their elemental nature connected them to spatiotemporal origins and ends. Unsurprisingly, within Christian discourse winds and angels became conflated as eschatological figures: Satan's fall inaugurates a new age; Gabriel's expulsion of Adam and Eve from Eden ends the first age of humans, while his Annunciation opens the age of Redemption; the end of time and the physical return of Christ to the earth are announced by angels. The key text here is the Revelation to John the Divine, the last in Christianity's canonical testament and a powerful source of cosmographic and cartographic global images.

As a cosmographic narrative, Revelation to John reverses the Apollonian perspective, opening with a terrestrial perspective on the heavens: "He cometh with clouds and every eye shall see him." Its key event is the opening of seven seals and consequent apocalyptic destruction—an earthquake, stars falling from the heavens, "every mountain and island . . . moved out of their places."[16] The opening of the final seal, which signifies ultimate destruction, is initiated with a vision of "four angels standing on the four corners of the earth, holding the four winds of the earth, that the wind should not blow on the earth, nor on the sea nor on any tree."[17] In the late-twelfth-century Duchy of Cornwall map fragment the trumpeting angel is also a wind head, located at the limits of both time and space. The passage raised the question of reconciling the rectangular earth it implied and the earth's known circular form. Graphically, placing the circle within a square frame resolved the problem and opened liminal spaces in the architraves where wind heads, angels, spirits, or the four beasts described in Revelation could mediate the geometries of physical and metaphysical space.[18]

Revelation reverses the order of universal history inherited from classical and Old Testament cosmogonies by moving from culture back to nature. The imperial city, Babylon, "clothed in fine linen, and purple, and scarlet," and rich through commerce and war, "in one hour is made desolate" (18:16). The geography of destruction expands "to deceive the nations that are in the four corners of the earth, and Gog and Magog," until "the first earth were passed away and there was no more sea" (18:19). The final verses describe "a new heaven and a new earth" (20:8) in a narrative that remaps the heavens in the image of classical social space (Fig. 3.1). The celestial Jerusalem is observed "from a great and high mountain" (21:10), foursquare, oriented to the cardinal points, and encompassed by angels, "according to the measure of a man" (21:17). A mineralogical description of the city is followed by one of a verdant "river of life" (22:2), replete with trees and fruits. The text projects the spatial perfection of initial creation and its paradise garden onto the New Jerusalem.

Revelation to John crystallizes Christian concerns with origins and ends and with the liminal characters located at the edges of time and space. It maps a spatiality and sense of boundary that differ from Roman imperialism's demarcated *limes*. The absence of enclosing, linear frontiers characterized thinking about the terrestrial globe itself, less actual and less known than the celestial spaces described in the testaments and patristic writings.

3.1. The angel of Apocalypse oversees the celestial city in a nineteenth-century Cretan icon, Preveli Monastery, Crete. Photograph by the author.

Centers, Edges, and Humanity on the Christian Globe

Augustus's and subsequent emperors' architectural and engineering projects in Rome had proclaimed the city's global centrality in marble and tufa. Acceptance of Christianity as the empire's official religion and the city's formal association with Peter as "Vicar of Christ" legitimated Roman claims to ecclesiastical seniority. Frankish success in establishing dominion in the Latin West and Charlemagne's claim to divine viceregency opened space for a papal claim to universal spiritual ascendancy. Rome's bishop declared spiritual dominion over the spaces of empire to the ends of the earth, *unus pastor et unum ovile.* The universal missionary *telos* implicit in Christian belief and the claim to imperial authority *ad termini orbis terrarum* coincided in Rome and were realized through conscious attempts to regulate autonomous churches at the edges of empire. But numerous political, moral, and physical obstacles stood in the way of this globalizing project. Rome's centrality may have been proclaimed by Pope Zacharias's mid-eighth-century world map on the wall of the Lateran Palace, but it was challenged by both Jerusalem and Byzantium. Jerusalem was the site of universal redemption, and its epiphanic centrality in a Christianized global geography accounts for its recurrence as the organizing center of T-O maps, so named because of their form of three rivers dividing the continents *(terrarum)* within a circular earth *(orbis).* Byzantium, to the East, was both politically and culturally a more vital imperial center, styling itself variously as the new Rome and the new Jerusalem. From the sixth century, Christian claims to global spiritual authority were undermined by an equally universalizing Islam, whose territorial control rapidly spread to include Jerusalem itself. Drawing upon Hellenistic and Alexandrine sources, supplemented with their own intensive observation, Islamic scholars from Toledo to Central Asia produced cosmographic images as aids both to *falsafah* (natural philosophy) and to contemplation in the esoteric, Neoplatonic tradition.[19] While these differed little from their Aristotelian and Ptolemaic precedents and would be the principal vehicle for the occidental transference of that tradition, *Ka'ba* images gave centrality to Islamic holy places, constructing an alternative spatiality to that of medieval Latins.

Medieval Rome's aspiration to global centrality was reflected in the city's landscape of fragments and memories scattered among fields, pastures, vines, and wasteland within the compass of the Antonine Wall and loosely coordinated by the cross described by its four basilicas. St. John Lateran, "mother and head church of the world," had been Constantine's imperial gift to the papacy on his departure for Byzantium. Reviled in St. Augustine's *City of*

God as a place of evil, much of the monumental evidence of pagan Rome was destroyed during the fifth-century invasions. As part of a global missionary strategy, it was Pope Gregory the Great (590–604) who adopted a syncretistic response to paganism, Christianizing rather than destroying pre-Christian remains. The domed Pantheon was thus rededicated to Mary and all the martyrs, its cosmic form unchanged and its universalizing sanctity brought into the Christian era. The severed hand and great bronze globe from the fifty-meter imperial colossus that had stood before the Colosseum were placed among other marvels, including the equestrian statue of Marcus Aurelius, outside the Lateran, thereby refiguring the papal palace as a temporal center of spiritual imperium. From the eighth century the "marvels" of Rome were described in guides for pilgrim visitors to a Christianized pagan metropolis, the site of martyrdom and sainthood, whose continuous history from antiquity served to substantiate the Carolingian and later Ottonian papacies' claims to a global inheritance. *Renovatio Roma* became a widely subscribed goal of both secular and religious leaders, culminating in Frederick II Hohenstaufen's powerful revival of imperial discourse in the mid-twelfth century.[20]

But even conceptually, papal claims to global centrality and authority faced critical contradictions. Two of the most influential texts inherited from late antiquity, Pomponius Mela's *De situ orbis libri tres* and Macrobius's *Commentary on the Dream of Scipio,* introduced to medieval Christians the possibility of other continents beyond a single ecumene.[21] Whether there was one southern or three hemispheric landmasses, the existence of *anti*podean, "other" continents raised significant theological issues. First, if they were inhabitable, as even a simplified conception of climatic zones (familiar from Latin encyclopedias popularizing Aristotelian lore such as Pliny's *Natural History* and Seneca's *Natural Questions*) suggested they could be, what creatures might they contain? Genesis made no mention of these, giving humanity a single parent, and even the three postdiluvian progenitors were allocated only three parts of the world island. If antipodean inhabitants were "sons of Adam," how could they have reached across the ocean and the torrid zone? Beatus of Liebana's *Commentary on the Apocalypse,* written in expectation of the world's ending in 800 C.E., narrates Christianity's universal apostolic mission. It is Beatus who allocates to each apostle a specific region for conversion, while to Paul, he claims, was given the world. Maps associated with Beatus's text appear to include a fourth continent beyond Europe, Africa, and Asia, reduced to minor proportions, and locate within it the *sciopedes,* shading themselves from the intense heat with their single gigantic foot (Fig.

3.2). It is unclear precisely what this space represents, as Isidore of Seville, one of Beatus's key sources, uses antipodes to denote both the inhabitants of a fourth continent and monstrous races in Africa, whom Augustine had explicitly proclaimed descendents of Adam.[22] Possessed of souls, such creatures demanded redemption. In the twelfth-century *Liber floridus* of Lambert de Saint-Omer, a clearly distinct antipodean continent having dimensions equal to those of the northern ecumene, is left blank, a terrifying absence in that part of the globe "unknown to the sons of Adam."

One response to the question of antipodean inhabitants space was to populate it with one or more of the "monstrous races." These owe their origins to early Greek myths, to the *kunokephaloi,* or dog-headed people of the Scythian regions, for example, familiar from the fourth century B.C.E. Like angels, in appearance they were part human and part nonhuman, and their location, at the geographical edge of chaos, was correspondingly liminal.[23] The first-century *Natural History* of Pliny the Elder was the secular text that most shaped medieval European images of the natural world.[24] Following a summary of Aristotelian cosmology and a geography of the ecumene, Pliny's seventh book deals with anthropology, describing a classification of semihuman creatures. Pliny associates these, like other natural wonders, with the East, and specifically India, the furthest latitudinal margin of Roman space. Later bestiaries, such as Julius Solinus's third-century *Collection of Remarkable Facts,* popularized the monstrous races, always locating them at the edges of terrestrial space. Mouthless *astomi* and dog-headed *cynocephali* belonged in India; cave-dwelling *troglodytes* and the *anthropophagi* who would come to dominate the European ethnographic response to discovery belonged in Africa and were also found in Scythia. The southern, fourth continent had its own race of *antipodeans,* whose feet pointed the opposite way to those of northern humans.[25]

The monstrous races recur in medieval universal histories, encyclopedias, and in *mappae mundi.* They are common too in the cathedral tympanum decoration, where they are placed beyond earshot of Christ's words. Corporeally, monsters were Other to the perfect microcosmic body of Christ and of his church, their deformities attaching them to the elemental earth rather than to the celestial heavens. Their participation in the scheme of universal salvation for humanity thus remained an open theological question, to become one of practical urgency with transoceanic European expansion in the late medieval period.

Hebraic testament and Aristotelian political philosophy provided authority for a Christian understanding of a single humanity within the *orbis ter-*

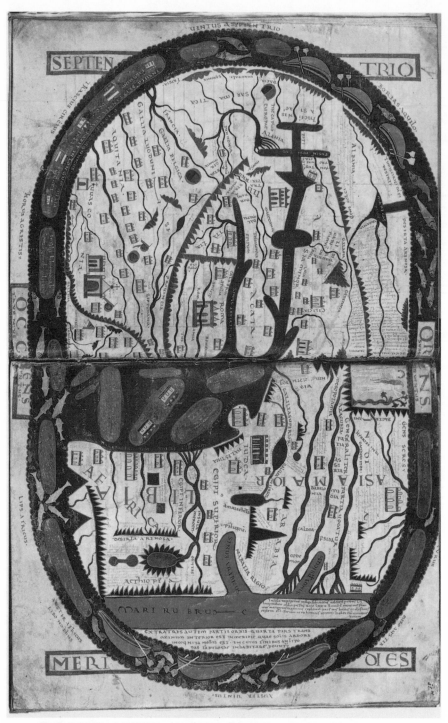

3.2. The habitable world and the antipodes according to Beatus of Liebana's *Commentary on the Apocalypse*. Photograph from Bibliothèque nationale de France, Paris.

rarum. Book 9 of Genesis deals precisely with the allocation of peoples to the various regions of the earth's surface after the universal flood, narrating ethnographic diversity through the story of Babel, a failed challenge to the ocular authority of a singular God, defeated by divine introduction of mutually incomprehensible tongues and the subsequent scattering of humanity over the earth.[26] The account in Genesis provided structure to a continental human geography within the ecumene and gave specific significance to the territory divinely allocated to the Jewish people, the *terra sacra* with its *axis mundi* at Jerusalem, for which the eleventh-century papacy launched two centuries of crusading warfare.

Medieval texts adopted the global division of a world island into three distinct spaces, a geographical expression of trinitarian thinking. In his *Ethymologiarum,* in the section titled "De orbe," Isidore of Seville (560–636) describes a circular terrestrial surface surrounded by a "great ocean." The world island is divided into three unequal parts—Europe, Asia, and Africa (Libya). Asia occupies half the total area and is separated from the other continents by the Nile and the Black Sea. Each continent is allocated to the descendants of one of Noah's sons: Asia to Sem, Africa to Cham, and Europe to Japhet. The specific linkage of continents to distinct human lineages, although implicit in Genesis, is essentially a creation of Christian thought. To Africa is allocated the progeny of Noah's sinful son, opening the way for the Christian association of African peoples with darkness and sin. Scholastic thought, with its penchant for elaborate speculative glosses, developed further associations from the spatial allocation of peoples. Honore d'Autun's *Imago mundi,* for example, reads the tripartite social divisions of feudal Europe onto the continental division of humanity: the Asian descendants of Sem are the world's kings and priests *(liberi),* Europeans descended from Japhet are its soldiers *(milites),* while Cham's people in Africa correspond to slaves *(servi).*[27]

The outer continental margins offered space for locating variously imagined, remembered, feared, and desired phenomena that would become familiar in the narratives of medieval and early modern travelers and romancers. In Africa they included the kingdom of Prester John, a trace in memory of Ethiopian Coptic Christianity separated from Rome and Byzantium by Islamic expansion; in the west, the mythic Golden River; and in the south, the Mountains of the Moon, source of the Nile. In Asia, they included Mount Ararat, resting place of Noah's Ark, various points of Alexandrine romance, and the land of the biblical giants Gog and Magog. Somewhere in Asia's furthest eastern reaches was located the terrestrial Paradise, whose four rivers—

the Tigris, the Euphrates, the Pishon, and the Gihon, flowed out in the cardinal directions. Paradise remained inaccessible to humans, however, surrounded since the Fall of Adam, according to Isidore's account, by fiery walls.[28] Theologically inspired phenomena shared these spaces with the fragments of a legendary antiquity—columns and altars from Alexander's eastern conquest, curious and marvelous animals from the bestiaries, and the various semihuman members of the monstrous races.

The margins of space and time coincided at the circuit of Ocean surrounding the ecumene. Strabo, Pliny, and Pomponius Mela, all known to early medieval scholars, suggested that Ocean might have been circumnavigated by the ancients. Accounts of heroic voyages into Ocean itself, from semimythical Greek and Carthaginian voyages beyond the Pillars of Hercules and supposed Roman visits to Thule, six days beyond Britain, to heroic Christian missions such as St. Brendan's passage into the Atlantic, decorated the medieval world continent with a necklace of actual and imagined islands, visible on such maps as the *Liber floridus.* Some of them, such as Brazil or the Island of Women, quite apparently located desires, anxieties, and fantasies to be confronted very directly in the historical process of European oceanic expansion. Descriptions of the Mediterranean archipelagos through which the Latin West, Byzantium, and Islam navigated, traded, and fought contained some of the most precise medieval mappings, transcultural productions that served to enhance the imaginative place of the island within the European geographical consciousness and gave rise in later medieval years to the island book, or *isolario.*[29]

From about the eighth Christian century two locations competed as centers of the Latin *orbis terrarum:* Rome was the physical site of a remembered, mythologized temporal authority; Jerusalem represented a transhistorical site of divine authority where events had occurred *in illo tempore.* Seen from Rome, the universal space of Christian salvation was tightly circumscribed; territory over which Christ's earthly vicariate held moral sway, when mapped onto the globe that his body embraced, was more circumscribed than that proclaimed by Augustus's geographers and poets. Christian space also lacked the intellectual coherence provided by Aristotelian natural philosophy, Pythagorean number, and Euclidean geometry. The slippage between geographical and temporal eschatology and the uncertain, anxious spatialities of medieval European globalism helps explain the pervasive power of apocalyptic vision. Commentators have identified this anxiety in medieval European travel writing, for example, in Egeria's *Peregrinatio ad terram sanctam,* with its visions of a timeless place where the sacred events of redemption and the promised celestial city coexist at the spiritual center of the world, and in St. Adamnan's seventh-century sensationalizing of the Holy Land in *De locum sanctis.*[30] The spatiality of such texts leaves traces in the travel writings of Marco Polo and even Columbus.[31] A center ruined, the greater world fragmented and surrounded by fantasy islands with monstrous chaos beyond, was "the self-image of a culture quite literally scared of its own shadow," constructing its normalizing core from fragment and memory and projecting alterity onto its assumed margins. "The alienation of Europe [that] is bespoken in almost all of the rhetorical strategies (and silences) that present themselves for analysis" is perfectly apparent in the global images, whose numbers and elaboration increase steadily from the Carolingian eighth century.[32] From this period we can begin to speak of a shared European *visualizing* of a global geography.

Global Maps: Texts and Images

In the mid-thirteenth century Paul of Venice claimed that "without a world map, I say freely that it is not merely difficult, but impossible to imagine or comprehend all that the texts, whether sacred or profane have to tell us about the sons of Noah, sons of their sons, the four monarchies and all the kingdoms and provinces."[33] To be understood, temporal processes had to be located in secular space; narrative alone was no longer sufficient. Paul signals an end to the long textual domination of Latin global discourse and a growing emphasis on visualization. The earliest texts whose exegesis was aided by world maps were either biblical, such as John's Revelation, or late classical compendia of knowledge emerging from the encyclopedic tradition of Pliny's *Natural History,* such as Macrobius's *Commentary on the Dream of Scipio,* Pomponius Mela's *Cosmographia geographia,* and Isidore of Seville's medieval texts *Etymologiae* and *De natura rerum.*

Macrobius had used Cicero's dream passage as the starting point for a detailed Neoplatonic reflection on the numerical and musical significance of classical planetary theory for an understanding of a "world soul," apprehended in the "universal dream." According to Macrobius's hugely influential dream theory, developed in the *Commentary on the Dream of Scipio* and in his *Saturnalia,* a set of allegorical interpretations of classical myth, the *somnium* is a "global" dream through which we come to know the nature of both the self and the world of things. Macrobius's world of things is Aristotelian, described by the thirty-six *klimata* of the frigid zone, the thirty of the temperate zone, and the twenty-four of the torrid. The known ecumene is but one of four continents, and Macrobius suggests, in the Stoic tradition, that since humanity is limited to only a tiny part of a fourth of creation, striving for glory in this world is foolish. The immortal soul merits our true concern, for through it we may attain divinity. Although almost certainly not Christian, Macrobius's Neoplatonism and invocations of humility attracted Christian attention. Neoplatonism was the only pagan religious system truly to survive into the Christian era, and its emphasis on a "world soul" has connected its texts quite consistently to global visions and images.[34] From the ninth century on, Macrobian manuscripts were commonly illustrated with a circular world map, its terraqueous space surrounded with as many as sixteen wind heads. Two continental islands are separated by the dual currents of the equatorial *alveus oceani,* flowing in opposing directions around the landmasses. Each hemisphere is divided into three climates; the whole of Europe, with the exception of insular Britain and Thule, is located entirely within the northern temperate zone. With the invention of printing, Macrobian maps would be among the earliest and most commonly reproduced, often incorporated into cosmographical texts such as Johannes Eschuid's fourteenth-century handbook of judicial astrology.[35]

An alternative image appears in illustrations of Isidore of Seville's seventh-century encyclopedic hybrid of pagan and Christian knowledge. Over a thousand manuscripts exist of his *De natura rerum* and his *Etymologiae.* Divided according to the system of liberal arts, Isidore's work summarizes human knowledge, including Aristotelian cosmography and geography. Within thirty years of Isidore's death the work was illustrated, and it was thereafter copied continuously into the age of print. Isidorian *terrarum-orbis* maps ignore global sphericity to concentrate upon the surface distribution of lands and seas upon a tripartite earth. Three continents, allocated textually or graphically to Noah's sons, are separated by the Mediterranean, the Nile, the Red Sea, and the Don and encircled by the Ocean Sea. The earthly paradise in eastern Asia and the Sea of Azov, regarded as a possible gulf of the eastern Ocean, reveal the growing influence of Islamic geography on Christian global knowledge. Sphericity and climatic zonation are dealt with in distinct sections of Isidore's text, commonly illustrated by cosmographic diagrams composed of connected circles.[36] A companion astrological text *De responsione mundi et de astorum ordinatione* develops Isidore's cosmology by means of an annular system of interlocked circles that illustrate the unity of creation and the relations between microcosm and macrocosm. Elements and humors intertwine within a single, harmoniously proportioned system of influences running through creation.

These eighth-century sources provide graphic foundations for later medieval *mappae mundi*, self-sufficient illustrations of the majestic order of creation consciously systematized in the work of the cathedral schools and universities that emerged throughout the Latin nations during the twelfth century. The information they conveyed synthesized Aristotelian natural philosophy and classical and biblical authority with the growing volume of empirical information brought back to Europe by crusaders and travelers. They picture the world as a marvel of creation, an optical wonder. Large-scale thirteenth-century *mappae mundi* such as the Ebstorf, Hereford, and Vercelli maps work as integral objects without accompanying text, *monstrances,* or visual proofs, of the variety and wonder of the world.[37] Their metric size renders these *imagini mundi* public sermons on the order of creation and the global reach of Christian redemption. Like the vibrant stained-glass windows, frescoes, mosaics, and statuary that decorated late medieval cathedrals with parallel cosmic images, later *mappae mundi* testify to an assertion of the power of the visual in relation to texts in shaping a Western geographical imagination.[38] All three of these *mappae mundi* place Jerusalem at the spatial axis of a circular world. At both center and margins the graphic *orbis terrarum* remained a place of miracles, remembered, promised, and desired more than known empirically.

Islam, Cosmography, and the Christian World Image

In the new millennium an increasingly political papal concept of Christendom, sharpened by both the 1054 schism with the Greek church and rivalry from secular claims to Roman territorial legacy in the West, increased the emphasis on Rome as Christianity's spiritual *axis mundi* over the claims of Jerusalem and Byzantium. The spatiality inherited from a patristic interpretation of Christian redemption for a single humanity, expanding globally, *ex urbe ad orbem,* confronted the crystallizing geography of a bisected ecumene across which East and West confronted each other in geospiritual, crusading conflict even as the flow of goods and ideas between them continued. Christendom's Eurasian border was a zone of active, often brutal contestation. In both Teutonic crusading in the Slav lands and the raids on settled Europe by Tartar horsemen, "the Eastern and Western limits of the *orbis terrarum* finally confronted each other in the flesh."[39] Marco Polo's travels between 1260 and 1295 reported an empire at the far outer edges of Asia; he confirmed in his *Milione* the marvelous and monstrous nature of ecumenical margins. His narrative was predicated upon a "belief in a cosmos where density, reality, proportion, and self-sufficient being clustered at the geographical 'center' while the margins sported the parodic or merely significant life whose image in art we call the grotesque," yet it claimed the authority of an eyewitness account.[40]

This discursive differentiation of East and West as opposing Others masked common inheritances. Not only do Islam's and Christianity's sacred texts draw upon the same Judaic sources and share a single Creator, if not divine incarnation in the material world, but Islamic scholars had sustained "an unprecedented movement of translation into Arabic of scientific and philosophical texts, either directly from Greek originals or from intermediate Syriac versions," elaborating the cosmographic knowledge of ancient Hellenism and transmitting it to medieval Europe in the transcultural centers of Islamic Spain.[41] Islamic science sustained the Ptolemaic image of the cosmos, firmly grounded in Aristotelian physics, and had shared the impact of late classical Neoplatonism, yielding intense philosophical speculation about the soul and the place of the individual within the cosmos. Islamic texts, elaborately illustrated with geometrical diagrams of celestial bodies and their movements, influenced cosmographic representation in the Latin West in the years following the Latin translation of Euclid's *Elements* in the early twelfth century, and by 1175 Ptolemy's astronomical *Almagest* was being translated into Latin in Toledo.[42]

Geometrical diagrams of the four terrestrial elements, the planets, fixed stars, and the crystalline sphere beyond illustrated a model of the universe whose validity was "never questioned by Muslim philosophers and scientists." They provided the graphic foundations of Western cosmography over three centuries, from John of Hollywood (Sacrobosco, d. 1244/56) and Albert Magnus to Petrus de Alliaco (Pierre d'Ailly, 1350–1420), Regiomontanus, and Julius Hygenus. The culturally hybrid nature of cosmography in the late medieval West, heavily dependent on astronomical and navigational science developed by Arabic and Sephardic Jewish authors such as Albumasar, Alcibiades, Alfarganus, Aben Ezra, and Zacharias Lilius, would subsequently be partially obscured by the historiography of the post-Copernican scientific "revolution." Islamic culture, drawing heavily on Neoplatonic Hellenistic sources to interpret cosmic influences, generated a wealth of complex geometrical diagrams to illustrate "the complex network of correspondences between the two major spheres of creation—the world of manifest, corporeal entities and the world of hidden, spiritual beings,"[43] intended to produce alchemical results and zodiacal prediction. Their introduction into the

West profoundly shaped the metaphysical significance of global thinking well into the seventeenth century.

Islamic images of the terrestrial globe identified seven latitudinal climatic zones in the Northern Hemisphere with unknown regions polewards of the seventh climate and below the equinoctial line. Repeated representation of this model with little concern for geometric accuracy of latitudinal measurement emphasized "the centrality, in the inhabited world, of the fourth region where the administrative center of the Islamic empire was situated."[44] An alternative system, of Persian origin, divided the world sphere into seven regions of equal size with a central circular zone surrounded by six others. It was based on ethnographic rather than physical considerations, the seven circles representing seven kingdoms into which the earth had been divided at the first rains, each characterized by a distinct people, language, and mores. Iran was the central *kishvar,* surrounded by India, China, Turkey and the lands of Gog and Magog, Asia Minor and the Slav lands, Egypt and Syria, and Arabia/Abyssinia, and the Latin West was placed at the outer margins of global space.

The period 1125–1250 saw a sustained translation of Arabic, Greek, and Hebrew texts into Latin, focusing the attention of late medieval thinkers on the sphere as a geometrical concept, a physical entity, and a metaphysical subject. In the newly founded universities at Padua, Bologna, Paris, and Oxford, as well as among the religious orders, intensive cosmographical study of an ordered creation placed emphasis on sight and light, combining astronomy, geometry, and optics within the scholastic curriculum. In the context of limited instrumentation, commentary and elaborate glossing of canonical texts was more significant than critical inquiry based on comparing texts with empirical observation. The fundamental authority remained Aristotelian.[45] Sacrobosco's textbook *De sphaera,* composed about 1220, was an immensely successful summary of Aristotelian cosmography, transmitting its principles through literally generations of European students and used instructionally into the seventeenth century. Manuscript and printed editions illustrate a rigidly concentric, geocentric universe with its seven planets, firmament of fixed stars, and sphere of the Prime Mover; at its center the elemental earth is marked with the seven Ptolemaic climates. For Sacrobosco, the spherical form of creation and its major parts can be attributed to three causes:

> likeness [*similitudo*], convenience [*commoditas*], and necessity [*necessitas*].
> *Likeness* because the sensible world is made in the likeness of the

archetype, in which there is neither end nor beginning; wherefore, in likeness to it the sensible world has a round shape in which beginning and end cannot be distinguished. *Convenience,* because of all isoperimetric bodies the sphere is the largest and of all round shapes the sphere is the most capacious. Wherefore, since the world is all-containing, this shape was useful and convenient for it. *Necessity,* because if the world were of other form than round—say trilateral, quadrilateral, or many-sided—it would follow that some space would be vacant and some body without a place, both of which are false, as is clear in the case of angles projecting and revolved.[46]

Within this spherical cosmos, Sacrobosco divides the earthly globe into four quarters by the equatorial line and a meridian joining the poles and dividing East from West, although its location is not specified. In his opinion only one of these quarters is habitable. The climatic zones are generally illustrated in Sacrobosco's diagrams as chords drawn across the circle, in the tradition of the text's Islamic sources, but a tricontinental land area centered on the Mediterranean may be roughly sketched into the northern ecumene, and in early printed editions this map may give way to a landscape view with cultivated fields, settlement, and church, representing the habitability of the temperate zone (Fig. 3.3).

Sacrobosco provided for scholars across Latin Europe a summary course in a science of cosmography, a science whose practical value lay in fixing the ecclesiastical calendar, especially for complex movable feasts such as Easter and for navigational purposes. *Ephemerides,* tables of astronomical positions, assisted navigation, astrology, and alchemy. Sacrobosco's text was the theoretical foundation for printed *ephemerides,* such as Regiomontanus's enormously successful almanac, the *Calendarium,* which tabulated decades of celestial positions for the late fifteenth century. A hand holding the astrolabe, the opening image in both manuscript and published copies of Sacrobosco's text, also illustrates editions of other authorities on the sphere—Ptolemy and Regiomontanus—and as light and optics emerged as the foundation of late medieval science, so the armillary came to signify philosophy itself, while the iconography of the astronomer was "orientalized." A book, an astrolabe, and perhaps an astrological chart accompany the sage, while the stereotype of the magus in European art (e.g., in paintings of the Epiphany) shares such Asiatic signifiers as a full-length coat of many colors, with tassels and jewels, with the personification of Asia herself. Physics and metaphysics shared a common emphasis on spherical geometry and vision. As

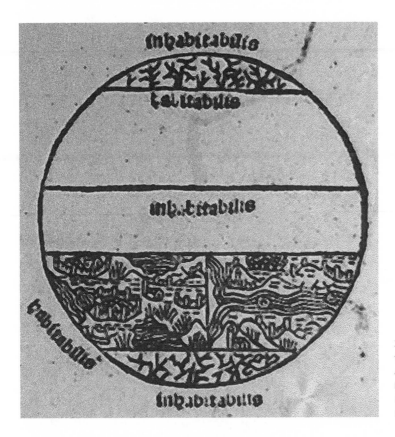

inhabitabilis

habitabilis

inhabitabilis

habitabilis

inhabitabilis

3.3. The habitability of the earthly globe according to Sacrobosco's *Sphaera mundi*. By permission of the British Library.

I discuss more fully in the next chapter, the Portuguese crown's fifteenth-century adoption of the armillary as its key heraldic device signified an act of transcultural globalization in which claims to both physical and metaphysical space combined with Portugal's very specific relations with the Islamic world.

Globe and Surface

Geographical representations in the Latin West until the fifteenth century illustrated a disk-shaped earth. In Nuremberg, Regiomontanus's mid-fifteenth-century text *Of Triangles*, heavily dependent on Arabic sources, discussed the techniques for constructing globes, anticipating their actual production by Martin Behaim and Albrecht Dürer at the turn of the new century. Projective mapping of the terrestrial sphere on a two-dimensional surface awaited the arrival in Florence of Ptolemy's *Geography*, but more than a century ear-

lier in England Roger Bacon (c. 1219–92) had described a coordinate system
for fixing locations by means of latitude and longitude, opening up, if not
realizing, the methodological possibilities for constructing a graticule. His
Opus maius, Opus minus, and *Opus tertium* together represent a scholastic attempt to reconcile his Christian intellectual inheritance with the newly available works by Aristotle and Euclid.[47] The ability to fix terrestrial locations
on a spherical globe or planisphere would eventually transform the uncertain spatiality of medieval Christianity, allowing qualitative knowledge of the
places that occupied those locations.

The absence of both constructed globes and the projective geometry necessary to construct a planisphere does not imply confusion in the medieval
West about the sphericity of the universe or any belief in a flat earth at its
center. The spherical form of the earth had been assumed as common knowledge by Latin encyclopedists such as Pomponius Mela and Macrobius. Among
the church fathers, only the fourth-century Lactantius had challenged this
conclusion. In the absence of direct access to Aristotle's texts, early medieval
images of the earth derived from commentaries long since disconnected
from their sources served as the knowledge base even for supposedly eyewitness accounts such as John Mandeville's *Travels.* Speaking of the Holy
Places, he points out: "That is not the EST that we clepe oure EST on this
half, where the sonne riseth to us, for whanne the sonne is EST in the par-
tys toward paradys terrestre, it is thanne mydnyght in oure parties on this
half for the roundness of the erthe."[48] Sacrobosco's *De sphaera* offered various theoretical proofs of sphericity, illustrating, for example, by means of
sightlines how a port may be visible to an observer atop the mast of a ship
at sea, while it remains out of sight to another on deck.[49] The perfection of
the earth's sphericity was a matter of debate; as a sublunar, elemental phenomenon it was, in Aristotelian terms, corruptible. The earth was often claimed
to be "round," *tending towards* the spherical but, unlike the heavens, not attain-
ing perfect sphericity. Its form was interrupted by the uneven appearance of
land and sea, so that the globe could assume fantastic shapes, revealing more
about the mind than about the earth; Christopher Columbus, for example,
figured it in the form of a breast complete with polar nipple.[50]

The globe's sphericity continued to raise theological questions long after
Islamic and Greek science had moderated Latin dependence upon late classical commentaries and encyclopedias.[51] Pierre d'Ailly, on whose *Ymago mundi,*
of 1410, Columbus and Martin Behaim both drew, was among those who
disputed the form of the outermost elemental sphere.[52] A number of the
wide range of *questiones* through which scholastic thinkers sought to bring

classical and biblical authorities into concordance directly concerned the
globe's form and geographical order.[53] One concerned the very existence
of a regular surface pattern of lands and waters. Medieval representations
consistently show the four elements as concentric spheres according to their
Aristotelian order: earth as the heaviest element is entirely enclosed by water,
water by air, and air by fire; fire, as the lightest and most refined of the elements, remained unseen at the sphere's outermost regions. Yet, by definition
lands were higher than the oceanic watery surface. They could thus be interpreted as imperfections natural to the sublunar sphere. Alternatively, God
had intervened to uncover land from waters on the Third Day specifically
to allow for human habitability. Among the more sophisticated answers was
that offered by the fourteenth-century thinkers Jean Buridan (1300–1358)
and Nicole Oresme (1320/25–82). Their answer turned on the proposition
that the centers of gravity of the two perfect spheres of earth and water did
not coincide, thereby leaving a part of the surface of the former exposed
above the latter: the area of the ecumene. Such a solution had the added attraction of removing the theological problems connected with a possible
antipodean ecumene. Noncentric spheres simply disposed of a southern continent and its complications for the lineage of the human race and the universality of Christian mission while reconciling Aristotelian physics with
Christian belief in an earth made for human habitability.

An equally knotty problem had to do with the globe's size. Although this
had been calculated with remarkable accuracy by Eratosthenes, later classical writers had differed in their estimations. Key medieval authorities, including both Roger Bacon and Pierre d'Ailly, argued for a relatively short diameter, with the implication that the oceanic distance from Europe to the
Chinese empire, which Marco Polo had reached overland, was short enough
for skilled navigators to cross by sailing west into the Ocean Sea. Such debates
assumed practical rather than merely scholastic significance with circumnavigation and the consequent explosion of globe making early in the sixteenth century.

Despite scholastic dispute, the image of a spherical earth at the stable center of concentric spheres, encompassed, penetrated, and animated by the
Trinity, enjoyed unchallenged acceptance in the Latin West. "All who learned
to read and write absorbed at least the skeletal frame of scholastic cosmology"; its principles were pictured in mosaics, stone carvings, and stained glass
in churches, abbeys, and cathedrals, the medieval world's centers of popular
culture.[54] The seven figures illustrating the 1483 edition of *Ymago mundi,*
printed at Louvain (Fig. 3.4), offer a compact graphic summary of this global

image in the years immediately preceding oceanic discovery. They illustrate d'Ailly's written summary:

> Immediately after [or below] the sphere of the moon, the philoso-
> phers place the sphere of fire, which is the most pure there and invisi-
> ble because of its rarity. Just as water is clearer than earth and air than
> water, so this fire is rarer and clearer than air, and so is the heaven [or

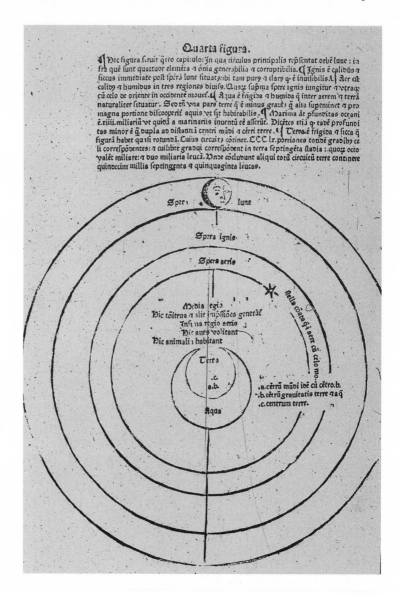

3.4. The globe of creation according to Pierre d'Ailly's *Ymago mundi*. By permission of the British Library.

sky] rarer and clearer than fire, except for the stars, which are thicker [or denser] parts of the sky so that the stars are lucid and visible. Afterwards is the sphere of air, which encloses water and earth. This is divided into three regions, one of which is the outermost (next to fire) where there is no wind, rain or thunder, nor any phenomenon of this kind, and where certain mountains, such as Olympus, are said to reach. Aristotle says that the starry comets appear and are made there and that the sphere of fire in this supreme region of air with its comets are moved simultaneously with the heaven [or sky] from east to west. The middle region [of air], however, is where the clouds are and where various phenomena occur, since it is always cold. The other [and third] region is the lowest, where the birds and beasts dwell. Then follow water *and* earth, for water does not surround the whole earth, but it leaves a part of it uncovered for the habitation of animals. Since one part of the earth is less heavy and weighty than another, it is, therefore, higher and more elevated from the center of the world. The remainder [of the earth], except for islands, is wholly covered by waters according to a common opinion of philosophers. Therefore, the earth, as the heaviest element, is in the center or middle of the world, so that the earth's center is the center of its gravity; or, according to some, the center of gravity of the earth and also of water is the center of the world. And although there are mountains and valleys on the earth, for which reason it is not perfectly round, it approximates very nearly to roundness. Thus it is that an eclipse of the moon, which is caused by a shadow of the earth, appears round. They say the earth is round, therefore, because it approximates to roundness.[55]

This dense verbal mapping emphasizes the significance in scholastic episte-mology of written authority compared with that of the visual image, despite scientific emphasis on geometry and optics. The fifteenth century saw a sharpening tension between these two modes of representation, signifying an emerging cultural division. The claim that "societal influences . . . did not shape the cosmological assumptions and judgements of scholastic natural philosophers," that "what did shape them was the bookish tradition they inherited and to which they responded,"[56] sits uneasily with evidence of an increasingly wealthy and acquisitive society in the courts and merchant cities of Latin Europe. Here a new elite attained and communicated status through both the material expressions of learning, employing scholars and humanists to search out ancient texts and commissioning paintings and illustrations of erudite subjects, and through the consumption and display of "worldly goods," thus encouraging the transcultural exchanges that accompanied merchant trading over greater global distances.[57] The Roman Church remained the unchallenged arbiter of intellectual truth, but the vast outpouring of printed editions of scholastic works in the first decades of printing began to place growing strain on the exercise of its authority.

Although the earthly globe was the central and most significant sphere in a spherical universe, and although optical science could map and perhaps even channel the forces passing between it and the unchanging supercelestial orbs within the sphere of fixed stars, direct communication with sacred space beyond the fixed stars was possible only by prayer and transcendence. The scale, mutability, and imperfection of the terrestrial sphere encouraged Christian humility and gratitude for divine redemption. For all their significance within creation, humans were confined in life to the earth's surface. Christ alone could attain the synoptic vision of the whole earth and see to its outermost boundaries, but his vicar in Rome was charged with the task of extending Christendom to those monstrous margins that bounded space and time. Commercial contact beyond peninsular Europe indicated that lucrative possibilities for acquiring and consuming the earth's exotic luxuries too lay at those margins. Armed with the theoretical and technical opportunities offered by Ptolemy's *Geography,* fifteenth-century cosmographers would pursue these graphically, navigators experientially. Both followed Apollo westward, promoting and transforming the Christian globe.

FOUR

Oceanic Globe

Between the Zone where Cancer bends his clutch
(To the bright Sun a Bound Septentrionall)
And that which for the Cold is shun'd as much,
As for the Heate the middle Zone of all,
Prowd EUROPE lyes: whose North, and all parts which touch
Upon the Occident, have for their Wall
    The Ocean; and, with unreturning Waves,
    Her South the Sea-Mediterranean laves

. . . .

The LUSITANIAN KINGDOM here survay,
Plac'd as the Crown upon fair EUROPE's Head:
Where (the Land finishing) begins the Sea,
And whence the Sun steps to his watry Bed.
This, first in Arms (by gracious Heav'n's decree)
Against the filthy Mauretanian sped:
    Throwing his out of Her to his old Nest
    In burning AFFRICK; nor there let him rest.[1]

The Lusiads, published in 1572 by the Portuguese poet Luis de Camões, narrates Vasco da Gama's exploratory voyage of 1497–98 from the Tagus River to the Cape of Good Hope and across the Indian Ocean to Calicut, opening an oceanic trading route between Europe and India.[2] Camões's literary model was Virgil's *Aeneid,* with da Gama cast in the role of its eponymous hero, departing, like Aeneas from Troy, to realize a heroic destiny by founding a maritime empire. Blending classical literary cosmography with geographical knowledge from a century of discovery, Camões retrospectively stages the globe as a maritime surface for Europe's *telos,* driven by lust for trade and the imperatives of Christian mission and structured imaginatively by recollection of antiquity.

In canto 3 Camões describes the Aristotelian ecumene, from frigid Hyperborea to the torrid zone of Libya and from the Sea of Azov to Portugal itself, on the very margins of both Ocean and Christendom. This is a *Euro-*centric rather than *Mediterra*nean construction, allocating the physical regions of the continental peninsula to the Latin *nationes.* This continental narrative generates a westward-driving geopolitical logic. Its eastern frontiers patrolled by Islamic empire, Christendom looks to its "crown," Portugal, which, located at the meeting point of antiquity's inner and outer seas, is driven by a cartographic logic to pursue empire westward into Ocean. Two graphic conceits are at work here: Iberia figures as Europe's head, crowned by Portugal, while the Atlantic figures as the greater Mediterranean, over which the Latins will reenact a classical narrative of conquest and colonization. Renaissance mapmakers commonly represented Europe as Christ's body, with Portugal as his divine crown. Such anthropomorphized maps also pictured the Mediterranean's northern and southern coasts as male and female, respectively, in a "marriage of the world," counterposing and uniting Orient and Occident. The idea of Ocean as a greater Mediterranean was more literary than cartographic and more Spanish than Portuguese, but it became a powerful trope in discovery narratives, gaining some cartographic support from Magellan's circumnavigation, whence Tierra del Fuego came to be represented as the northern tip of a great antipodean continent closing the southern Atlantic (see Fig. 1.2).[3]

Camões's Europe is engaged in a world-historical struggle for the light of Christian truth against the darkness of unbelief. Portugal's destiny follows Apollo west to illuminate the globe. The material effects masked within this discourse, constructing modernity, transforming and destroying "other" worlds than Europe's totalizing globalism, are familiar.[4] Here I trace the rhetorical construction of a European imperial discourse with specific attention to the idea of an oceanic globe. The voyages that Camões celebrates were commercial, and, rather than a colonial empire, they strung a necklace of trading enclaves along the continental coastlines. Camões's own *periplus* narrative structure reflects this coastal perspective on the globe more than a celestial and synoptic view. Practically, oceanic globalism altered both Europe's imperial vision and its constructions of humanity.

Christian Empire

The worst news of all for Christendom spread west across Europe in the summer of 1453. After decades of evident weakness, Byzantium had fallen to Mehmed II's Ottoman army, echoing for readers of Virgil—including probably Mehmed himself—the fall of Troy, which had launched Aeneas's wanderings. Byzantium held a unique claim to Rome's imperial heritage, with a direct lineage from Constantine and the cultural and material worlds of antiquity. Since the Arab sack of Rome in 846 Christianity's universalist claims had never appeared so weak. Crusading dreams of Latin control over Jerusalem and the Holy Places faded irrecoverably, while confessional divisions within Christianity sharpened after the 1439 Council of Florence failed to resolve its great schism. For Rome, these events signified a *translatio imperii,* a transfer of imperial heritage, to the West. Despite papal injunctions, contact with the eastern Mediterranean continued, even increased, trade fueled by competition, intellectual exchange by the movement of intellectuals and the new circulation of printed materials. A transcultural circulation of merchandise, gifts, and intellectual and artistic treasures included both Islamic sultan and Latin prince in competition for the patrimony of antiquity and the creative talent of their day.[5] But the Aegean trading colonies of Genoa and Venice, including Rhodes, Cyprus, and eventually Crete, fell under Turkish rule, so that a European spatiality, in the making since the division of Rome's Eastern and Western Empires, intensified from the mid-fifteenth century. Latin Christendom's doctrinal boundary with the "Orient" actually bisected the Mediterranean basin from east to west. Granada, western Europe's last Islamic state, fell to Spain in the month that Columbus sailed; in the same decade Portugal's attempt to defeat Islam in Morocco was routed. Geopolitically, Ottoman and Habsburg contested a shared imperial heritage; conjuring Alexander and Julius Caesar, respectively, as heroic models, their rivalry was symbolized in sumptuous palaces at the Mediterranean's gateways, Topkapi Saray on the Golden Horn and Charles V's imperial additions to the Alhambra at Granada. Charles V revised inscription from the Pillars of Hercules, *Plus Ultra,* emblazoned everywhere on his palace, celebrated his authority over oceanic waters but not those separating Europe and Africa.

In terms of a tricontinental world, rounding Africa might still be described as coasting. It did, of course, involve oceanic navigational maneuvers in the southern mid-latitudes in order to reach the Cape, namely, the arcing swing, or *volta,* away from land into open ocean that Columbus would put to such effect in the North Atlantic.[6] Circumnavigating Africa certainly revolutionized European global thinking, accessing by sea an Indian Ocean deemed closed by the ancients. But it recast rather than fundamentally transformed the image of a world island. To be sure, the *alveus oceani* indicated on Macrobian maps disappeared as a geographical phenomenon (although it was long reproduced in printed editions of Macrobius), and the torrid zone was shown to be passable. A southern ecumene could no longer be conceived as a mirror image of the northern. But as an extension of a known landmass, its coasts connecting to known parts of Asia, and as a morbid physical environment holding little material attraction for European penetration and colonization, Africa did not revolutionize the Western geographical imagination as a "new world" in the way that the Americas and Australia did.[7]

Oceanic navigation differed from coasting, not only in calling upon specific technical skills of shipbuilding and navigation—the skills of the *marinheiros*[8]—but also imaginatively. It was explicitly global and surficial. Oceanic navigators placed faith in the theoretically predicted size and surface of the sphere, as Columbus did in the writings of d'Ailly and Regiomontanus and in his own study of ancient and modern texts and maps. *Marinheiros* and navigators came to recognize a surface geography of the oceans themselves. No longer undifferentiated matter, the oceans were inscribed by unique patterns of winds and currents that did not correspond to the simple compass winds of Mediterranean culture. The oceanic globe connected mathematically to the heavens by means of ocular instruments, establishing continuities with received knowledge, although stars and constellations scarcely known to antiquity rose higher as the mariner sailed south. The classical spheres were familiar to educated seamen from simple cosmographies, and to break the enclosing frame offered the joint prospect of material wealth and worldly fame.

European discovery in the Atlantic dates from the fourteenth century, pioneered by Genoese and Portuguese seamen. The Canary Islands, antiquity's Hesperides and Ptolemy's prime meridian, were appropriated in 1336, followed by Madeira and the Azores by 1400. Such discoveries were readily accommodated into the image of a world continent surrounded by a scattering of islands. Despite sometimes strong resistance, enslavement and demographic collapse of their indigenous populations followed, anticipating the experience of all the oceanic regions beyond the ancient world continent. While they did not fully conform to received images of monstrous races, the cultural and linguistic differences of their native peoples, their insularity, their lack of Christian faith, even their tactics of warfare, raised questions in Latin minds about the humanity of these island peoples and the implications of their conversion.[9]

Late-fifteenth-century Atlantic landfalls—by Christopher Columbus in the Caribbean, by John Cabot in 1497 on the Labrador coast, by Pedro

Álvares Cabral in 1500 on the Brazilian coast, and by Cortez Real in 1501 on Newfoundland—were initially recorded as islands in the Ocean Sea. Indeed, as late as 1523 an informed correspondent could refer paradoxically to "those islands in the Oceanic Sea which are called Terra Ferma."[10] The imaginative grip of a single landmass surrounded by archipelagos was strong. Landfalls could be presumed to be on new-found islands fringing the ecumene's eastern edges. The earliest surviving terrestrial globe, commissioned in 1490 from Martin Behaim by Nuremberg traders wishing to participate in the emerging African trade, perfectly illustrates this image of a tricontinental main encircled by a satellite belt of islands. Acknowledgment of a fourth continental landmass was slow and contested. Columbus himself touched the mainland of Central America only in 1498, and he died insisting on its Asian identity. Writing in 1507, Martin Waldseemüller, while indicating a western landmass separating Europe from Asia with an unbroken American coastline, seems ambiguous about its status: "The earth is now known to be divided into four parts. The first three parts are continents, while the fourth part is an island, inasmuch as it is found to be surrounded on all sides by the ocean."[11] Before Magellan's 1520–22 circumnavigation, a world map such as Francesco Roselli's 1506 conical projection could exploit the convenient fact of having to cut global space along a meridian to leave ambiguous the question of America's continental identity. The De Bure globe of about 1528 makes the Americas a peninsula of China, while the Englishman Edward Wright's 1599 global maritime chart still implies a single ecumene.[12] America's physical separation from Eurasia was not fully confirmed to Europe until Vitus Bering's second expedition, in 1741. The extended process of Europe's mental embrace of a fourth continent allowed the idea and image of the oceanic island to play a powerful role in its global imaginings.[13]

In the cultural context of theocratic universalism the competitive speed of oceanic discovery and intra-Iberian trading rivalry in the late fifteenth century required papal intervention in order to determine a spatial and anthropological framework for the expanding spaces of Christendom. Oceanic discovery coincided with a changing spatiality *within* Europe as visions of a continental polity under emperor and pontiff faded before an emerging system of absolute, territorial sovereignties culturally defined through print as much as through any other medium, for example, by Protestant bibles translated into national languages. But while a papal imperium faded within Europe, the Catholic rulers of Spain and Portugal offered the opportunity for its imaginative extension across oceanic space. Two papal declara-

tions defined the geopolitics of global space and humanity, although neither
was implemented as intended. The 1494 Treaty of Tordesillas sought to determine the spatial order of Christian empire, while Pope Paul III's 1537 encyclical *Sublimis Deus* sought to fix the boundary between human and animal natures by determining whether the inhabitants of newly discovered
lands fell within the scope of divine salvation.

Recognizing Portugal's African trading claims, the 1481 papal bull *Aeterni
Regis* designated global space south of the Canaries and west of Africa as
exclusively Portuguese. In his report, Columbus falsified the location of his
initial discoveries, placing Española and Cuba at 34° and 43° north, respectively, to suggest that they were extensions of the Canaries and thus
Spanish. The Atlantic was now a European geopolitical space, and the pope
claimed Christ's delegated authority over it. Alexander VI thus issued bulls
in 1493 granting Spain sovereignty over the territories discovered or to be
discovered by Columbus, fixing a line of longitude 100 leagues west of the
Azores, accepting Castillian claims west of this line, but leaving the status of
undiscovered parts of Africa and India unclear. In the Treaty of Tordesillas,
signed in 1494 by Portugal and Castille, the papal meridian was relocated to
370 leagues west of the Azores, thereby allocating Africa and Asia to Portugal and the western oceanic island world to Spain. The treaty was the first
geopolitical instrument of global imperialism since Augustus's claim to empire *ad termini orbis terrarum*. In the years immediately preceding the Lutheran
challenge to Catholic unity, Europe's world map—not, significantly, a spherical globe—was divided between the two crowns whose territorial struggle
with Islam had been most immediate and recent and whose location lay at
the juncture of Europe's inland and oceanic middle seas.

Continental distributions do not obey geometrical regulation; the South
American landmass projected deep into Portuguese space. Discovered by
Cabral in 1500, this space was given the crusading name Santa Cruz,[14] its
terra ferma filled on Portuguese maps with the most terrifying of the monstrous races, the *anthropophagi*.[15] By the 1520s, circumnavigation required extension of the Tordesillas line around the sphere, forcing the issues of where
East and West originate and how the line should be drawn among the scattered Spice Islands. A treaty negotiated between Spain and Portugal at Saragossa in 1529 undertook the task, this time using globes but, in a tacit acknowledgment of spatial relativity on an oceanic surface wherein longitudes could
not be fixed over sea, allowed the highly profitable Spice Islands to shift arbi-
trarily to either side of the line in future negotiations.[16] However inadequate,

these geo-cartographic acts, sanctioned for Europeans by spiritual authority, marked the origins of a modern globalism.

Representing the Oceanic Globe

The lack of an agreed, global geopolitical space in the opening years of the sixteenth century was a consequence of the mode of its representation. Since the thirteenth century, mariners, trading across the suture of Christian and Islamic worlds from the Catalan, Provençal, and Ligurian ports and using Arabic compasses and astrolabes had developed a mode of charting marine space from coastal traverses, which in the enclosed Mediterranean basins could themselves be closed. From the earliest surviving example of these portolanos, the Carta Pisana,[17] through gilded collections such as the twelve-leaf, bound Catalan Atlas of 1375, presented to Charles VI of France, to mid-seventeenth-century highly decorative charts, these works evolved into a sophisticated mode of mapping that would be stretched to its technical and conceptual limits by oceanic navigation. The portolanos, which show conceptual if not historical affinities to ancient *periploi* descriptions of coastal navigation, similarly charted the edges of known space, and their origins were equally practical.[18] Thus, while this mode of mapping eventually came to represent the global scale, it did so as a logical extension of deckboard observation and recording. Portolanos originated as written lists of locations and compass bearings. Translation from text to graphic image left its mark in the characteristic listing of names that follows the meandering line of coast, given the appearance of mathematical regulation by superimposition of one or more circles marked by wind or compass points, from which a net of rhumb lines is constructed. Such charting in fact attains global perspective only by linear extension and cumulative observation, not by theorizing a synoptic view over the earth's surface.

As a graphic record of practical wayfinding, the portolano decenters the conceptual and imperial spatiality of the Apollonian vision of the globe. Its shipboard perspective, centered on a moving individual from whose eye the curving oceanic surface extends, remains even after solar observations complemented the use of portolanos. Johannes Galle's engravings illustrating Theodore de Bry's *Historia americae,* which celebrates Columbus and Magellan, picture this experience of oceanic space.[19] *Triumph of Magellan* shows the circumnavigator skirting the perilous coastline of Tierra del Fuego, caught—like Ulysses between Scylla and Charybdis—between the arrow-

4.1. Johannes Galle, *Triumph of Magellan.* The circumnavigator passes Tierra del Fuego in a late-sixteenth-century engraving. In *Pars Quarta* of Theodore de Bry's *Historia americae* (Frankfurt, 1594), pl. 15. Photograph from Bibliothèque nationale de France, Paris.

eating Patagonian giant and the fires of the antipodean continent (Fig. 4.1). He holds the compass against his astrolabe as the sun god, Apollo, places a protective hand on the ship's prow. Significantly, Apollo has here descended to the level of the waves. In *Apotheosis of Columbus* the Genoese, costumed as a chivalric knight, holds a flag bearing the crusaders' icon, also found on portolanos: the crucified Christ (Fig. 4.2). Guided by Neptune and his Tritons and protected by Diana and the Nereids, Columbus gazes over islands toward the curving main, which extends to the hemispheric horizon. The dove on the prow, surmounted by the cross, signifies both the guiding Spirit and Noah's sign that land had emerged from the Deluge, thus associating Columbus directly with the Genesis narrative of a new world emerging from the waters. While these engravings date from a century after the events they romanticize, they capture a truth about the oceanic globe. The heroic navigator is also a vulnerable mortal on a flimsy craft, unable alone to grasp the full scope of global space. On portolanos the alignment of words and symbols assumes an immobile eye and a revolving image whose orientation is continuously adjusted to the fixed viewing position. Mobility and extension by way of fixed points follow promontories, river mouths, bays, and harbors named for trading or penetration into a blank interior. Territoriality and boundary are subordinated to mobility and circulation, while possession is marked by the decorative flags that signify control of points along the coast.

Conventions of representation on portolanos show some consistency, whereas scales and units of measure vary locally between ports. The coast appears as a scalloped line, broken to indicate access points such as rivers and harbors. Dominating the image is the boundary between land and sea, signaled by the tight list of names inscribed on its landward side. Symbolic conventions can be recognized, such as red dots signifying hazardous shallows and sandbanks, blue dashes for chains of coastal inlets, small crosses indicating reefs or promontories, gold and blue areas illustrating river deltas. Larger

4.2. Johannes Galle, *Apotheosis of Columbus*. The discoverer of America as crusader in a late-sixteenth-century engraving.In Galle's *Speculum diversarum imaginum speculativarum* (Antwerp, 1638). Print Collection, Miriam and Ira D. Wallach Division of Art, Prints and Photographs, The New York Public Library, Astor, Lenox and Tilden Foundations.

cities may appear topographically as towers tightly grouped within walls. The emblazoned flags of large-scale sixteenth-century portolanos, made for presentation rather than practical navigation, record the coastwise extension and competition of Europeans' oceanic navigation, their disinterest in interior land spaces other than as resource hinterlands reflecting the reality of initial globalization. These empty areas leave enormous scope for rhetorical decoration on presentational maps.

Latinized names, Christian symbols, and drawings of spices, precious metals and beautifully plumed birds for trade or plunder, the tents and palaces of potentates, monsters carried from medieval *mappae mundi*—such decorations trace memories, desires, and fears onto the shifting spaces of European cartography. The 1519 Miller Atlas composes a single global image—painted cosmography—from a series of large-scale regional portolanos (Fig. 4.3). Oceanic space teems with carefully rendered Christian and Islamic trading vessels; named continents contain distinct vegetation, flora, fauna, ethnography, and individually realized cities. In Santa Cruz (Brazil) these images are coordinated into genuine landscapes, inviting a systematic reading of the New World. On either side of the sharply delineated coastlines distinct spaces stand in iconographic opposition: at sea, Christian ships, their white sails aggressively billowing the red crusading cross as if on the swelling chest of a Frankish knight, push toward the shore. Continental space, by contrast, is a riot of promiscuous forms and color: gorgeous birds flit among parkland trees, while monkeys and dragons observe naked Indians building fires under the guidance of chiefs clad in feathers and sporting bows. Violently contrasting worlds are brought within a single spatial frame by the latitudinal lines of climates connecting this particular mapped image to a more theoretical globalism.[20]

Individual portolanos were inscribed onto parchments, which accounts for their characteristic rectangular shape tapering to a narrow neck. This neck area served as a place for information such as the title, the name of the artist, and, commonly, a sacred image dedicating the map, and thus the spaces it represents and the journey it guides, to the crucified Christ or Our Lady Star of the Sea, the Virgin holding the infant Jesus. In a faint echoing of the *mappa mundi,* the protecting—and projecting[21]—presence of the Redeemer is placed beyond the space of cartographic representation but clearly connected to it. The parchment shape, together with the sacred image at its neck, was often maintained in the framing line of maps drawn on other materials than parchment and bound into a volume.

The portolano initially constructed a land-girt marine space, one or more

4.3. European navigators and new world *anthropophagai:* "Terra Brasilis" from an early-sixteenth-century map, Miller Atlas, 1519. Photograph from Bibliothèque nationale de France, Paris.

of the eight separate basins and seas that make up the Mediterranean and Black Seas.[22] As the Atlantic coasts of Europe, Africa, and America came to be represented in the same mode, Ocean shifted from the earth's margin into a greater Mediterranean, bounded by named and familiar coasts. Both Portugal and Spain adopted the portolano as the form representing the information flowing into Lisbon and Seville from oceanic navigation. The "first maps of empire" were the Padreo Real and the Padron Real, the master charts of the Casa de Mina and the Casa de Contratación, respectively, through which the two crowns sought "to keep knowledge of new discoveries within the control of the state and to ensure the standardization of knowledge, so that errors and inconsistencies among charts could be eliminated and they could be revised and updated as new discoveries were made."[23] Despite the problems of regulating and updating the master map because of inconsistent data collection and recording, the portolano, in the form of Spain's Padron Real, came to cover the globe itself, only fully replaced by Ptolemaic mapping with Mercator's 1569 mathematical hybrid of grid and loxodrome. Earlier maps had sought to combine rhumb lines and graticule, for example, Pedro Reinel's 1522 azimuthal projection from the South Pole, showing the ultimate limits of oceanic space. In one of the earliest antipodean images of the globe an oceanic surface dominates, its circular horizon a pattern of coasts whose implied linear connection would render the southern Ocean the third and largest Mediterranean.[24] Following Waldseemüller's prototype *carta marina* of 1516, sixteenth-century editions of Ptolemy's *Geography* acknowledged the distinctive perspective of the navigator, regularly incorporating alongside their graticular planispheres nonprojective hydrographic world maps modeled on the structure of portolanos, constructed with consistent orientation to cardinal points, compass roses, and rhumb lines. A popular example was Laurent Fries's successful woodcut *Orbis Typus Universalis Iuxta,* published in Strasbourg in 1525, with its decorative frame of marine knots linking together the winds and written instructions for lay readers.[25] Appended to collections of Ptolemaic maps, marine maps served to emphasize the epistemological and representational differences between a territorial mode of global representation, with its attention to boundaries, and an expanding, boundless oceanic perspective in which landmasses are primarily islands.

An Insular Globe

The island book *(isolario* or *insulaire),* often produced by makers of portolanos, gave fullest expression to the oceanic globe. Island books too grew out of navigational experience in the medieval Mediterranean, specifically in the Aegean Sea, where Genoese and Venetian sailors vied among the islands of the archipelago with Greek and Arab fleets for the lucrative Byzantine and Black Sea trades. A late-sixteenth-century manuscript by the Venetian Antonio Millo preserved at the Museo Correr at Venice binds together seventy-six watercolor island maps. Designed for practical use, it summarizes a Mediterranean navigator's picture of a world composed of islands surrounded by three continental coasts, finally opening into Ocean. Its preface reads:

> Island book and portolano of the whole Mediterranean Sea by
> Antonio Millo, which orders all the islands of the said Sea with their
> ports and measured distances and directions between each, their size,
> length and breadth, together with a portolano which begins at the
> Strait of Gibraltar continuing along the entire coast of Europe as far
> as the city of Constantinople, and then the coast of Asia as far as the
> Nile river, and the coast of Africa as far as the Strait / with a porto-
> lano of the ocean sea beginning at the Strait of Gibraltar as far as the
> coast of Flanders with ports, carefully recorded distances between
> points, soundings of depths, and all tidal currents.[26]

Each island in the *isolario* is distinctive in shape and is presented as a disconnected entity, a self-contained world (Fig. 4.4). Initially these works simply described individual islands in words, prompting recognition as they emerged out of undifferentiated aquatic space. With oceanic navigation the contents were extended to encompass the greater sphere, capturing a global vision that sought to extend rather than rupture the world-island geography inherited from antiquity and signaling the imaginative appeal of islands.

Island books were produced more for educated consultation on land than for use at sea, responding to a vision of the island deeply colored by classical narratives of the marvelous and the monstrous. They had a classical precedent in the work of the Greek Diodorus, who gave the title *Nesiotikè* (Islands) to the fifth book of his *Historical Library,* in which he described a hierarchy of Mediterranean islands according to size.[27] In its late medieval form the earliest island book was by the Florentine antiquarian Cristoforo Buondelmonti. His *Description of the Islands of the Archipelago* was the product of a humanist's journey through the Aegean in the second decade of the fifteenth century in search of antiquities, its islands described in the language of the classical *locus amoenus*.[28] While drawing on Islamic and Greek sources, Buondelmonti's book is also indebted to the medieval convention of pilgrimage guides, describing the routes to Byzantium and the Holy Land.

4.4. An island world: map of Crete from Cristoforo Buondelmonti's *Isolario*. Courtesy Biblioteca Nazionale Marciana, Venice.

Appropriately for the inhabitants of an island city whose empire was made up of islands and coastal enclaves, most importantly Crete (the subject of a separate geographical treatise by Buondelmonti), Venetians made the *isolario* distinctly their own. The first printed example, complete with forty-nine woodcut maps, is a text composed in rhyming couplets between 1478 and 1485 by a Venetian captain, Bartolommeo dalli Sonetti. Its poetic structure testifies to a humanist interest in the locations and landscapes of Greek antiquity that deepened the connections between the island and the exotic.[29] Bartolommeo describes the Aegean archipelago, illustrating his text with maps that indicate topography, contemporary settlement, and ancient ruins.[30] The form was copied by Benedetto Bordone (1528) and Tomasso Porcacchi da Castiglione (1572). In Venetian hands the island book became a way of imagining the whole earth. Porcacchi figures the world as an island—"in guisa d'insola." Bordone and the greatest island books, including André Thevet's unfinished *Grande insulaire* (1586–87), pushed its contents far beyond the Mediterranean to include not only the large Atlantic islands such as Britain, Ireland, and Thule but also newly discovered oceanic lands. Boschino's 1658 island book continued this Venetian tradition into the Baroque period of lavish cartography, and the tradition culminates with Vincenzo Coronelli's spectacular *Isolario* of 1696, occupying two volumes of the thirteen-volume *Atlante veneto.*

The island book reflects the distinctive spatiality of its principal subject matter. From classical antiquity a canonical distinction of geographical form has been made between continent and island.[31] The distinction was fundamental to the picture of a tricontinental landmass encircled by minor fragments of land. For the navigator, however, the distinction is less apparent; land emerges slowly from marine space, only incrementally differentiated into headlands, beaches, bays, lagoons, and estuaries. Only the Apollonian perspective grasps scale and form instantaneously. Island book images adopted a portolano style of representation, coasts appearing as scalloped embayments and angular headlands to render in approximate form the overall shape of the island. Interior space is either blank or occupied with pictograms of mountains, forests, rivers, cities, towns, and fortifications. The diffusion of this mode of representation is apparent from islands pictured in "world" landscape paintings by artists such as Altdorfer and Bruegel.[32] The Mediterranean island offered the perfect scale for the map form known from Ptolemy as chorography, which specifically disconnected the mapped from its coordinate geographical position and emphasized its qualitative characteristics of *locus.*

Lacking a means of fixing longitude at sea, the precise location of oceanic islands remained virtually impossible to determine.[33] In a sense they literally floated on the earth's surface (some believed that Aristotle had professed this as a physical truth). Thus islands could be moved on maps to gain commercial or diplomatic political advantage, as in the case of the Moluccas in the Saragossa negotiations. Their geographical indeterminacy also increased their imaginative resonance. Part of Diodorus's intention in writing his island book had been to claim islands as the places chosen by the gods to deliver the benefits of civilization to the earliest Greeks.[34] Individual islands appear in *isolarii* as vacant space, disconnected from all spatial markers except for a scale and a compass rose. Bordone sets each island directly over the compass, a design device that provides minimal orientation but gives the impression that the island is a self-contained, centralized world, a persistent cultural assumption: Ulysses' wanderings after the fall of Troy are conducted through islands, construed environmentally and morally as distinct worlds, for example, Polyphemus's and Circe's realms. It was on Patmos that St. John's Revelation had been composed, making this tiny Aegean speck of land a place of epiphany in island books. The exotic spice trade connected Europe with a distant archipelago; Thomas More's imaginary island world of *Utopia* (1516) and Shakespeare's magical island of Prospero in *The Tempest* worked the lore of islands into imaginary worlds that served as social and moral prisms for viewing the actual one. The illustrated Louvain edition of More's island and Holbein's allegorical map for the 1518 Basel edition both borrow the style of contemporary island books.[35]

In the West's figuring of the globe, images of self-contained island worlds have supported much of the imaginative burden of Europeans' shift from passive contemplation to active possession of the marvelous.[36] Neither Bordone nor André Thevet ever questioned the reality of islands occupied solely by demons, troglodytes, satyrs, or Amazons. Bordone illustrates an island in the Antilles chain "inhabited solely by women," who at a chosen moment during the year copulate with cannibals living on a nearby island, the resulting male offspring departing for the cannibal island after three years' infancy. The name of Isla Mujeres, an island off Mexico's Yucatan coast, still records this fantasy. Another island of women, with rather different reproductive arrangements, is located by Bordone in the Indian Ocean.[37] A complex mixture of sources, including Marco Polo's *Milione,* classical myth, and reports from oceanic travelers such as Antonio Pigafetta, lie behind these images. It is easy to recognize the sexual fantasies being played out in a geographical space that offered opportunities for their endless deferral to its undiscovered margins.[38] In this sense Renaissance islands complemented continental interiors as liminal spaces where the European social order could be inverted.[39] André Thevet also reported an island pair gendered according to patriarchal conventions: Imaugle, the women's island, a peaceful and fertile home of gardeners and fishers, and Inébile, a wild and sterile male space of hunters and warriors, regularly decimated by warfare and repopulated by an annual orgy held to reestablish the demographic balance. At the confines of time, in both a mythic past and a fantasized future, the island drifts at the edges of space. Such wild and savage places, even the island of demons located by Thevet in the Canadian waters of French colonial initiative, would be sanctified by the arrival of the cultivating Christian planter.[40]

Cosmography as Global Image

A discursive play between scales of island and world extends from Bordone, publisher of both *isolario* and "apamondo in forma rotonda de balla," to today's photographic images of an insular globe floating in space.[41] As lists of European names edged around continental capes and crept into new-found bays, the expanded scale of Europe's knowledge space was managed conceptually by picturing analogies to more familiar closed spaces of the known world, in a kind of intellectual equivalent to the portolano. If Columbus's Caribbean archipelago inherited something of Ulysses' Aegean or the Aeolian Isles, the sixteenth-century "South Seas" became Ocean at the world's edge, accessed through the Straits of Magellan, the new Pillars of Hercules.[42] As Francesco Chierigato said of Magellan's triumph, "Surely this has thrown the deeds of the Argonauts into the shade."[43]

The task of collecting, coordinating, and regulating oceanic knowledge in the first century of European globalism fell to cosmography, a broad and ill-defined field. Literally, cosmography was the description of the Aristotelian *kosmos,* the entire physical universe. Its application to the study of the terraqueous globe stemmed in part from a titular mistranslation of Ptolemy's *Geography;* more significantly, it represented the continued authority of Aristotelian physical science, inflected for some with Platonic and Neoplatonic cosmology. Cosmography "supposed a full, global world with no other limits than the celestial orb that, projected onto it, formed its poles, regions and zones."[44] Fifteenth-century cosmography was a transcultural product of printing, which increased the circulation of Latin, Greek, and Islamic texts and commentaries and navigators' reports. Cosmography was the science of merchants, princes, scholars, and a growing number of literate lay people across Europe, so that among cosmographers were counted skilled mathematicians and court cartographers such as Diego Ribero, Bartolomeo Velho, and Oronce Fine, publicists and popularizers such as Martin Waldseemüller and Peter Apian, and barely literate nautical technicians such as the Portuguese Jean Alphonse. Regiomontanus of Nuremberg exemplifies the potential scope and intellectual range of Renaissance cosmography. Instructed by the Greek scholar Bessarion, who had left a collapsing Byzantium for Venice and donated to his adopted city his unmatched library of ancient knowledge, Regiomontanus had himself studied Ottoman and Greek texts during a spell at the Hungarian court. In Nuremberg in the 1470s he worked at the center of a network of merchants, scholars, and publishers that included Martin Behaim and Hartmann Schedel, constructing scientific instruments, composing, publishing and circulating mathematical texts and epitomes of Ptolemy, and supplying all Europe with printed ephemerides and astronomical tables. Regiomontanus was at the hub of an international cosmographic discourse that expanded rapidly in the new century.[45] Martin Waldseemüller's *Cosmographiae introductio* of 1507, with its accompanying planisphere and inclusion of Amerigo Vespucci's *Voyages*,[46] and Peter Apian's 1533 *Cosmographicus liber*, constantly republished throughout the sixteenth century and translated into Europe's vernacular languages, register a widespread demand for popular mathematical cosmography.[47]

While Waldseemüller and Apian subordinated the descriptive part of cosmography to the mathematical, texts such as Sebastian Münster's *Cosmographia* of 1552 and the French cosmographer André Thevet's *Cosmographie universelle* (1584) passed rapidly over technical aspects to make the cosmography an amalgam of the medieval encyclopedia, the bestiary, and the universal history, with the addition of contemporary geographical discovery. For them, cosmography's task was to present the universe to the reader's eye as a marvel, a visual spectacle. Thevet recalls the etymology of the Greek word *kosmos,* "ornament," suggesting that the cosmographer's task, made urgent by oceanic discovery, was to reassemble the world's fragments in order to display to the human eye the perfection of the single jewel-like sphere that the Creator had fabricated.[48]

Connections forged within cosmography over the course of the fifteenth century between the practical interests of merchants, navigators, and rulers and the interests of the scholars and humanists who served them can be illustrated by the work of two cosmographers, one at each end of the century. Leonardo Dati was an early-fifteenth-century Florentine merchant residing in Valencia and trading across the sea lanes of the western Mediterranean. His work *La sphera* was widely available across Europe, in manuscript and then in print.[49] It is often taken as a popular summary of how the globe was viewed in the late medieval period. Martin Waldseemüller's *Cosmographiae introductio* was an influential summary of cosmographic knowledge at the turn of the next century, remembered for its recognition of a fourth part of the terrestrial globe beyond the tricontinental world island. A comparison of the two texts signals the issues raised by the parturition into European consciousness of an oceanic globe.

Leonardo Dati represents a growing class of educated citizens in such merchant cities as Florence, Nuremberg, and Venice with an interest in trade, commerce and politics, and natural philosophy.[50] *La sphera,* like Sonetti's island book, is composed in verse. It celebrates the merchant's life, acknowledging the vagaries of Fortune, atop her globe, who can make or break a mariner's success. The text is transcultural: Dati drew on an Arab original for its structure, and some copies actually reproduce portolano-style maps. It opens with a description of the Aristotelian universe and its parts, following the model set by Albert Magnus or Sacrobosco. Dati's theoretical description of the earth's surface contrasts sharply with his outline of trade routes to the Levant and North Africa. While the former superimposes the T-O form of world image found in Isidorian texts over the theoretical climatic belts, marking continental spaces with biblical locations such as Ararat and Babel, the latter is detailed, based on mathematical measurements, and practical. The tension between a theologically authorized geography and a linear, fragmentary coastal world remains unresolved in the text. While the latter derived from Dati's navigational experience, much of the former undoubtedly derived from Dati's brother, a high-ranking Dominican priest. In Anthony Grafton's words, "Merchants and navigators on the one hand, scholars and philosophers on the other inhabited much the same cosmos, imagined much the same history, and saw no necessary conflict between the lessons of experience and those of books."[51] Their distinct epistemologies would with time become irreconcilable.

By Waldseemüller's time the intellectual challenges were becoming sharper, and the authority of received texts more open to challenge. His cosmography covers much of the same ground as Dati's, although in a more sophisticated Latin prose. Its nine chapters open with a summary of the elements of geometry necessary to understand spherical form. The terms *horizon, sphere, axis,* and *pole* are defined, and such phenomena as the heavenly circles, the ecliptic, the five celestial zones and their application to the earthly sphere, and the parallels of latitude are explained. The last three chapters deal with the earth's *klimata,* the winds and the division of the earth's surface into lands and waters. Antiquity's continued cosmographic authority is clear not only in Waldseemüller's scientific sources but in his repeated quotation of Ovid and Virgil to support his definitions and descriptions.[52] But Waldseemüller also drew upon the modern writings on spherical geometry and perspective by Nicholas Cusanus and Regiomontanus, upon Latin translations and commentaries on Ptolemy, and upon the advantages for critical comparison offered by printing, as well as the empirical evidence of navigation. The final chapter of *Cosmographiae introductio* deals with descriptive cosmography: patterns and distributions on the earth's surface. Here Waldseemüller makes his observation about a fourth part of the world.[53] Yet his caution toward this altered globe is striking. He acknowledges the scale and significance of the Atlantic discovery, devoting more space to Vespucci's narrative than to his own geographic survey. But immediately he bows to antiquity, refusing to call America a continent. It is surrounded by ocean, and although he does not refer explicitly to the encirclement of the tricontinental Old World, Waldseemüller insists that "there is only one ocean" and that all authorities accept a sea-girt earth. After Magellan's 1522 circumnavigation such a compromise could no longer easily be sustained.

Such tensions between knowledge based on texts and knowledge based on eyewitness accounts were also becoming apparent in fifteenth-century *mappae mundi*. The most elaborate of these, and effectively the last, are those produced in the Venetian workshops of Fra Mauro and Giovanni Leardo.[54] These Venetian world maps illustrated the influence of portolano modes of representation, for example, in the scalloped outline of their seacoasts and their network of rhumb lines.[55] In these Venetian maps such lines are decorative rather than practical, and they cover both lands and seas. Like sight or light rays, they seem to gesture toward the growing status attributed to observation over contemplation. Fra Mauro's world map of 1459, produced for King Alfonso V of Portugal, retains the characteristic circular form but places the earthly disk within a rectangle, using the corners for secondary circles—the heavenly spheres, the constellations, and the earthly paradise. Their liminal location beyond the terrestrial surface implies a degree of unease with the conflation of sacred and secular space. The world is pictured as a sanctified place of wonders, the graphic equivalent of Hartmann Schedel's combination of empirical evidence and legend. Indeed, Fra Mauro and Leardo reproduce long descriptive texts in the blank continental space of Asia and Africa, some almost direct plagiarisms of the *Milione,* by Marco Polo, who is given the status of eyewitness and guarantor of truth. However, the Mauro map, like Dati's text, is powerfully unitary, closing its single landmass tightly within the circular frame and reducing oceanic space to a narrow belt scattered with islands at the earth's edges. By comparison, Martin Behaim's globe, which followed thirty years later and was not vastly different in information content, is oceanic. Of the two characteristically Venetian modes of representing geographical space—the *mappa mundi* and the *isolario*—the latter was clearly more suited to accommodate oceanic expansion. Indeed, it was the *isolario* that survived and continued to evolve in Venice into the seventeenth century; the *mappa mundi* died at the close of the fifteenth century, overtaken by the globe and the mariner's planispheres, themselves foreshadowed by Fra Mauro's maps.

The sixteenth-century French cosmographer André Thevet made perhaps the most ambitious attempt to unify expanding oceanic knowledge into a single image of the terraqueous globe. A Franciscan monk, Thevet was Henry III's royal cosmographer. His intellectual goals were encyclopedic: to organize and make coherent sense of the stream of varied, discontinuous, and sometimes contradictory information about the world's contents pouring into France by the mid-sixteenth century. By this time cosmography had necessarily become a collective exercise, requiring the author's collaboration with not only explorers and navigators but also a team of writers and engravers, placing the cosmographer at the hub of a network that he was obliged to command and coordinate. Thevet's "center of calculation" was Paris, and the success of his cosmography depended on his determining the accuracy and reliability of the information he was handling, as well as choosing a fundamental ordering principle, a convincing mode of collation and classification. Among the models he chose was the island book. The *insulaire* reflected a nonhierarchical conception of the world, "a landscape made up of drawing and writing." Inevitably, given this indeterminate character, Thevet's *Grand insulaire et pilotage,* containing more than three hundred charts and commentaries on the islands of the globe, was neither completed nor published.[56]

Thevet's island maps were collected into two volumes, devoted to the Mediterranean Sea and the world Ocean, respectively, signaling the continuing imaginative hold of antiquity's "Middle Sea" and allowing Thevet, alongside his appeal to eyewitness truth and experience rather than textual authority, to sustain the humanist parallel of ancient and modern worlds. Presenting a globe composed entirely of islands "resolved in its way—the most elementary possible—one of the major difficulties posed by the cartographic construction of the globe at the time. The fragmentation and dichotomization it authorized provided a way of overcoming the hiatus that existed between a science of mathematical projection and the art of placing fragmentary empirical data on a canvas."[57] Thevet's montage treatment of the global surface allowed scraps of empirical information to be attached to individual islands, each presented as a distinct "world." As Frank Lestringant points out, however, this apparent resolution was achieved only at the expense of both the logic of the voyage narrative, which formed the space-time connection between the discovered and the known while guaranteeing the authenticity of the former for the latter, and any pretense at practical navigational value for his work (a matter of some importance to Thevet). Thus "the great Atlantic archipelagos share their drifting fragments with each other in the universal brew of this disordered maritime encyclopaedia."[58]

Thevet's only graphic attempt to integrate his assemblage of individual island maps and commentaries was a set of two polar-projection hemispheric maps placed at the beginning of his work. But their coordinates bear no consistent relation to those of individual island maps, so they offer no possibility of reconstructing, even theoretically, a single global surface. Although the individual island maps have graticule coordinates and a partial set of rhumb lines, apparently proving Thevet's aptitude for both theoretical cosmography and practical navigation, he actually ended up confusing both and signally failed to adopt Gerardus Mercator's technical solution, available for nearly two decades, using a conformal cylindrical projection, on which graticule and rhumb lines could be coordinated. Consequently, on Thevet's insular globe elements *drift* in space, allowing whole degrees of latitude and longitude to be filled by the imaginative work of memory and desire. That drift appropriately reflects the mariner's perspective on an oceanic globe. As a cosmographer, Thevet makes no Icarian or Scipian ascent over the terra-queous sphere to cast an Apollonian eye across its surface. Rather, as in Galle's engraving, Apollo is brought down to the level of the sea, and the global surface fills with a scattered assortment of marvels and monsters, drifting across the greater Mediterranean and available for Christian trade and empire.

As a global project, cosmography spanned physical and metaphysical, empirical and theoretical knowledge. Sixteenth-century cosmographers found themselves stretched epistemologically between the fixity of an Aristotelian cosmos and the rapidly differentiating terrestrial surface, forced to negotiate and regulate disparate and disconnected strands of text, commentary, and eyewitness accounts that moved across information networks controlled by powerful political and commercial interests. An imaginative expression of European desires as much as an intellectual engagement with cognitive knowledge, cosmography continued to articulate an imperial sense of global unity, as Waldseemüller's dedication of *Cosmographiae introductio* to the Habsburg Maximilian makes clear:

> Since thy Majesty is sacred throughout the vast world
> Maximilian Caesar, in the furthest lands,
> Where Phoebus Apollo raises his golden head from eastern waves
> And seeks the straits called by Hercules' name,
> Where midday glows under his burning rays,
> Where the Great Bear freezes the surface of Ocean;
> And since thou, mightiest of mighty kings, dost order
> That mild laws should prevail according to thy will;
> Therefore to thee in spirit of loyalty this world map is dedicated
> By him who has prepared it with such wondrous skill.[59]

Here the Apollonian eye imperially resolves the dilemma that obscured Thevet's oceanic vision. Theories and practices of vision had been of growing epistemological significance in Europe since the twelfth century.[60] By the fifteenth century *perspectiva,* a Euclidian description of light, was central to natural philosophy, underpinning both Sacrobosco's *Sphaera mundi* and Albert Magnus's *De caelo et mundo,* the principal scholastic texts of Aristotelian physics. The optical measurements that produced the portolano and the eyewitness integrity upon which navigators' information relied differed from the inward, or imaginative, eye that witnessed a visionary globe.

FIVE *Visionary Globe*

They marcht not long, when of the arduous *Hill*
They gain the top; where an inameld *Flat*
(In a *Field Em'rauld*) *powdred Rubies* fill,
Making them think old PARADICE was *That.*
Heer, in the Ayre a GLOBE (by wondrous skill
So fram'd that *Thorough Lights*) they contemplat,
That th'unresisted Eye the *Center* sees,
As plainly as the *superficies.*

—Luis Vaz de Camões, *The Lusiads*

The first-century corpus of global knowledge by Claudius Ptolemy (c. 90–168) reentered Latin knowledge space as complex transcultural texts transcribed and redacted by Arab and Byzantine scholars, as the title *Almagest* for Ptolemy's astronomical work indicates. Two versions of the *Geography* in Byzantium had come from Alexandria via Arabic scholars in Baghdad and Samarkand.[1] Ptolemy's texts summarized and synthesized Roman and Hellenistic terrestrial knowledge, offering instructions for calculating and representing the earth's spherical form and for recording locations on its surface. They were prized equally by Ottoman and Latin Renaissance rulers, as much for their antiquity and symbolic authority as for their scientific or practical value.[2] The full Ptolemaic corpus came to the attention of Western scholars at separate moments. The *Almagest* and the *Planisphaerium,* describing the earth-centered cosmos and concentric planetary spheres, had been translated into Latin in the twelfth century. A Greek manuscript of the *Geography,* studied in Byzantium since 1300, was delivered to the Florentine humanist Pallo Strozzi by Emanuel Chrysoloras in 1397; a Latin translation by Jacobus Angelus to which he gave the title *Cosmographia* was available within a decade, and the work circulated rapidly within humanist circles on both sides of the Alps.

Beyond the prestige of an ancient text to be decorated, presented, and possessed as an item of cultural capital, three specific features of Ptolemy's *Geography* helped alter the West's global image: Ptolemy's scientific hierarchy of spatial representation (book 1), his methodological descriptions for plotting the sphere on the plane (books 1 and 8), and the eight thousand coordinates for places across the ancient ecumene (books 2–7). From this gazetteer a set of world and regional maps *(tabulae)* could be constructed. These graphic images, initially in manuscript form but by the final quarter of the fifteenth century cut and engraved on wood or copper plates for printing, offered a visual bridge to the imperial spaces of antiquity. The agency of print expanded their circulation and speeded the processes of criticism and emendation. The *Geography* offered a model for the *atlas,* the mathematically coordinated and systematically scaled representation of terrestrial spatiality that connected Western global representation to the book. What the globe illustrated in three dimensions, Ptolemy's opening map, *Typus orbis terrarum* (see Fig. 1.2), projected onto two, the controlling image for the systematic spatial subdivision that followed.

Ptolemaic spatial order would be extended in the sixteenth century to a threefold hierarchy. *Cosmography* represented the totality of a spherical cosmos, *geography* provided geometrically exact images of the earth's spherical surface and its major divisions, and *chorography* pictured the form and character of localized spaces and places. This was a different spatiality from the navigator's or the chartmaker's. Ptolemaic science fitted a theoretically unlimited set of spatial data onto a geometrically predetermined surface. Fitting empirical (oceanic) information within the new spatial framework was a long, critical, and very public process, enabled largely by print culture.[3] Incunable editions of the *Geography* were printed in Rome, Bologna, Ulm, and Vicenza. By 1500 new *tabulae,* or maps illustrating differences between ancient and modern geography, were being bound into the text, a process of hybridizing the classical work that opened up the possibility for progressive addition of maps showing spaces unknown to the ancients and the ultimate erasure of the received world image.[4] Within a century of its printing the *Geography* had become a historical item only, a structural trace in Abraham Ortelius's *Theatrum orbis terrarum* of 1570 and the pedigree of atlases that followed.

But the Ptolemaic contribution to an Apollonian imperial discourse of global sovereignty in the early modern West and its implications for conceptions of a global humanity were vital. Through the graticule, globalism emerged as a conception that connected discourses of humanity and civi-

lization. In the maritime and intellectual centers of Europe, Abraham Ortelius, Gerard Mercator, and Jodocus Hondius followed earlier cosmographers—Martin Waldseemüller, Oronce Fine, Peter Apian, and others—in
elaborating ways of picturing a habitable earth, not a restricted ecumene, by
means of illustrated globes and atlases. These pictured in detail a great theater of the world wherein dreams of Christian redemption, global imperium,
and a singular humanity might be projected and rhetorically secured.

Ptolemaic Perspective

The figure of Claudius Ptolemy is widely represented in art of the late Middle Ages and the Renaissance, grasping the armillary sphere or astrolabe that
represents the cosmic scope of his work. He is clothed as an Arab or oriental seer, not uncommonly a magus, frequently sharing the iconography of
the Wise Men, whose cosmographic skills foretold the birth of the Christ-
Apollo (Fig. 5.1).[5] His *Almagest* catalogued 1,028 visible stars and twenty-
one constellations, and its calculation of planetary movements was the
empirical foundation for astrology, connecting the Alexandrian astronomer
to ancient Egyptian esotericism. The most striking feature of the carved
wood Ptolemy in the choir stalls at Ulm, the location for an early printing
of the *Geography,* is his narrowed eyes staring intently at the armillary sphere.
The astronomer's vision extends over space and time. While the mariner
scans the horizon and interrogates the compass, the astronomer's vision rises
conceptually over the surface, escaping the contingencies of location and
moment in order to grasp a cosmic order and regularity. The cosmographic
language is geometry, in the view of Platonizing Renaissance cosmographers
the divine language of creation.[6]

This conceptual capacity to grasp the earthly sphere from a cosmic location is essential to Ptolemaic mapping. In the *Almagest* Ptolemy details constructional techniques for making a celestial globe, emphasizing its value in
displaying the heavens to the eye.[7] The *Geography* is driven by the idea of
rendering visible the form and pattern of the earthly sphere. Geography
is founded on geometry, "the art of delineating solid objects upon a plane
surface so that the drawing produces the same impression of apparent relative positions and magnitudes, or of distance, as do the actual objects when
viewed from a particular point." Geographical science draws on geometry
to make "an imitation and description of the whole of the known world and
all the things which are almost universally related to it."[8] The globe is best
represented by a graticule of 360 lines of longitude converging at the poles

5.1. Claudius Ptolemy examines the heavens and the terrestrial globe in a fifteenth-century manuscript copy of the *Geography*. Photograph from Bibliothèque nationale de France, Paris.

and 180 degrees of parallel latitudinal lines measured from the equator. Co-ordinating these numbered lines allows point location to be precisely determined on the conceptual surface. The implications of representing earth space through an infinite array of fixed points are more than merely instrumental. The graticule flattens and equalizes as it universalizes space, privileging no specific point and allowing a frictionless extension of the spatial plot. At the same time it territorializes locations by fixing their relative positions across a uniformly scaled surface. Its geometry is centric only at the poles, which, practically speaking, are the least accessible points on its surface; otherwise it extends a nonhierarchic net across the sphere. Geometric projection allows the spherical surface to be transformed and molded on the plane while retaining consistency of locational relationships. What David Woodward has called the route-enhancing properties of the portolano and the center-enhancing structure of a Jerusalem-centered *mappa mundi* are displaced in favor of the space-equalizing and area-fixing properties of the graticule.[9]

Ptolemy offers four techniques for transforming the sphere into a two-dimensional surface, each involving areal and directional distortion. The simplest inscribes parallels and meridians as a rectangular grid, rapidly distorting shape and scale as one moves away from the equator. Two "conic" transformations seek to accommodate spherical curvature, one by radiating straight meridians from the poles to the equator across curving parallels, the other by maintaining a straight central meridian while progressively curving the longitudinal lines to its right and left, thereby maintaining their true position at both pole and equator and giving a visual impression of sphericity. In book 8 Ptolemy describes a fourth alternative, placing the observer's eye at a position where the visual axis is in a latitudinal plane; the earth is seen frontally, and mathematical calculations are made from a distance point located within the sphere. Ptolemy effectively represents the earth as a transparent sphere, similar to the astronomer's hand-held armillary. While this fourth "projection" cannot be produced mathematically, unlike the stereographic projection described in Ptolemy's *Planisphaerium,* it conjures an image of how the earth might look from space—the Apollonian perspective. The close historical and cultural affinities between the reappearance and dissemination of Ptolemy's fourth model and the pictorial method of linear perspective demonstrated by Filippo Brunelleschi in 1425 and theorized by Leon Battista Alberti a decade later have been widely remarked.[10] Representational space was subjected to the conceptual logic of ocular vision in the opening years of the West's historic encounter with transoceanic global space. Given the increased mobility of reproduced images promoted by movable type, printmaking, and small-scale oil painting on canvas, the imaginative and intellectual possibilities opened up by Ptolemaic mapping were considerable. It entered and enhanced a culture in which visual images of terrestrial, architectural, and natural spaces at a variety of scales circulated widely.[11]

Ptolemy's graticule covers 360 degrees, and his transformations can generate flat maps of the sphere. But the specific locations coordinated in his text are limited to the classical ecumene. Conical sections easily accommodate this curving rectangle, largely confined to the Northern Hemisphere. Ptolemy's text thus confirmed the continental image on Macrobian maps while more precisely elaborating it.[12] It described the same coasts as portolanos did, reinforcing a sense of an unmediated inheritance of classical empire, while the graticule offered the flexibility of assimilating and integrating ancient authority with empirical discovery. Thus Giovanni Leardo's and Fra Mauro's mid-fifteenth-century circular *mappae mundi* combine Ptolemaic and maritime modes of representing global space. For the scholars who introduced the *Geography* to the West, and for those who financed translations and reproductions of its text and maps—the Medici in Florence, the Montefeltri in Urbino, the Este in Ferrara, Aeneas Silvius Piccolomini (Pope Pius II) in Rome, and Mehmed II in Istanbul—accommodating new knowledge into its picture of the earth was not a primary concern. Its image of ecumenical space assumed significance in the context of other ancient texts and of competing claims to the imperial inheritance of Rome more than that of oceanic navigation. As the Greek scholar Trepuzuntios put it to Mehmed II, "You are the emperor of the Romans. Whoever holds by right the centre of the Empire is Emperor and the centre of the Empire is Constantinople."[13] Ptolemy's *tabulae* demonstrated such claims graphically. To add new maps that covered areas unknown to the ancients and to "correct" errors in the *Geography*'s image of the ecumene (proclaimed as the victory of experience over authority) reflected an intense geopolitical and moral debate over the patrimony of classical empire in a modern world.

Like manuscript copies of the *Geography,* incunable editions were lavishly illuminated and dedicated to competitively acquisitive patrons. In the sixteenth century, the number of editions and commentaries rapidly increased as it was translated into Italian and German and became a standard library work. The Greek manuscript from which Chrysoloras and d'Angelo had worked, together with an illustrated copy of their Latin translation, was secured by Duke Federigo da Montefeltro at Urbino, part of a 1482 collection of more than eighteen hundred classical manuscripts.[14] Vitruvian architectural proportion and geometry are the governing spatial principles of Federigo's palace at Urbino and its managed views over ducal territory. The sumptuous decoration of the bound manuscript of the *Geography* closely mirrors that of the palace. Like Ambrogio Lorenzetti's fourteenth-century disk map at Siena, Federigo's *Geography* is more than simply an item of cultural capital; its images of terrestrial space participate in a cosmographic rhetoric connecting Montefeltro's tiny territory across time to the classical empire and across space to the great orb itself.[15] The imperial conceit is pursued in the marginal decoration, which includes Federigo's stemma intertwined with portraits of Roman emperors and set against the landscape of the Italian Marches, over which the duke actually ruled. The decoration of this codex is especially extravagant, but printed editions of the *Geography* were similarly lavish, testimony to the *Geography*'s cultural significance in a Renaissance visual discourse of territorial authority.

In 1482 a poetic rendering of the *Geography* was dedicated to Federigo. The work, by the Florentine Francesco Berlinghieri (1440–1500), offers an insight into the cultural context into which the Ptolemaic perspective on the globe was precipitated. Berlinghieri, trained in Greek, rhetoric, and poetics, pursued his humanist career as an orator at the Este court in Mantua and later at Lorenzo di Medici's Florentine court.[16] He was a member of the group that gathered around the Greek scholar and Platonist Marsilio Ficino, with whom Berlinghieri debated Ptolemy's intellectual significance.[17] Ficino's own Latin translation and commentary on Plato's works, to which Berlinghieri contributed financially, was a defining achievement of Renaissance humanism, a task interrupted briefly in 1463 by Lorenzo's demand that Ficino translate Hermes Trismegistus's writings, supposedly the pre-Mosaic teachings of an Egyptian priest, philosopher, and king.[18] Berlinghieri's poetic rendering of Ptolemy's *Geography* was begun in 1464, making it precisely contemporary with Ficino's work. It was prepared in two huge, magnificently decorated codices, and the printed edition by the German Nicolaus Laurentii was illustrated with copper engraved maps.[19] The dedicatory lines, composed by Marsilio Ficino himself, make explicit reference to the Hermetic theme of spiritual ascension through the spheres. In this context Berlinghieri's poetic rendering of the Apollonian perspective and its elegant visual realization in copper engraving clothe Ptolemy's *Geography* in the quasi-spiritual habits of Ficinian philosophy. The dedication to Federigo in the printed edition links this Platonic embrace of the sphere to the more secular conceit of imperial subjection of the globe.

This Neoplatonic rendering of the *Geography* into Dantean verse was published in 1482 as *Septe giornate della geografia di Francesco Berlinghieri Fiorentino,* a title suggestive of the Creation narrative in Genesis. Berlinghieri follows the structure of Ptolemy's text but incorporates commentaries on humanist themes such as the mythical and poetic origins of classical places, producing a book that is more philological than scientific. In book 5, a long dilation on ancient Egyptian burial and other sacred rituals draws upon

Strabo, Pliny, and other ancient sources as well as on medieval texts, contemporary nautical maps, and chorographies of Britain, France, Spain, Italy,
and the Holy Land, implying that Ptolemy offers a foundation for cosmographic knowledge. Berlinghieri draws upon Buondelmonti's *Liber insularum
archipelagi*,[20] whose author was familiar to the Florentine circle as the discoverer of Horapollo's *Hieroglyphica,* the foundational text on Egyptian hieroglyphics as the original universal language.[21] Appropriately, therefore, Berlinghieri's opening paean to geography places its study at the heart of all
learning:"How many [disciplines] are affected by the delay of this great work,
which takes into full view the whole earth. It feeds not only military art but
also philosophy, scripture, history, and poetry. The sweet life of agriculture,
medicine, and art that animates the love of nature in the human breast. In
sum, no greater need have our faculties than knowledge of the earth."[22]

Ptolemy is the "light and ample glory of the world" who "raises us above
the limits of an earth obscured by clouds, which hide our view of the surface." Berlinghieri appeals directly to Apollo and the angelic choirs to raise
his poetic imagination to the task of presenting the whole universe to the
marveling human eye. His song will follow Apollo's curving path over the
earth to offer a conspectus, a vision of its surface, without yielding to the
temptation of Icarus. His achievement betrays the imaginative power of
geography, which permits an *intellectual* vision of the globe:"It offers divine
intellect to human ingenuity, as if it were by nature celestial, demonstrating
how with true discipline, we can leap up within ourselves, without the aid
of wings, so that we may view the earth through an image marked on a
parchment. Its truth and greatness declared, we may circle all or part of it,
pilgrims through the colors of a flat parchment, around which the heavens
and the stars revolve."[23] Imaginative vision across celestial space was a commonplace of the Ficinian discourse. The ascent of the soul through the
stages of sensual, emotional, and intellectual love, corresponding to the terrestrial, celestial, and supercelestial realms of the Ptolemaic cosmos, respectively, was achieved by spiritual purification and contemplation. Its ultimate
attainment, a beatific vision of universal love, was a key theme in Ficino's
attempted synthesis of Hellenistic philosophy and Christian theology. Berlinghieri's "human genius," "divine intellect," and "rising up within ourselves" echo the language of the *prisca theologia,* the supposed pre-Mosaic
anticipation of the Christ-Apollo's global redemption and renewal.[24] The
recovery of the Ptolemaic perspective restimulated the poetic association
between global vision and Apollonian ascent.

The actual itinerary that Berlinghieri constructs across the Ptolemaic ecumene is similar to a portolano *periplus,* a circuit of the Mediterranean coasts, closely dependent upon Buondelmonti's island book. But the copper engraved maps in the printed edition do not reflect the navigators' perspective, nor does their selection and arrangement follow the logic of the *isolario* (see Fig. 4.4). Their sequence is governed by the Ptolemaic *tabulae,* to which they add modern plates of Iberia, Britain, France, and Italy. Berlinghieri's is a hybrid globe, mapping the territorial logic of the classical empire from its initial *orbis terrarum* dominated by the land ecumene through the *tabulae* marked with the altars and columns of Alexander, Pompey, and Caesar, while narrating a maritime sequence more appropriate to the emerging spatiality of modern maritime empire.

From Ptolemaic Ecumene to Global Space

No Ptolemaic map survived from antiquity. All had to be reconstructed from manuscript tables of names and coordinates. Each version contained materials for a map of the ecumene plus either twenty-six or sixty-five regional maps. Most manuscript and early printed editions of the *Geography* in the West contained twenty-seven images: a world map plus ten European, four African, and twelve Asian regional maps. Berlinghieri's Latin codex included bird's-eye chorographic maps of eleven cities. The opening world map, the key to the succeeding regional images, could be constructed according to any one of Ptolemy's methods. Early Western mappers invariably adopted one of the conical sections, illustrating a rectangular area stretching from a western meridian through the "Fortunate Islands" across 180 degrees to the shores of an enclosed eastern sea into which trans-Gangetic India and the island of Taprobane extend as if into a greater Mediterranean.

Early Western reconstructions of the Ptolemaic world map have been extensively classified and studied.[25] They illustrated an unambiguously spherical space, of which the geometrically framed ecumene, although occupying most of the area displayed, was clearly a limited section. A graded and numbered frame contained a gridded conceptual space of meridians and parallels. Both latitudinal and longitudinal lines constituted a purely terrestrial geometry disconnected from cosmographically determined circles of equator, tropic, ecliptic, climates. Potential expansion of the mapped space through 360 degrees was implicit in the image. Wind heads, still often angelic in their features, are awkwardly positioned between rectangular page and ecumene. They blow from the meteorological spaces of air and fire assigned them by Aristotle, disconnected from the two-dimensional surface described by the graticule, their gaze mirroring that of the human spectator. Framing, shape, and measure rather than the pattern of lands and seas are novel. Thus the world map in Nicolaus Germanus's 1466 Ulm edition of the *Geography* quite dramatically diminishes the space of Ocean, presenting a land-dominated world.[26] Ptolemy's African east coast extended 25 degrees below the equator, necessitating a corresponding Atlantic coastline, currently the object of Portuguese navigation. Nicolaus extends Africa westward from the "Gulf of Hesperus" into oceanic space and eastward to connect with Asia, through *terra incognita secundum Ptolemeum,* reinforcing the graphic argument on all early Ptolemaic maps of a land-dominated global surface.

The distinction between Ptolemaic and oceanic images is epistemological rather than consistently representational. The former assumed a retrospective spatiality of territorial empire easily appropriated to the Mediterranean geopolitical context after 1453 and reinforced by its incidental contents of classical triumphs, columns, and altars. Both genres still peppered the spaces beyond the ecumene with *anthropophagi* and *ichthyophagi* and other inherited "marvels" and "monsters." From their first appearance, however, Ptolemaic and oceanic perspectives were brought into dialogue in shaping Western globalism. This is apparent, for example, in the form of coasts on the maps accompanying Berlinghieri's text. They reveal the typical scalloped form of island-book maps. And the frame of the Ptolemaic ecumene was immediately placed under pressure from oceanic knowledge, literally apparent in Nicolaus Germanus's Ulm world map, where, at the northwest corner of the ecumene, the 65° north parallel breaks to incorporate the *mare glaciale,* the Scandinavian lands and various north Atlantic islands. By 1500, cosmographers such as Martin Waldseemüller could locate oceanic knowledge within the Ptolemaic conceptual frame and challenge its territorial emphasis, although the addition of a distinct *carta marina* endured in sixteenth-century editions of the *Geography* alongside a world map on a projected graticule until Gerardus Mercator's 1569 graphic marriage of the two spatialities.

Fusion of the territorial and oceanic perspectives was a northern European rather than a Mediterranean achievement, as was the introduction of the new cosmography to a wide reading public. By 1500 the Grüninger workshop in Strasbourg was reproducing large-scale woodcut maps in significant numbers; it printed five editions of Waldseemüller's *Cosmographia introductio* between 1507 and 1509 and two Ptolemy editions in 1522 and 1525.[27] Lorenz Fries's 1525 German-language edition of the *Geography,* based on Angelus's Latin translation and using Waldseemüller's maps, competed with Waldseemüller's text as an accessible guide to cosmography. It was accompanied by fifty folded maps and a twelve-sheet *carta marina,* the first indexed world map in a vernacular language. Although it explained how to coordinate a projection, Fries's map was not itself a projection; it reduced the known oceanic space, reinforcing the image of a Mediterranean-centered terrestrial world. In Nuremberg, where Regiomontanus's mathematical legacy combined with the patrician humanism of Willibald Pirckheimer's circle, interest in Ptolemy was part of a broader cultural interest in space and its representation, facilitated by the exchange of letters, books, and instruments between European centers, from navigators in Lisbon to Copernicus in Frauenburg. "Inquire into the hidden and powerful workings of the earth," proclaimed Pirckheimer, a translator of Ptolemy whose study of German place names, *Germaniae explicatio,* appeared in 1530.[28] The planetary spaces of the cosmos and the territorial spaces of painted landscape concerned this group as much as the oceanic space of navigation, although like those Venetians who saw in the Aztec capital, Tenochtitlan, a New World reflection of their own city, Nurembergers imagined a mirror of their own city at Refugio, on the New England coast.[29] Pirckheimer's close friend Albrecht Dürer, who contributed along with Michael Wöhlgemut and Wilhelm Pleydenwurff to Hartmann Schedel's *Liber chronicarum,* in 1515 designed a widely reproduced globe woodcut illustrating Ptolemy's fourth projection, while Nuremberg merchants commissioned the instrument makers Martin Behaim and Johann Schöner to produce some of the earliest terrestrial globes. Alongside his advanced mathematical texts, Peter Apian published the popular cosmographic handbook *Cosmographicus liber* in 1524, the first to distinguish and illustrate cosmography, geography, and chorography, whose clear descriptions and cleverly inserted vovelles allowed ordinary readers to understand the calendrical and seasonal implications of planetary motion and a spherical earth.[30] The Latin work was translated into French, Flemish, Spanish, and Italian, and it was revised and republished into the 1570s.[31]

In northern Europe, Ptolemaic science entered a context of intense moral and theological as well as commercial debate. Ottoman expansion into the lower Danube and the Lutheran challenge to Rome cast the idea of a Christian ecumene in serious doubt. The simple, uncompromising black lines of German woodcut prints are less seductive as images of global harmony than lavish courtly images of Italy and Iberia. Lutherans saw the Ottoman expansion as divine judgment on a corrupt Christendom; some responded with a quietist and familist idea of concord that in some respects paralleled Catholic Neoplatonism in Italy.[32] The cordiform map projection, pioneered by

Oronce Fine for François I of France in 1519 and printed for the first time by Apian as *Tabula orbis cogniti* in 1530, would become a specific object of contemplation among northern familists, its heart shape signifying a world ruled by love in an age of intensifying doctrinal hate.[33]

Gerardus Mercator's mathematical resolution of oceanic and Ptolemaic images by means of a cylindrical projection, which retained the Ptolemaic grid's geographical coordinates while incorporating the mariner's loxo-dromes, has been criticized as a Eurocentric image, enlarging the temperate regions at the expense of the tropical regions of European imperial expansion. At the time of its making the map diminished the areal significance of Europe, and its relations with empire are subtle. Blending the practical, surface spatiality of the navigator with the conceptual, Apollonian eye of Ptolemaic mapping, Mercator's world map offered a graphic solution to the sixteenth century's growing tension between textual authority and empirical evidence as foundations of secure knowledge. At the same time, Mercator's profound interest in hermeticism, the power of terrestrial magnetism, and Pythagorean metaphysics, apparent in his *Typus vel symbolum universitatis,* suggests that for Mercator resolving different modes of vision through the mathematics of spherical projection was as much a spiritual as a geopolitical act.[34]

Certainly, the technical combination of oceanic and Ptolemaic visions depended on the capacity to think globally, to break the frame of the classical ecumene and imagine the world as a spherical surface. This epistemological break owed more to Ptolemy's textual descriptions of a spherical earth within a mathematically structured cosmos and its subjection to geometrical manipulation than to Ptolemaic maps.[35] As the *Geography* circulated in Europe's courts, terrestrial globes began to be constructed. The earliest extant example is Martin Behaim's *Erdapfel* of 1492, commissioned by Nuremberg merchants the previous year, although a Vienna-Klosterneuburg manuscript of the 1430s refers to the manufacture of spherical world maps, and in 1444 Guillaume Hobit, astronomer to the duke of Burgundy, constructed a "round world map in the form of an apple" based on Ptolemy's description. In 1477 Pope Sixtus IV obtained from the Ptolemaic mapmaker Nicolaus Germanus a terrestrial and a celestial globe for the new Vatican Library.[36] Bramante's *Heraclitus and Democritus,* dating from the same years as Behaim's commission, clearly intends to show a spherical earth. These examples all point to a widespread interest in the fifteenth century in representing a spherical earth.

The Magellan circumnavigation prompted an explosion of globe making, both engraved metallic spheres and, more commonly, printed gores attached to wooden globes. Globes now became essential instruments for political strategy, academic study, and trade, if not practical navigation. Association with the spherical globe signified the social, intellectual, and moral status of "a man of vision" in the sixteenth century. One of the two tapestries called *The Spheres,* a set commissioned by João III of Portugal in the 1520s, shows "Earth under the protection of Jupiter and Juno": the Portuguese sovereigns gesture toward a terrestrial globe, contained by the graticule, the Mediterranean ecumene entirely diminished in scale by the spaces of Portuguese navigation (Fig. 5.2). In Holbein's *Ambassadors* globes accompany an array of astronomical and navigational instruments that underline the moral stature of the Frenchmen de Dinteville and de Selve, signal the authority and fortune of England's Henry VIII, and respond to Pirckheimer's injunction to "Inquire into the hidden and powerful workings of the earth."[37]

Ptolemy's graticule, so emphatically represented in these images, allows spaces to be "envisioned" before they are encountered and tied logically into uniform global space. Globes and manuscript maps were the means of achieving this. In the late 1480s Henricus Martellus extended Ptolemy's 180-degree ecumene by nearly 100 degrees of longitude; in his *Cosmographia* Martin Waldseemüller pushed it to 360 degrees. By 1508 a "whole-earth" map covering 360 degrees of latitude and extending to both poles had been constructed by Francesco Roselli, "drawn on an oval projection into which every point on earth could be theoretically plotted and on which every potential route for exploration could be shown."[38]

The Cosmographic Globe

The Apollonian gaze assumed by Behaim's globe, Roselli's map, and João III's tapestry is a powerful theme in sixteenth-century art and literature. Cosmography became the discourse that brought together celestial and geographic exploration, represented space and scale, and theorized the place of humans within nature.[39] This globalism was underscored by moral pressures resulting from both reformation and discovery to bring within a single intellectual frame the emerging Renaissance sense of the sovereign individual, so powerfully explored in Dürer's self-portraits, and an expanding world of humanity.[40] This required simultaneously engaging with the globe's materiality and rising above it into the celestial spaces of imaginative vision. In Bernard van Orley's *Spheres* image sun, moon, planets, and stars have no logical location other than that determined by design considerations. Their disks

5.2. The imperial globe presented and protected by the Portuguese monarchs, as Jupiter and Juno. Tapestry by Bernard van Orley, "Earth under the protection of Jupiter and Juno." © Patrimonio Nacional, Madrid.

are subordinated to a dominating global sphere, to which the Portuguese monarchs gesture, at once recognizable individuals and personifications of Jupiter and Juno. They share the firmament with angelic wind heads at the margins of celestial space. The image contrasts with the cosmos in Hartman Schedel's 1493 *Liber chronicarum,* an encyclopedia, bestiary, and wonder book combined, published in Nuremberg, in which a seated patriarchal God accompanied by fixed ranks of angels and saints oversees a tightly bound cosmos whose central earth with its continents is the merest button (see Fig. 1.3).[41] The *Chronicle* fits its geography into a Christianized cosmographic narrative divided into seven ages, an amalgam of biblical and classical learning. The image of the cosmos succeeds six woodcuts illustrating the Creation and is followed by a simplified Ptolemaic ecumene and a chorography of Eden.[42] There follow a synopsis of biblical history, lives of the saints, tales from Homer, Ovid, and other classical writers, a geographical description of the parts of the earth, and illustrations of its cities. The sixth, or current, age is followed by blank pages, to be filled in the immediate future, before the seventh age, which describes the Apocalypse, judgment, and the dance of death. It is this linear and hierarchical cosmography that van Orley's globalism unconsciously subverts.

Schedel's terrestrial globe is firmly contained within the enclosing spheres of a greater cosmos (although his brief reference to Martin Behaim's "discovery" of islands in the Ocean Sea indicates Nuremberg's close contacts with Iberian events). The noncentric arrangement of supercelestial and temporal spheres illustrates the Aristotelian separation of incorruptible, "simple" heavens from the corruptible, "composed" earth. South of the Alps, the Thomist conviction that no direct interaction was possible between these spheres was under challenge from Platonic thinkers who emphasized the philosophic implications of spherical harmony and universal unity. Sixteenth-century cosmographic images such as van Orley's reflect a simultaneous erosion of physical and metaphysical boundaries upon and between globes, the philosophical legacy of thinkers from the previous century, such as Marsilio Ficino and Nicholas of Cusa (1401–64). Cusanus (his humanist name), who had been a student at Heidelberg and Padua, would play a key role in those attempts to reunify Christianity before the fall of Byzantium, which had brought Greek and Latin scholarship into such close contact.[43] A key figure in the Platonic critique of scholastic Aristotelianism, Cusanus was also closely involved with Ptolemaic science, purchasing cartographic materials from the Klosterneuburg monastery in 1444, discussing with German scholars at Rome in 1450 possible new *tabulae* to supplement the *Geography,* and producing his own map of central Europe, the source for that in Schedel's *Chronicle*.[44] Cusanus was a *global* thinker, both theoretically and practically, whose driving belief was unity. Spherical geometry offered him intellectual evidence of divinity, "propounding the idea of an infinitely open universe, whose center was everywhere and whose circumference nowhere. As an infinite being, God transcended all limits and overcame every opposition. As the diameter of a circle [or sphere] increased, its curvature diminished; so at its limit its circumference became a straight line of infinite length. Likewise, in God all opposites coincide . . . in the universe God is both centre and circumference."[45]

Cusanus was equally committed to the unification of faiths and peoples, proposing "a comprehensive vision of all reality, God, the world and man."[46] Developing the Majorcan Ramon Lull's argument that Christ's incarnation was necessary to unite divine and human natures and based on his own knowledge of Greek sources, Cusanus published *De docta ignorantia* in 1440. According to Cusanus, Christ is the necessary union between the cosmic spheres, which Thomism had seen as unable to communicate. Consulting *Timaeus* in the original Greek rather than its Macrobian Latin summary, Cusanus argued for a created world animated by divine love, whose expression is optical geometry. The subcelestial realm moves toward a perfect order of the supercelestial sphere, which may itself therefore be imag(in)ed through visual observations received by the globe of the human eye from the material cosmos.[47] Cusanus lectured on optics at Padua, where one of his students was Paolo dal Pozzo Toscanelli, who later corresponded with Columbus about the size of the globe. Cusanus's thinking gave an almost divine significance to the sun as the source of light, reinforcing the imaginative appeal of the Christ-Apollo, an indirect stimulus to Copernicanism, whose author had also studied at Padua.[48] Cusanus's surveying and mapping concerns complement his interest in spherical geometry and optics, as well as his interest in Christendom's geopolitics. In the aftermath of 1453 he even flirted with the idea that Mehmed II might be the agent of global religious and political unity.[49]

Among the theological implications of Cusanus's work was a renewed and distinctive emphasis on the medieval motif of the human microcosm, connecting it to the Platonic ascent of the soul, uniting the individual and the cosmos. For Cusanus, "man is a microcosm not because he comprises in himself all the different degrees of reality and thus is subject to all its conflicting forces, but rather because—situated at the centre of creation, at the horizon of time and eternity—he unites in himself the lowest level of intellectual reality and the highest reach of sensible nature and is thus a bond which holds creation together."[50] Cusanus offers a mapping conception of the microcosm: the human individual is literally relocated at the center of the cosmos, capable of rising imaginatively and spiritually above the globe as an eyewitness to the beauty and harmony of creation. Physical light connects the spheres of earth and eye within a spherical cosmos, while the divine illumination of universal love is its metaphysical equivalent. Imaginative mapping of the human individual within the cosmos was an insistent theme among Ficino's Florentine group. According to Paul Kristeller, Ficino's commentary on Plato's *Symposium* "treats love as a cosmological principle of the unity of things, as a *viriculum mundi,* while his *Platonic Theology* gives the same role to the rational soul."[51]

Ficino's brilliant young colleague Giovanni Pico della Mirandola (1463–94) pushed the autonomy of the microcosm even further in his 1486 *Oration on the Dignity of Man,* detaching the human individual from any fixed cosmic location, freeing it to "observe whatever is in the world," determining its own place in creation: "Thou canst grow downward into the lower natures which are brutes. Thou canst again grow upward from thy soul's reason into the higher natures which are divine."[52] Pico owned a personal copy of Berlinghieri's poetic translation of the *Geography,* which offers a reading of Ptolemy's work as a realization of the cosmographic ascension.[53] The elevated perspective is given moral superiority over any surface view. Correspondence between microcosm and macrocosm is figured through physical and imaginative *vision:* "Without some voyage of the soul, there can be no instantaneous point of view over the cosmos. The kinship between cosmography and sacred poetry was . . . essential and primary."[54]

Terrestrial and ocular globes connect in the metaphor of the mirror, the *speculum,* an insistent trope in sixteenth-century culture. Apian's most sophisticated vovelle, based on an azimuthal world map, is called a cosmographic *speculum,* or "mirror," illustrating the motion of the globe to the eye. The French moral philosopher Charles de Bovelles (1479–1567) refers to the human individual as a "mirror who stands outside and opposite the rest of creation in order to observe and reflect the world. He is thus the focal point of the universe in which all degrees of reality converge."[55] Cosmography thus laid claim to being "the most fundamental science" in the sixteenth century, dealing with the implications of this revolutionary relocation of a human individual liberated to soar imaginatively above the earth through the spheres of creation.[56] Uniting moral and natural philosophy within the universal scope of cosmography yielded both Copernicus's reordering of the planetary spheres in *De revolutionibus orbium caelestium* (1543) and the speculative hermeticism of such figures as Cornelius Agrippa.[57]

Sixteenth-century cosmography was a pan-European project. By 1500 printing presses competed to produce texts such as Apian's *Cosmographicus* and, from 1532, Oronce Fine's *De mundi sphaerae, sive cosmographia, libri V* in large numbers for a mass market.[58] Broader in conception were texts such as Sebastian Franck's *World Book: The Mirror and Portrait of the Earth* (1534) and Sebastian Münster's *Cosmographiae universalis,* which appeared in 1550 and was republished into the 1590s. Part universal history, part contemporary encyclopedia and news journal, part geographical gazetteer or even atlas, these works continued the tradition established by Schedel, Rollewink, and others, complex publishing ventures in both Latin and vernacular languages. Münster, a professor of Hebrew at Basel, had produced his own edition of Ptolemy in 1542. His *Cosmographia* was structured through the conventional microcosms of *annus, mundus,* and *homo,* dealing, respectively, with astronomical knowledge, global geography, and the human body. Its scope and the continuous incorporation of new information produced a textual collection of marvels. Common to such texts was a recognition of the illustrative and explanatory attraction of the visual images that engraving and printing techniques allowed. But the coherence of cosmographic texts that sought to retain the medieval encyclopedic model of universal history was increasingly undermined by the desire to incorporate a totality of knowledge, producing the disconnected, fragmentary, and even contradictory collections of texts and images into which the popular cosmographic project eventually collapsed.[59]

Cosmography was both instrumental and intellectual, increasingly required in the sixteenth century to accommodate sectional interests within its universalizing premise as its various activities became harnessed to a competitive imperialism among the Christian nations. Cosmographers such as André Thevet and Guillaume Postel in France or the Hakluyts and John Dee in England promoted and recorded the course of their nations' discoveries. Spain's Alonso de Santa Cruz wrote a *Historia universal* in 1536, a translation of Aristotle in 1545, and an *Isolario general* in 1542, as well as a geography of Peru and a brief introduction to the sphere.[60] Pedro de Medina, official cosmographer to the Casa de Contratación from 1552, included among his tasks examining ships' pilots and navigational instruments. Nicolas Nicolai, the official cosmographer to Henry II of France, had similar duties, while in England William Digges and William Cuningham titled themselves cosmographers and competed for influence in the court of Elizabeth I.[61] In Venice the cartographer Giacomo Gastaldo was given the title *cosmografo*, while Pope Gregory's official cosmographer, Egnazio Danti, was centrally involved in the 1582 reform of the Christian calendar.[62]

Portuguese cosmography offers an example of how these competitive demands on cosmography spread its influence widely over the sciences and the arts. One of João II's negotiators at Tordesillas, Duarte Pacheco Pereira, published his *Esmeraldo de situ orbis* in 1505. Principally a mathematical treatise on the sphere, it also includes a universal history and geography.[63] Bernard van Orley's tapestries, as we have seen, appropriate cosmography's graphic elements to signify the Portuguese crown's claim to universal empire. By the late sixteenth century Luis de Camões was deploying cosmography to construct a national rather than a universal history and geography. *The Lusiads,* whose lines open this chapter, concludes with its hero Vasco da Gama, here a heroic literary hybrid of Ulysses and Aeneas, offered a vision by the goddess Tethys from the mountainous peak of her island realm.[64] He sees a crystalline globe, "infinite, perfect, uniform, self-poised," a model of the fabric of creation, made "by that *All-wisdome,* that *All-eye.*" Vasco's own "unresisted eye" penetrates to its very center, piercing the transparent orbs of individual planets and stars to reach the earthly sphere at its center:

> . . . . the seat of MAN:
> Who, not content in his presumptuous pride
> T'expose to all *Earth's* Mischiefs his life's span,
> Trusts it to the unconstant *Ocean* wide.[65]

In the tradition of Cicero's Scipio, da Gama is literally seized with wonder and desire at his Ptolemaic vision. Tethys describes the earth's oceans and continents, bequeathing them to imperial Portugal, ruled by a Christian Alexander, who "through the WORLD shall spread." An oceanic empire, its detailed geography narrated coastwise like a portolano itinerary, is marked upon the cosmographic sphere.[66] Vasco's ascent and vision leads to Aeneas's choice at Dido's Carthage between an imperial vision *ad termini orbis terrarum* and insular confinement in the arms of a desirable and available woman. Similar episodes recur throughout Renaissance literary cosmography, for example, in the lunar episode of cantos 34–36 of Ariosto's *Orlando furioso,* whose printed editions included Ptolemaic maps of Roland's African and Asian adventures.[67]

In a series of illustrations to the Genesis narrative, Camões's contemporary Francisco de Holanda (1518–84), the Portuguese humanist and artist, seems almost to illustrate Vasco's vision and Lusitanian cosmography's asso-

5.3. The origins of the elemental sphere: Day One of *Genesis* according to Francisco de Holanda. Gouache on paper, from a set of six images, 1545–47. Courtesy Biblioteca Nacional, Madrid.

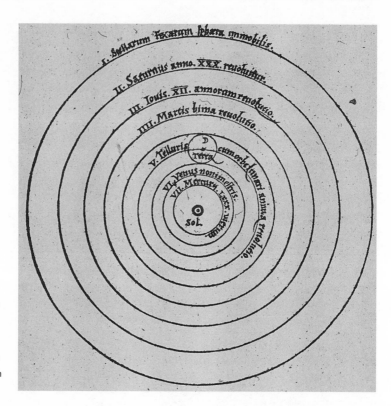

5.4. Copernicus's helio-
centric cosmos, from
De revolutionibus
orbium coelestium
(Nuremberg, 1543). By
permission of the British
Library.

ciation of spherical geometry and Neoplatonic speculation.[68] In 1573 he
completed work on 154 gouache illustrations titled *De aetatibus mundi imag-*
ines. These apply the sacred geometry of sphere and triangle, light, and
vision as cosmogonic principles to the hexaemeron, drawing upon sources
including Cusanus and Sacrobosco.[69] De Holanda represents the first day of
Creation by means of a transparent heavenly globe, dividing light from dark
through a trinity of triangles, simultaneously penetrating and containing
with fire the other three elements of corruptible matter. The words *Fiat Lux*
transform these from chaos into a spinning volumetric globe of Earth (Fig.
5.3). The image, which owes its geometrical construction to illustrations of
Cusanus's *De coniecturis,* allows the Father, represented by the innermost,
equilateral triangle, literally to "compass" the created cosmos.[70] Ficino had
included the circle and triangle in his translation of the *Timaeus,* the sole
illustration in his *Complete Works* of Plato. *A* and Ω in de Holanda's image
represent the divine closure of time as well as space, the ends of creation. In
the words of de Holanda's contemporary the English cosmographer William
Cuningham, "Whatsoever is betwixt the seate of the almighty governour of

all lyving creatures, and the centre of the earth: is called the world, and it is compared to a round ball and globe."[71] De Holanda's second day illustrates the Christ-Apollo as human microcosm creating the concentric transparent orbs of the firmament, again recalling Camões's image. The breath of life is illustrated by a cartographic wind head. The separation of lands and waters on the third day stretches scalloped coastline and islands over the curving surface of the globe under the eye of the Trinity and choirs of angels. The illumination of the planetary firmament on the fourth day draws directly from illustrations of the lunar eclipse in Sacrobosco's text, its tiny central earth turned to reveal a distinctly Portuguese global space: Western Europe, Africa, the Indian Ocean, and the Brazilian land of Santa Cruz.

These cosmogonic images replace an anthropomorphized Father with pure geometry, while the process of creation becomes a play of triangles and globes. Cosmography, "a Science which considereth and describeth the magnitudes and motions of the celestiall or superior bodies,"[72] is here rendered as mystical geometry. The mystery that drew cosmography repeatedly toward speculation was the paradox of planetary motion within an immutable cosmic order. Copernicus's contribution to that question made his simple image of 1543 a pure cosmographic vision (Fig. 5.4).

The Geographic Globe

To geography is given "knowledge teaching to describe the whole earth and all the places contained therein, whereby universall maps and Cardes of the earth and sea are made."[73] Sixteenth-century geography was a hybrid of textual authority and empirical observation that opened a broad space for imagination and invention in representing the global surface.[74] As for Camões's Vasco da Gama, the Apollonian gaze authorized both an individualist, imperial quest for *Fama* (Fame) and a more structured metaphysics of global order and harmony.[75] Global geographic space is examined here as it appeared in three significant locations in sixteenth-century culture: the princely map gallery, the world landscape painting, and the printed atlas.

The Map Gallery

More than twenty sixteenth-century galleries of painted map cycles are scattered across Europe, notably in the Italian courts. Popes and princes centered themselves imaginatively on the global stage by commissioning images of cosmography, geography, and chorography for their palaces. The tradition traces back at least to Augustus's mausoleum complex in imperial Rome. Lorenzetti's work at Siena included the now lost disc referred to by Ghiberti as a "cosmografia," possibly an astronomical or calendrical device but equally likely an image of the ecumene, integrated into the chorographic panoramas of Siena's territories. At Windsor and Winchester in England world maps and tapestries were hung in direct line of sight from the throne as signs of monarchy.

Ptolemaic study, oceanic navigation, and mathematical representation altered the content and form of such schemes, while astrological and allegorical conceits such as those connecting Cosimo de Medici's name to celestial images and Medicean iconographic use of Galileo's discovery of Jupiter's moons further elaborated their meanings.[76] Two of the most elaborate cycles were the inspiration of Egnazio Danti (1536–86), cosmographer to Pope Gregory XIII and adviser on the reform of the Western calendar. Danti held a chair of mathematics at Bologna, where he designed the great gnomon on the floor of San Petronio; he edited Sacrobosco, surveyed, mapped, and engineered fortifications and watercourses for various Italian states, and in 1586 raised the great obelisk in St. Peter's Square as a Christianized gnomon around which global time and space revolved.[77] Danti's first map gallery was a room of fifty-seven Ptolemaic maps commissioned for the Uffizi Palace in 1562 by Cosimo de Medici and based on Girolamo Ruscelli's Ptolemy. Below the maps were paintings of the plants and animals of each region, and above them were portraits of their rulers. From the ceiling, which was decorated with zodiacal signs, a great terrestrial globe could be lowered by windlass, "so that, when fixed, all the pictures and the maps on the cabinet will be reflected therein, each part being thus readily found on the sphere."[78] Cosimo could stand at the very center of space surveying heavenly and earthly globes.

Even grander in conception was Danti's work between 1577 and 1583 for the Bolognese Gregory XIII, undertaken in the Vatican's Terza Loggia and on the upper floors of the Belvedere, a scheme including mechanical globes, painted planispheres, Ptolemaic tables, and forty geographic maps showing the papal dominions in Italy. A connection between cosmos and globe is made at the Tower of the Winds, whose astronomical instruments fix the equinox and solstice for the longitude of Rome, as towers in Athens and Rome had done in antiquity. Its interior decorative scheme demonstrates the latest theories of light and perspective. An astrolabe depicting the cosmos and a terrestrial globe were incorporated into the scheme, so that the Ptolemaic wall maps in the Terza Loggia were related to the spherical earth—cosmography to geography.[79] The vaulted Gallery of Geographical Maps is concerned less with the expansion of faith than with the affirmation of Rome's continued global centrality. Thus Danti's map of Italia Antiqua quotes Pliny: "Italy is a land which is at once the foster child and the parent of other lands, chosen by the providence of God to make heaven itself more glorious, to unite scattered empires, to make manners gentle and to become throughout the world the single fatherland of all peoples."[80] The decorative scheme comprises scenes from Old Testament and church history connected to forty regional wall maps of classical and papal Italy in a geohistorical defense of Catholic orthodoxy.

Danti's images lead the eye from a vertical perspective over the territory, through high-angled bird's-eye views, into intimate landscape scenes: the verdant slopes of Etna, the monuments of Rome, the vine-hung terraces of Campania, and then upward toward the infinite emptiness of cosmic horizons. In the Liguria panel, for example, the eye ascends from the two pilgrims following their winding road through an Apennine valley to a panorama of the Gulf of Genoa and the cities along its coast. In the depth of the image the alchemical symbol of angler and stream allegorizes territorial fortune. Passing over the sea toward the Corsican coast, the eye is arrested by two images. On the left a Christian captive aboard a Berber ship signifies Christendom's threats to south and east, while on the right Neptune drives a chariot toward the western Ocean, its gilded seat occupied by Christopher Columbus, son of Genoa, whose fame in discovering a new orb is proclaimed by Neptune's banner (Fig. 5.5). The vignette captures papal globalism a century after Pius II's cosmography pleaded for a crusade against the Ottoman conquest of eastern Christendom.

World Landscape Painting

Danti's images reflect a connection between painting and global mapping found in the key centers of printing and map production in the Renaissance, notably southern Germany, Antwerp, and Venice. It was also in these centers that a genre of "world landscapes" became popular in both the opening and the closing decades of the sixteenth century.[81] Panel painting was held up by humanists such as Pirckheimer as a more appropriate format for cosmographic description than language, and world landscapes appear on panels, small, framed canvases, or etchings. They are portable images, responding to the demands of a bourgeois clientele that might own or be familiar with Apian's handbook or Münster's encyclopedia. As Juan Vives wrote in 1531,

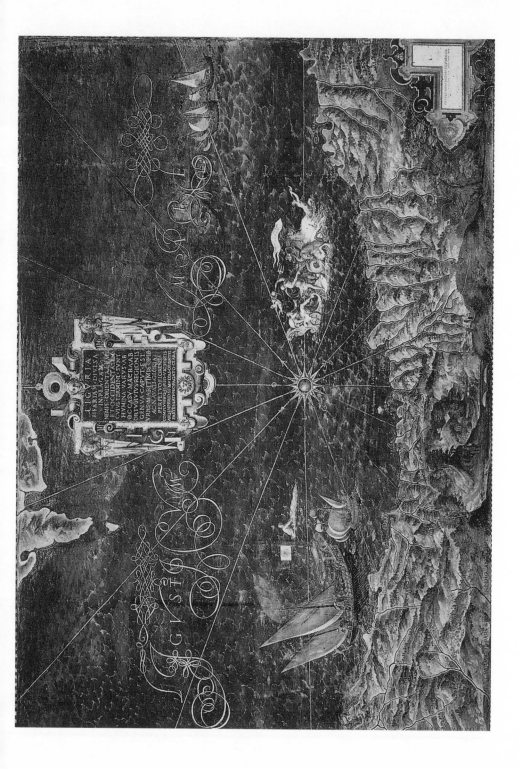

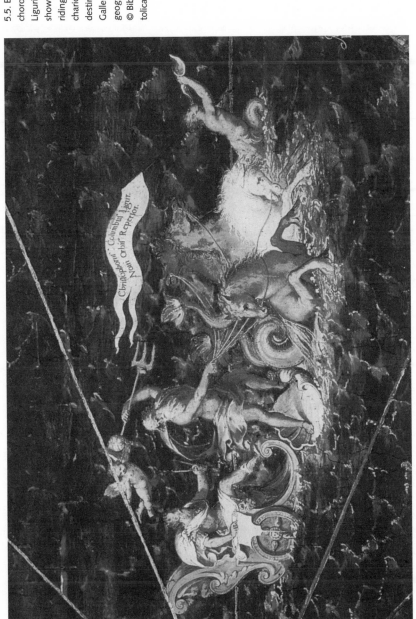

5.5. Egnazio Danti's chorographic map of Liguria, with a detail showing Columbus riding Neptune's chariot to an imperial destiny in the west. Galleria delle mappe geografiche, Vaticana. © Biblioteca Apostolica Vaticana.

"The whole globe is opened up to the human race, so that no one is so ignorant of events as to think that the wanderings of the ancients (whose fame reached to heaven) are to be compared with the journeys of these travellers [contemporary explorers], either in the magnitude of their journeyings, or in the difficulty of their routes."[82]

World landscapes appear to illustrate a substantial segment of the globe's surface, a vast panorama with multiple viewpoints, "dazzling the spectator with a rich profusion of natural scenery: mountains, plains, and valleys, rivers and seas, harbours and castles, and villages. . . . The sense of vastness is enhanced by the unnaturally elevated horizons . . . and the clarity with which even remote regions are often shown."[83] The earliest examples are by Joachim Patinir of Antwerp, Lucas Cranach of Vienna, and Albrecht Altdorfer of Regensburg in the years immediately preceding Magellan's circumnavigation. Patinir's *Martyrdom of St. Catherine* typically removes the narrative content to one side of the frame, allowing the eye to sweep out over a wide estuary with its portolano-style scalloped bays. Like the angel who hovers over the dying saint, the viewpoint is high above an incidental world of locations and events. Patinir's round earth appears as a marvelous, glittering jewel, recalling Erasmus's words, "What spectacle can be more splendid than the sight of this world?"[84]

Albrecht Altdorfer achieved even more dramatic effects in a series of heroic scenes from antiquity commissioned by Wilhelm IV of Bavaria. *The Battle of Issus* of 1529 illustrates Alexander's defeat of Darius in 334 B.C.E., which opened Asia to European empire (Fig. 5.6). From the foreground details of combat, the eye passes mountains and cities, ascending across the Levantine coast and a foreshortened Cyprus to a vision of three continents meeting in the eastern Mediterranean, the Isthmus of Suez, and the Nile Delta. The Red Sea stretches to the horizon, while above the cosmic swirl of cloud, sun and moon occupy opposite corners of the firmament.[85] In *The Conquest of Tunis,* a set of tapestries commissioned to celebrate Charles V's African campaign of 1535, a similar perspective over the Mediterranean reworks Rome's defeat of Carthage in a much more explicitly cartographic rendering of a transhistorical struggle for global empire.[86]

World landscape painting peaked in the work of Pieter Bruegel the Elder (1528–69) in the 1560s, losing appeal by the time of Peter Paul Rubens, the last great painter to follow the genre. Bruegel's world landscapes, such as *The Tower of Babel* and *Magpie on the Gallows,* are characteristically vast in conception yet jewel-like in detail, visual equivalents of contemporary cosmographies in their search to frame both the scale of an ordered globe and the

5.6. *The Battle of Issus,* oil painting by Albrecht Altdorfer, 1529, a world landscape view of a foundational moment of the European imperial imagination. Alte Pinakothek, Munich.

fragmentary and contingent nature of its contents. *Landscape with the Fall of Icarus* (1558) adopts the most appropriate of all themes for representing global landscape, incorporating Ovid's story within a single, curving horizon. The image captures the moment when Icarus, his waxen wings melted by Apollonian fire, disappears below the waves, falling unnoticed as the globe turns, the plowman marks the changing seasons, and the galleon slips out into the gulf.[87] The scalloped lines of promontories and bays are echoed in the forms of the setting sun, the plow lines, and the lateen rigging of the ship. And the cycle of cultural time is captured within the frame by the movement from fisherman to shepherd to plowman to mercantile city.[88]

Printed Atlas

Among Pieter Bruegel's closest associates was Abraham Ortelius (1527–98), an Antwerp antiquarian and map publisher at the center of a European "republic of letters."[89] Ortelius's *Theatrum orbis terrarum* of 1570, a collection of fifty-three maps engraved to a uniform format by Franz Hogenberg, synthesized the printed cosmography and Ptolemaic *Geography* to produce a "theater," *conspectus,* or mirror of the whole earth. Its inspirational sources were Ptolemy's *Geography,* now thoroughly superseded as an image of known geography, popular cosmographies such as Schedel's and Münster's, and the bound but unsystematic collections of printed maps sold by printers such as Bertelli, Forlani, and Camoccio.[90] The *Theatrum* was predominantly graphic, Latin text being located to the left of the maps, whose order followed a rigorous geographical logic: from globe to continents to countries and regions. It was hugely successful commercially, reprinted four times in its first year, regularly updated and translated into six languages by 1612, and continued by Mercator; its lineage is traceable through the seventeenth-century Dutch atlases of Jodocus Hondius and the Blaeu family. Ortelius's work encouraged the idea of private, vicarious enjoyment of geographic discovery, which had become a common feature of mapping rhetoric.[91] The individual could master the globe at a single glance. The Apollonian dream was domesticated, a point emphasized in the laudatory poems that introduce Ortelius's venture. Adolphus Mekerchus, for example, describes the editor seated with Phoebus Apollo, witness to the world ("qui conspicit omni"), himself a discoverer of unknown places and peoples hidden in the remotenesses of the globe:

> Ortelius, whom Phoebus Apollo has allowed to be conveyed with
> himself in the four-horsed chariot over the air, from where he may

circle the lands that lie below and the [Ocean] depths that flow around. Let men say that to him [Ortelius], Phoebus—who sees all things—has shown regions utterly unknown and situated far below the vault of Heaven, formerly known only to natives, and disclosed a new orb, and races and peoples and the secrets of a new world.[92]

The totality of Ortelius's vision is illustrated in his frontispiece. A Doric arch supports a crowned Europa robed in imperial purple and seated below a vine-entangled pergola. She holds the orb and scepter of universal rule.[93] At her sides are celestial and terrestrial globes, while below stand female personifications of the other two ancient continents and their value to Europe: Asia in bejeweled silks, holding an urn of smoking myrrh, and black Africa, seminaked under the sun of the torrid zone, holding a branch of balsam. Below and between the columns of this arched entrance to the printed earth sits nude America, decorated only with feathers, holding bow and arrows in one hand and the medicinal wood *guanacum,* supposedly a cure for syphilis. She carries a severed European head, signifying the monstrous nature of her peoples. A fifth head emerges from a block of stone: *terra incognita* of the undiscovered southern continent, heir to the classical antipodes, which Ortelius's world map marks as a vast extension of Tierra del Fuego (Fig 5.7). Ortelius explains that the map shows the earth given by God to the sons of Noah, as well as a fourth continent and a fifth yet to be fully known. The ethnological assumptions of his sexualized iconography reveal both the continued resonance of humanist, "Mediterranean" spatiality and the desires and fears of a patriarchal culture. This image and the opening *Typus orbis terrarum* (see Fig. 1.2) construct a spatial dialogue between a narrative of surface discovery—through a proscenium arch or the Pillars of Hercules—and a cosmographic synopsis from which the eye swoops to examine selected locations in greater detail.[94]

If Ortelius's frontispiece emphasizes a European global territoriality, *Typus orbis terrarum,* which colors continents distinctly, makes Europe's diminished continental size and northerly location on a globe of 360 degrees instantly apparent, prompting for the critical reader those moral questions of European normalcy and superiority raised in Montaigne's precisely contemporary essay on cannibals.[95] The cylindrical globe projects from encircling clouds within a rectangular frame. It is a predominantly terrestrial space; the ship-filled ocean occupies no more than a quarter of its surface. The Stoic theme so consistent in humanist global images is stressed by Cicero's epigrammatic question, "What in human affairs can appear great to him who

5.7. Frontispiece to the German edition of Abraham Ortelius's *Theatrum orbis terrarum* with figural images of the continents. Harry Ransom Humanities Research Center, The University of Texas at Austin.

is familiar with all eternity and the size of the whole world?"[96] The revised *Typus* of 1587 carries further epigrams from Cicero and Seneca dignifying contemplation of the earth as a defining feature of humanity while mocking human pride and folly, such as the following from Seneca: "Is this that pinpoint divided by sword and fire between so many peoples? How ridiculous are the boundaries of mortals."[97]

Ortelius's humanist commitment structures the *Parergon,* a companion volume to the *Theatrum,* which maps an antiquity now disconnected in textual space and time from the modern globe but still its guiding moral influence. This volume constructs an image of ancient empire, Ulysses' and St. Paul's Mediterranean odysseys, and the Holy Land. Together Ortelius's two collections consummate a global vision of Christendom, transcending profound religious and political divides in the republic of images produced by Catholic, Lutheran, and Calvinist mapmakers. Like Oronce Fine's cordiform world projection, overseen in Apian's 1539 *Cosmographicus* by Apollo and the Habsburg emperor (Fig. 5.8), Ortelius's Christian humanism deploys the authority of antiquity to create a moralized geopolitical globe. Mercator's 1595 *Atlas* further developed Ortelius's concept into his cosmographic *oculus mundi.* Mercator's preface outlines a "planetary strategy" for a Europe diminished only in geographical scale to realize its historical *telos* in imperial conquest: "Here [Europe] wee have the right of Lawes, the dignity of the Christian religion, the forces of Armes . . . Moreover, Europe manageth all Arts and sciences with such dexterity, that for the invention of manie things shee may be truely called a Mother . . . she hath . . . all manner of learning, whereas other Countries are all of them, overspread with Barbarisme."[98] Despite this triumphalist rhetoric, however, sixteenth-century atlas makers— those of Flanders above all—could scarcely ignore Christian Europe's own doctrinal barbarism. The Stoic marginalia to their maps should be read in the light of Ortelius's pietism and Mercator's hermeticism and his 1544 arrest for heresy. The vision of a unitary globe, graphically revealed by the map and atlas makers, was intellectually appealing to humanist scholars in the context of Calvinist and Tridentine Catholic fundamentalism. Jodocus Hondius's heart-shaped projection, based on Ortelius's 1564 map and suspended from the hand of God, became a common emblem of universal love at the end of the century (Fig. 5.9).[99]

The global territoriality envisioned by Ortelius and Mercator was complemented by the six-volume *Civitates orbis terrarum* (1572–1617), edited by Georg Braun and Franz Hogenberg, whose selection of city images mapped Europe's claims as the location of *civis* and thus civilization. While Braun

5.8. Oroncé Fine's cordiform world map, 1534. Photograph from Bibliothèque nationale de France, Paris.

and Hogenberg's collection, like Ortelius's, offers ample graphic testimony to trans-Danubian, non-Latin territorialities, Mediterranean antiquity is still privileged as the source of civilization. Thus *Civitates orbis terrarum* includes a set of images by Georg Hoefnagel (1542–1600) illustrating his journey with Abraham Ortelius from Salzburg to southern Italy.[100] The northern humanists visit landscapes, ruins, and natural wonders of antiquity. From Rome the Appian Way leads them to the Bay of Naples and the Straits of Messina with views of Etna. Wide panoramas record Hadrian's Tivoli, the festooned vineyards of the Campania, Scylla and Charybdis. Ortelius's presence as a traveler in these images provides eyewitness authentication for the vicarious explorer in library or study of a historical narrative that reaches into the heart of the Mediterranean, paralleling the geographical narrative of Ortelius's *Theatrum* images, extending the eye to *terra incognita* at the edges of the earth.[101]

Envisioning Global Humanity

Petrus de Turre's *De locis et mirabilis mundi,* a medieval corruption of Herodotus and Pliny summarizing the marvels and monsters to be found at the edges of the earth was bound into many early printed editions of Ptolemy's *Geography.*[102] Petrus described the monstrous races, the earthly paradise with its four rivers and angelic guards, Taprobane's elephants, gold, jewels, and trees that never shed their leaves, India's variety of serpents, Arabia's phoenix, and Central Asia's cannibals. Descriptions of the Fortunate Isles to the west anticipate the ambiguities of New World landscapes, at once Edenic gardens and howling wildernesses.[103] As the true dimensions of the globe, America's continental size, and oceanic space became recognized, Europe's spatiotemporal centrality, implicit in Ortelius's frontispiece, figured Asia in terms of an exotic past and the transoceanic West as an unformed future.[104] Hybrid human-animal creatures are more explicitly territorialized. Some, like the *kunokephaloi* and the *sciopedes,* remained outside empirical known space, while others, most dramatically the *anthropophagi,* threateningly within it.

The expansion of Ptolemy's ecumene raised immediate and insistent questions about how those creatures—who were physically human and capable of speech but whose appearance, cultures, languages, and practices appeared utterly alien—might fit within the scheme of salvation through which Europeans had normalized their humanity. From the time of Aristotle language and evidence of a settled "community" had signified humanity, while such signs as nakedness, body hair, "nomadism," and lack of agriculture or per-

5.9. Emblematic use of the cordiform map: map by Jodocus Hondius in Pieter de Hondt's *Album amicorum*, 1589 (KBR, II 2254). Copyright Brussels, Royal Library.

manent settlement revealed proximity to animal nature. Christianity had added monotheism and rejected divinity in nature as pagan. Renaissance concepts of selfhood and human dignity emphasized conscious distancing from "animal nature," shaping responses to other peoples. At the extreme, New World observers such as the Spanish colonizer Genesium Sepulveda argued that Americans had no souls, conveniently excluding them from redemption and placing them in the category of animals. Others, such as Bartolomé Las Casas, whose observations were eagerly seized upon in Protestant Europe as evidence of Spanish brutality, adopted the inclusive view of humanity adopted by the papal bull *Sublimis Deus,* of 1537, which pronounced all such peoples capable of redemption, guaranteeing their humanity even as it legitimated for some the infliction of unspeakable brutalities on fallen bodies for the sake of saving souls.[105]

Attitudes toward the humanity of newly discovered peoples were neither simple nor consistent. The inheritance of Greek ethnocentrism was pervasive. No longer were Europeans presented only with images of other beings against whom to define their own humanity; in American and later oceanic discoveries they were presented with physical people. Resolving the complex tensions between image and embodiment has since remained a central ethical issue for Western culture. "The problem, then," according to Anthony Pagden, "was precisely how to understand 'otherness' in terms which made sense both as an account of 'their' lives or beliefs, *and* as an account of the lives and beliefs of beings who were still sufficiently like 'us' to be clearly recognizable as part of what all contemporary Europeans understood to be the 'brotherhood of man.'"[106] The Stoic epigrams commonly attached to sixteenth-century global images normalized brotherhood along Christianized Platonic lines. Inevitably, globe images neglect the atomistic Epicurean tradition associated with Lucretius's *De rerum naturum* (actually condemned by Ficino in 1470), with its historical account of chance human separation from animals. Amerigo Vespucci himself referred to South Americans as Epicureans rather than Stoics, thus placing them *ex leges,* sharing the natural morality of beasts. Confronting the moral challenge of cultural difference in new worlds, writers such as Montaigne embraced the relativist position, refusing to pass judgment on unfamiliar conduct and accepting the domestic primitivism of New World peoples.

In practice, the European response to global difference was to domesticate nature by colonization and cultivation. Even though Columbus reported his discoveries in the language of marvel and monster, the seventeen ships of his second voyage bore people, plants, and animals to settle Hispaniola.[107] His own name plays are instructive: he signed himself "Christian ferrens" (Christ-bearer) and acknowledged *Colón,* his family name, as meaning "someone who settles a land for the first time." By his fourth voyage, the eschatological themes of Columbus's letters reveal his self-image as embodying the *telos* of Christendom.[108] By 1513 Spain had codified the *Requirimiento,* an authenticating sign of linguistic colonization to be delivered within the hearing of aboriginal inhabitants. Like Schedel's chronicle, the *Requirimiento* narrated the history of the world up to the time of the first pope and required the listeners' submission to the king of Spain. It was a "cosmogonic" speech, bringing new space into Christendom's temporal framework and justifying a physical colonization that would translate nature to culture and redeem fallen peoples. For Las Casas, it would introduce a second Fall to the globe's other Eden.[109]

The confusion of experience and representation that shaped the European global vision in the sixteenth century is graphically evident in the Frankfurt engraver and publisher Theodore de Bry's fourteen-volume *Grands voyages—historia Americae* (1590–1634).[110] Giambattista Ramusio's earlier *Navigazioni et viaggi* narratives of the Atlantic discoveries had lacked illustrations, partly because their maps were destroyed by fire, and the impact of de Bry's copper-engraved illustrations was considerable. Opening with an image of Adam and Eve under the tree of knowledge, de Bry locates the New World within Christian space and time. His characteristic mode of geographic representation resembles Danti's Vatican chorographies in its synoptic view over a mapped coastline, sweeping down to details of topography, ecology, and ethnography. Recognizably human and far removed from the monstrous races, the peoples of the New World—naked, exotic, ornamented, dignified in posture and conduct—often assume the pose and stature of classical figures. A questioning of normative European humanity similar to Montaigne's is suggested by de Bry's inclusion of ancient peoples drawn from beyond the classical Mediterranean ecumene, for example, Picts and ancient Britons. Their savage posture and painted bodies place a question mark over the European reader's own cultural inheritance.[111]

Not only Europe's past offered challenges to the image of global harmony implicit in the Apollonian vision. An extraordinary mid-sixteenth-century engraving by Pierre Eskrich accompanying a cosmographic text by Jean Baptiste Trento uses the Ptolemaic picture in a devastating Calvinist parody of papal claims to global mission. The *Mappe-monde nouvelle papistique* illustrates the whole earth contained within Satan's gaping maw, the wind heads not angels but fire-spitting devils.[112] Papal imperialism is a lust for power, the New World's spices are the opium of Catholic superstition, the Mass itself a conflation of cannibalism and theophagy. Habsburg geopolitics circle the globe in a belt of human destruction. The map adopts even as it subverts the principles of cosmographic science, its fanaticism shaping the image of the world into a labyrinthine representation of Rome's urban topography.

The papist world map's emblematic use of the global image exemplifies an important feature of late Renaissance cosmographic discourse. In his *New Atlantis* of 1637 Francis Bacon wrote that "the end of our foundation is the knowledge of causes, and secret motions of things; and the enlarging of the bounds of human empire."[113] Bacon's triad would find graphic expression in the emblematic globe.

SIX *Emblematic Globe*

Yee noblest sprights, that with the bird of Jove,
have learned to leave, and loath, this baser earth,
and mount, by your inspired thoughts above,
To heaven-ward, home-ward, whence you had your birth

. . . .

And smile to see a multitude of Antes
upon this circle, striving here and there.[1]

Henry Peacham's emblem shows an eagle rising in a cloud over the globe's continents and oceans toward the rising sun. Image, motto, and rhyme combine to generate an object of moral contemplation, the ascent of the soul, escaping its temporary terrestrial prison for celestial enlightenment. On both sides of a doctrinally divided but deeply pious Europe emblems were popular devices, simultaneously communicating and obscuring precepts. Cheap printed texts and illustrations spread moral and religious ideas well beyond the confines of aristocratic and scholarly culture. The emblem's ambiguities served as a social marker for those who could interpret it and as a guard against charges of religious unorthodoxy (Fig. 6.1).

The globe's sphericity made it a common emblematic motif. Peacham's connects the spheres of earth, heaven, and eye. At once clear and obscure, emblems characterized mannerist and baroque culture, including its globalism. Navigation's continuous disclosure of an ever vaster and more varied globe filled the *Wunderkammeren,* or cabinets of curiosities, of seventeenth-century scholars and princes, the richly decorated globes and atlases that adorned studies and reception rooms offered and accepted as prestigious gifts between monarchs and states. But much of the globe's geography remained obscure, hidden, esoteric; whole continents, peninsulas, and islands connected and separated, expanded and contracted,[2] drifting across a sphere

H E A R E what's the reaſon why a man we call
A little world? and what the wiſer ment
By this new name? two lights Cœleſtiall
Are in his head, as in the Element:
Eke as the wearied Sunne at night is ſpent,
    So ſeemeth but the life of man a day,
    At morne hee's borne, at night he flits away.

Of heate and cold as is the Aire compoſed,
So likewiſe man we ſee breath's whot and cold,
His bodie's earthy: in his lunges incloſed,
Remaines the Aire: his braine doth moiſture hold,
His heart and liver, doe the heate infold:
    Of Earth, Fire, Water, Man thus framed is,
    Of Elements the threefold Qualities.

D d 1 .   And

6.1. Henry Peacham's emblem 28 from *Minerva Britannia* . . . (London: Dight, 1612). By permission of the British Library.

whose conceptual and empirical spaces could not finally be coordinated until a way was found to fix a longitude at sea.[3] The private meetings of humanists, merchants, and mechanicians gave place over time to the royally patronized academy, where the secrets of natural philosophy and geography could be systematically investigated. The cosmographic search for a unity and harmony across the apparently haphazard terrestrial surface took on an air of desperation as dreams of Christian unity drowned in thirty years of internecine bloodshed and the sanguinary struggles of competing mercantilist empires.[4]

The decorative richness of baroque maps and globes made them desirable items for collectors of geographic art; they were designed to satisfy the lusts of the eye. Vision was a matter of profound intellectual concern in early modern Europe both to the exoteric, empirical experimentation and discovery we associate with "scientific revolution" and to the still vital tradition of imaginative, esoteric speculation.[5] Ocular evidence authorized an experimental science increasingly assisted by telescopic and microscopic lenses. Imaginative vision, itself working through the fabrication and contemplation of visual images, sustained the speculative tradition. These were not distinct activities: thinkers as diverse in their conclusions as Johannes Kepler and Athanasius Kircher adopted both forms of vision. As Peacham's cloud of unknowing suggests, there was an intense awareness of how much remained to be disclosed: the southern continent, the globe's magnetic field, the form and process of celestial mechanics, the meaning of Egyptian hieroglyphics, the calculatory possibilities of the Tetragrammaton.[6] Any one of these might provide a master key to a creation still widely believed to be governed by cosmic harmony.[7] The globe's aesthetic and metaphorical representation remained a legitimate mode of its exploration.

Belief in a "hypercoded universe," to use Fernand Hallyn's telling phrase, could support the pietism of Protestants and reforming Catholics alike. An academy of natural philosophy such as Andrea Navagero's on the Giudecca in Venice could be a place for evangelicals to discuss natural philosophy.[8] Meetings opened with music, the audible expression of the cosmic resonance touched by the elevated spirit.[9] On the same island and with similar intent, choral harmonies echoed through the geometrical spaces of Andrea Palladio's Tridentine churches.[10] If, after Galileo's trial, Copernican heliocentricity was anathematized by Rome and only slowly accepted even among Rome's opponents, the *poetic* appeal of Copernicus's underlying conviction "that man can know the world in its reality and totality," together with heliocentrism's recentering of the cosmos, signaled a rational order wherein nothing was arbitrary or accidental and where aesthetics, logic, theology, and calculation should coincide in a unified space. The homology with the political and imperial discourse of absolute monarchs such as Louis XIV of France or Pope Innocent X is direct. The aesthetic embrace of emblematic, complex, but mutually reinforcing phenomena found expression in various representations of terrestrial space, not only the globes, atlases, and maps already mentioned but the giddy spaces of theatrical church ceilings, the architectural and garden complexities of royal palaces, and the construction of mechanical models of planetary motion. Global illumination became a general trope for scientific discovery. In the emblematic frontispiece of Francis Bacon's *Sylva Sylvarum*, of 1627, the *mundus intellectus* is shown as a great sphere, complete with graticule and the outline of continents and seas, half illuminated by divine light. It waits beyond the *non plus ultra* of ancient knowledge to be disclosed through observation: geographical and intellectual discovery have become one.[11]

Metaphorical and physical exploration converged in *geographia sacra,* the belief that the "imperfections" in which the earth's physical surface diverged from spherical purity, harmony, and order might symbolize deeper truths. Some imperfections were newly discovered, for example, the irregular continental pattern, although the size of the southern continent (Magellanica) on pre-Enlightenment globes and maps indicates a continued desire for symmetry. Other excrescences had bemused Europeans since antiquity. If Galileo's telescope revealed mountains *on* the moon, the Mountains *of* the Moon, the supposed location for the Nile's source, remained undiscovered. *De locis et mirabilis mundi,* the text describing marvels that was so often bound into Ptolemy's *Geography,* in addition to presenting ethnographic and animal oddities, speculated upon unresolved physiographic mysteries inherited from Mediterranean antiquity that global exploration might answer for the cosmographers: the sources of the annual Nile flood, the regular movement of the ocean tides, and the causes of volcanic fire.[12]

The simultaneous drive for empirical accuracy and decorative elaboration that characterizes the operatic globes and atlases of seventeenth-century Europeans such as the Dutch William and Joan Blaeu and the Venetian Vincenzo Coronelli connects the worlds of commercial competition, political rhetoric, and emblematic conceit. Over time the idea of rational illumination would come to predominate over more shadowy themes of occult speculation.

Globe and Culture in Late Renaissance Venice

> In regard to what you write me about M. Paolo, I thoroughly approve
> of his taking up the sacred study of astrology and geography, subjects
> of study for every learned gentleman and nobleman, as he would have
> as his guide and teacher the well-known Piedmontese to whom we
> owe so many excellent things, but first I should advise you to have
> M. Paolo construct two solid spheres. On one of these there should
> be represented all the celestial constellations, and the circles should
> all have their place, that is to say, not as Ptolemy represents the stars as
> they were located in his time, but according to the investigations of
> our own times, that is, about twenty degrees further east.[13]

This advice on the place of globes in a gentleman's education was given to
Gianbattista Ramusio, secretary to the Venetian Senate, by the physician and
natural philosopher Girolamo Fracastoro (1483–1553). The armillary sphere
that Fracastoro holds in his portrait signals the unity of his scientific interests
in the greater and lesser worlds, macrocosm and microcosm. Ramusio, Fracastoro, and two of the latter's close associates, Giacomo Gastaldi, the Republic's cosmographer, and Pietro Bembo, philologist, poet, and Venice's official
historiographer, together constituted an informal group devoted among other
things to global study. All four owned and exchanged actual globes, and their
correspondence provides an insight into a geographical culture in which the
globe stood as the very figure of poetic and scientific unity. While Bembo
and Fracastoro had died by the date of its foundation, Ramusio and Gastaldi
both joined the Venetian Accademia della Fama (1557–61), whose project
reflected the four friends' earlier interests and whose members included the
diplomat Federico Badoer and the senator Nicolò Zeno.[14] The academy's
declared goal was *renovatio,* renewal of Venice's civic and material fabric in
the light of changing commercial and political circumstances. The academy
maintained a publication program of three hundred titles, including Plato's
Timaeus, Ovid's *Metamorphoses,* ancient geographical works by both Strabo
and Ptolemy, and modern writings of Regiomontanus and Federico Del-
fino, the Paduan mathematics professor whose *De fluxu et refluxu maris* sought
to explain oceanic tides.[15] The academy's motto, *Io volo al cielo per risposarmi
in Dio,* "I can fly to the heavens to rest in God," connects its goal of civic
renewal through scientific study to personal spirituality;[16] it might equally
have served as the motto for Peacham's emblem.

The West's most cosmopolitan city in the sixteenth century, Venice was a perfect location for such a cosmographic academy. Editions of Ptolemy had been printed there since the incunable period, and Venice dominated European map engraving and publishing until the plague years of the 1570s.[17] The Venetian workshops of Camoccio, Pagano, Forlani, and Bertelli were challenged only by Lafreri in Rome for the quality and comprehensiveness of their map stocks. Gerardus Mercator deemed it necessary to take out privileges in Venice to protect his Flemish work from the promiscuous piracy of the city's publishers.[18] Venice's comprehensive diplomatic service and its vast mercantile fleet made it Europe's principal public clearinghouse for reports of geographical discovery at a time when such knowledge was tightly controlled in those European states competing more directly for the spoils of oceanic navigation.

Ramusio's post gave him unrivaled access to the best geographical information in Venice, supplemented by official contacts and the frequent exchange of letters with Venetian ambassadors, merchants, and learned correspondents across Europe. The three volumes of his *Navigations and Voyages,* published between 1563 and 1606 used these sources to report the disclosure of the globe as a continuous process beginning in Venice with Marco Polo's *Milione.*[19] Ramusio added his own commentaries, raising more general geographical questions, for example, about the relations between patterns of human occupancy, global form, and motion.[20] Were it not for a fire that destroyed the premises of Giunti, Ramusio's publishers, Giacomo Gastaldi's world and continental maps would have made Ramusio's three volumes a textual and graphic monument of early modern cosmography. Gastaldi was Europe's most accomplished designer of global and continental maps, in which oceans and continents shared equal visual significance. Commissioned to paint two great hemispheric canvases, one of Africa and South America, the other of Asia and North America, in the Sala dello Scudo of the Ducal Palace, Gastaldi drew upon Ramusio's unrivaled access to the most recent geographical information, and Gastaldi's own printed cosmography synthesizes learning from Ramusio, Bembo, Fracastoro, and the Accademia della Fama, explaining horology, the calendrical effects of circumnavigation, and calculation of longitude at sea.[21]

The cosmography reflects Gastaldi's and Ramusio's earlier conversations with Girolamo Fracastoro, who had studied the astronomical inconsistencies in Ptolemy's *Almagest,* which eventually led Copernicus, Fracastoro's fellow graduate of the University of Padua, to heliocentricity. Tycho Brahe drew upon Fracastoro's study in developing his own system,[22] while the Venetian's verse explanation of the origins of syphilis supplied the name by which the

first disease of globalization has since been known.[23] The group's philological interests are best represented by Pietro Bembo, a patrician humanist and
Venice's official historian. *Gli asolani,* Bembo's Neoplatonic pastoral, addresses
the themes of ascent, illumination, and love in the Renaissance global vision:
"The very fabric of the world which is so large and fair, and which needs
our minds rather than our eyes to understand, comprises everything; yet if
it were not so full of love, which binds it altogether with the chain of its own
discordant elements, it would never have lasted so long, nor would it be here
now."[24]

In his youth Bembo had climbed Mount Etna. More than a Petrarchian
literary exercise, his poem *De Aetna* (1496) connects Bembo's exploratory
interest to both Pliny and Columbus, Mediterranean antiquity and Atlantic
contemporaneity. Bembo speculates on the origins of volcanoes as "natural
prodigies," challenging philosophical belief in natural perfection.[25] In 1527
he acquired the famous *Tabula bembina,* a signal discovery of ancient arcana,
whose hieroglyphic engravings promised the possibility of translating the
world's original language, drawn upon by Plato for the *Timaeus.* The sixth
book of Bembo's history of Venice is a geographic and ethnographic description of the New World, pictured as a golden-age landscape offering Europe
the prospect of *renovatio.*[26] The history opens with an outline of cosmography in which Bembo acknowledges that the earth's habitability, formerly
thought to be limited to a single hemisphere and a single climatic zone, is
now revealed to be global. A Christian Platonist should have anticipated this,
for otherwise "it would be almost necessary to believe that God had been
imprudent, having fabricated the world in such a manner, that the greater
part of the earth, through its surpassing intemperance [should be] void of
men, itself having no utility."[27]

In both publications and private correspondence the Venetian group
brought a prodigious, if promiscuous, knowledge to bear on a range of cosmographic concerns, reflecting the totalizing sweep of the *furor geographicus*
in early modern Europe.[28] In one dialogue we find Fracastoro debating with
Ramusio about the nature of the universal soul, arguing from the *Timaeus*
that everything in the universe is governed by the celestial bodies, that animal and plant souls cannot act outside the *anima mundi,* that only humans
possess such freedom. The two men consider the modes of human communication with God and the intervening agency of angels.[29] In another
dialogue they debate the cause of the Nile's flood, raised elsewhere by Ramusio in the first volume of his *Navigations.* Fracastoro claimed that its cause
was equatorial rainfall produced by the perpendicular rays of the sun at

low latitude, challenging Ramusio's suggestion of permanent snow in the Ethiopian mountains. Ramusio's *Navigations* comprehensively reviews the writings of the major classical authors on this geographical mystery, concluding that the question can only be resolved empirically.[30] The friends also discuss tide times at Venice and Seville and the tidal oscillation every six hours. Fracastoro wonders whether the whole body of the ocean moves at once, or only its surface, leaving lower waters to flow in a compensatory opposite direction. Terrestrial magnetism, another global mystery, interested Fracastoro and Gastaldi. In *De sympathia et antipathea rerum* (1546) Fracastoro offers an entirely Platonic answer based on mutual love between elements, controlled by the same *anima* that governs attraction and repulsion between humans. Ramusio, drawing on navigators' reports of magnetic deviation in high latitudes, asks why these forces operate given the different degrees of predictability in different parts of the globe.[31]

Academic debates and the meetings of the Accademia della Fama took place around an actual globe. In the presence of the *balla del mondo* speculation could turn imperial, considering the colonization of the globe as a renewal and extension of Roman civilization into uninhabited regions such as Madagascar and the poles now that old ideas of limitations to human occupancy had been swept aside, or the possibilities for interoceanic canals at Suez and Panama. Venice did not directly participate in the new imperialism, but its citizens did so imaginatively; thus, in Tintoretto's *Paradise,* painted in the Ducal Palace in the years that Ramusio and Gastaldi worked there, both the Father and the Son hold globes.

Venetian geographical culture was broad and socially all-encompassing. Livio Sanuto, cosmographer at the Accademia della Fama, and his brother Giulio produced the sixteenth century's most detailed description of Africa and one of the century's largest terrestrial globes, respectively. Venetians collected and displayed globes, printed maps, and geographical memorabilia in the public rooms of palaces and villas, and this was a feature of seventeenth-century metropolitan culture across Europe.[32] Giuseppe Rosaccio (1530–1620) supplied a wider market with cheap pocket cosmographies.[33] His *Sei età del mondo* mapped a universal history across four continents, while his *Teatro del cielo e della terra* describes the Ptolemaic universe, explaining universal sphericity through a Christianized Aristotelian argument. This is complemented by a theologically inspired description of the spheres. Rosaccio locates *inferno* below the surface of the terrestrial globe, the goal of human life being to rise above the earth toward a celestial beatific vision. Pietism is justified by navigation: in the past, hell had been placed in the torrid or frigid zones—in Norway, for example, or even in the crater of Etna—but navigation has disproved such claims and Rosaccio draws upon the Dantean model of three elemental spheres surrounding a central hell: purgatory, limbo, and the surface world of the living. Navigable and measurable, Rosaccio's earth remains a metaphysical construct, an alembic for the transmutation of souls, above whose surface the spheres ascend 999,995,500 miles to the heavenly throne. Rosaccio's *Teatro* world map adheres to Ptolemy's *Geography* rather than to Ortelius's *Typus* and includes an image of Adam and Eve gesturing toward the Tetragrammaton that shines from the ethereal spheres.

The *Sei età* is a historical and anthropological companion to the *Teatro's* natural philosophy. Modeled on Münster's cosmography, it recounts a universal history of the six eras of Creation, corresponding to the six days of God's original working week, Christ's death and resurrection initiating the sixth imperial era of Christendom. This formula is common on seventeenth-century world maps. Predictably, Rosaccio's other popular texts include a description of the human body as microcosm, travel guides, and illustrated world maps.[34]

Girolamo Ruscelli and the Emblematic Globe

Gianbattista Ramusio's complaint that his European contemporaries, unconcerned with ideas, were preoccupied with "applying themselves to the contents of the whole round earth in order to satisfy their immense greed and avarice" might reflect the sour grapes of a once dominant trading city obliged to watch the oceanic scramble from the sidelines.[35] But the integrity and influence of a metaphysical geography in late Renaissance culture cannot be discounted. The Venetian philologist and academician Girolamo Ruscelli connects the science of globes directly to the moral and metaphysical discourse of emblem and allegory. Ruscelli's 1558 publication of Plato's *Timaeus* in Italian for the Accademia della Fama was followed in 1561 by his Italian translation of Ptolemy's *Geography*.[36] Its twenty-six ancient and thirty-six modern maps were drawn by Gastaldi and engraved by Giulio Sanuto. The work included Ruscelli's commentary and a treatise on the practical fabrication of small globes by Galileo's tutor and professor of applied mathematics at Padua, Giuseppe Moletto.

The verities of vision and representation are a recurrent theme in Ruscelli's Ptolemaic commentary. Wishing to describe the whole earth on sphere or flat map, Ptolemy had been obliged to use cartographic signs rather than mimetic images for cities, countries, and events. Ruscelli focuses on Ptolemy's distinction between geography and chorography, whose distinct modes of spatial representation turn upon relationships between eye and mind as tools for acquiring knowledge. Globes and world maps offer the means for humans to "see" the earth as a unity and thus to know it directly.[37] But the visual image alone is insufficient. A text is also required; the globe maker produces an *iconotext*,[38] and Ruscelli compares Ptolemy unfavorably with Strabo as a writer of regional description. At the global scale, the earth's curvature makes for technical difficulties in bringing together image and printed text, necessitating the production of gores. These representational problems are compounded by the increased size of the ecumene, which modern discovery has extended far beyond Ptolemy's single-hemisphere rectangle. Ruscelli recommends the double-hemisphere map as "the most rational, the truest, and the best way of representing our modern world, that is, the whole globe of the habitable earth, in the plane."[39] This *tavola universale* has the advantage of allowing the viewer to "see" the globe's sphericity. It is no longer adequate simply to locate places; we need a visual impression of the earth's rotundity now that both hemispheres have been shown to be habitable. The double-hemisphere map was not Ruscelli's invention, but he was the first to use and justify it in an atlas.[40] His text also offers the first substantive technical discussion of globe construction. Linking it directly to the techniques of linear perspective, Ruscelli also uses spherical analogies, for example, the surface of a melon held on a skewer as axis.

Ruscelli's interest in the logics of vision and relations between picture, text, and number recur in his illustrated edition of Ariosto's *Orlando furioso* and in his emblem book.[41] Ruscelli illustrated each canto of Ariosto's crusading romance, which takes as its canvas the Ptolemaic ecumene, with a woodcut, a brief verse synopsis, and scholarly annotation in the manner of the emblem. Many of his images set Orlando's adventures against a world landscape or Ptolemaic map offering an Apollonian view of the dramatic action. Canto 2 shows the North Sea littoral; canto 20, the Mediterranean; and canto 15, the ecumene from the mouth of the Ganges to the Mountains of the Moon. The images reflect Ariosto's theme of crusading Christian imperialism. In his preface Ruscelli claims that his combination of image and text increases the reader's understanding of the narrative, and he stresses the significance of perspective in achieving a realistic vision. This discussion was plagiarized by John Harrington in his own preface to the English translation of *Orlando furioso:*

The use of the picture is evident, which is, that (having read over the booke) you may read it (as it were againe) in the very picture, and one thing is to be noted, which every one (haply) will not observe, namely the perspective in every figure. For the personages of men, the shapes of horses, and such like, are made large at the bottome, and lesser upward, as if you were to behold all the same in a plaine, that which is nearest seemes greatest, and the fardest, shewes smallest, which is the chiefe art in picture.[42]

In his collection of emblems, *Le imprese illustri,* Ruscelli uses the globe recurrently. The first emblem book to introduce human figures, it became a standard text. As in his Ptolemy, Ruscelli offered technical instructions on making such devices. The idea of correspondence is fundamental to the concept of the emblem, whose combination of visual image and printed word, Ruscelli points out, simultaneously reveals and obscures meaning.[43] The emblem's purpose was "to feed at once both the minde, and eye,"[44] typifying a concern for intellectual and spiritual illumination in the shades of esoteric allegory, the *sfumatura* so favored among mannerist and baroque painters. The world's deepest truths—the unity and interconnectedness of nature—lay beyond and below surface appearances, revealed to humans through physical, intellectual, and spiritual exploration, and often only indirectly.[45] The emblem fulfilled the epistemological goal Ernst Cassirer attributed to all symbols: continuous and necessary ascent from the bonds of the sensory toward purely intellectual abstraction.[46] Visual image, textual legend, and verse combined in an allegorical, moral, or intellectual argument, revealed to the cognoscenti while hidden from the vulgar and uneducated.[47] The principle of the emblem connects directly to the idea of a *higher* knowledge and its attendant ambitions and dangers, amply captured in Bruegel's emblematic *Landscape with the Fall of Icarus.*[48] It also expressed a "critical commonplace of the time . . . that both poetry and painting signify something beyond what they actually are, and beyond what they seem to represent."[49]

As a cultural phenomenon, the emblem remained significant from the mid-sixteenth century until well into the eighteenth, emblem books being translated, plagiarized, popularized, and pillaged for images and ideas in all European languages.[50] The foundational collection was by the Italian Achille Bocchi; others were those by Claude Paradin in France, Andreas Alciato and Michael Maier in Germany, and Henry Peacham in England.[51] Cesare Ripa's *Iconologia,* of 1593, gave the emblem a classical pedigree, classifying and regulating the allegorical meanings of visual images.[52] An emblem, claims Ripa, "can represent a visible thing and also something different from it, yet having conformity with it, because if the former often persuades by means of the eye, the latter moves the will by means of words, and by means of the former [the image] the metaphor of things is seen that, when conjoined with the latter [the text], states the essential."[53]

Emblems dealt primarily with natural philosophy and secondarily with human concepts, virtues, and habits, in each case extending naive vision. In the emblem every element and gesture is meaningful; nothing is any more accidental than in the world itself. Reading the world for hidden truths found intellectual justification in both Plato's image of the shadowy cave and Christ's parables: the most profound or elevated things cannot be stated directly. By Ripa's time the emblem had become a genuinely popular cultural form, an element of devotional practice in both Catholic and Protestant communities, "tapping and creating a large area of the common consciousness," and it would remain so well into the eighteenth century.[54]

In attributing meaning to visible form the emblem intentionally blurs any distinction between sight and insight. Thus, the sphere as form and object of vision held particular attractions for emblem makers. Galileo's Copernicanism and discovery of Jupiter's moons cannot be disconnected from the emblematic significance attributed to globes in the Medicean court,[55] and the globe's imperial associations make it a recurrent device in Ruscelli's *imprese*. For Philip II of Spain, he places Apollo's chariot between terrestrial and celestial globes with the motto *Iam illustrabit omnia*. For Ferdinand of Austria, flags at the cardinal points of a globe signify both Christ's universal authority and the Habsburg role in opening the globe to Christian faith. For Henry II of France, Ruscelli gives the globe a more Ficinian interpretation: "Thus we see through the ordering of the scale of Nature how man is placed at the center, supreme over all created things, closest to the angels" (Fig. 6.2).[56] The terrestrial globe figures in Ruscelli's personal emblem supporting History, Poetry, and Music to reveal how the man of science illuminates the world in the manner of the Sun and Moon.

The emblematic meanings of the globe were complex and varied; it could signify, for example, a bubble (or bauble) and thus nothing *(nihil)* or at most a distracting trinket, a display of *vanitas* and the transience of material acquisition. Thus the cordiform map is placed within the fool's cap, while the gorgeously dressed Lady World in Jodocus Hondius's 1597 *Christian Knight Map of the World* is crowned with the globe but gestures toward Death, and Magnus Jörgensen's *Vanitas* makes a similar point by means of the bubbles bursting over the globe (Fig. 6.3). Alternatively, the globe could signify en-

6.2. Girolamo Ruscelli's emblem for Henry II of France (1580). In Ruscelli's *Le imprese illustri con espositioni, et discorsi del S.or Ieronimo Ruscelli al serenissimo et sempre felicissimo re Catolico Filippo d'Austria* (Venice: Francesco de Franceschi Senese, 1566), 29r–v. Harry Ransom Humanities Research Center, The University of Texas at Austin.

6.3. The globe as an emblem of worldly vanity: Magnus Jörgensen's *Vanitas* (Copenhagen, 1709). Statens Museum for Kunst, Copenhagen.

lightenment, the creation of a beneficent God, as in the frontispiece to Du Bartas's *Deuine Weekes.*[57] Emblematic globes draw upon geographical discovery in Francis Bacon's frontispiece, connecting contemporary imperial expansion to the teleology of universal history. The arch that decorates the title page of Walter Ralegh's *History of the World* (1614) is supported by Corinthian columns inscribed with hieroglyphics.[58] History, its central figure, holds aloft a globe with a loosely Ortelian image of continents and a southern landmass. A battle is being fought in the mid-Atlantic between English and Spanish fleets. The globe is supported by Good and Evil Fame, represented by healthy and plague-ridden angels, illuminated and clouded, respectively. Angelic trumpets and the skeletal figures of Death and Oblivion indicate the destinies of victor and vanquished, respectively, under the impartial eye of Providence. The female figures of Experience and Virtue stand between the columns.

Globe and world map were insistent Elizabethan and Jacobean allegories and literary conceits, from Maria's caustic comment on Malvolio in *Twelfth Night* that "he does smile his face into more lines than is in the new map, with the augmentation of the Indes," to John Donne's and Milton's constant global and cartographic metaphors for the human body. The medieval connection between microcosm and macrocosm was renewed in the parallel exploration and mapping of globe and human body by navigation and anatomical science. John Donne's lines "At the round earths imagin'd corners blow / Your trumpets, Angels. . . , and arise, arise / From death you numberlesse infinities / Of soules . . ." work the eschatological globe, mapped within the frame to contemplate the fate of the poet's own body.[59]

Emblematic Mapping: Seeing and Reading

Map and emblem demanded similar skills of printing and engraving, so that publishers such as Fernando Bertelli in Venice and Theodore de Bry in Frankfurt traded in both products.[60] There was a shared belief that graphic representation might overcome the relations between words and things that underlay doctrinal division and ethnographic difference and thus help reunify mankind.[61] As Ruscelli pointed out, neither the globe's sphericity nor the larger geographical patterns on its surface could be seen, yet it had a definite material reality. In this it resembled those other "unseen realit[ies] upon which the visible world was based,"[62] such as intellectual concepts or moral propositions, which required the medium of graphic expression to be rendered visible. Emblem, globe, or map could serve for moral commentary or reflection.[63] The principles of correspondence inherent in an allegorical epistemology dictated that if the globe and its major parts were available for use in emblems, then global and continental cartography could be read emblematically. This is explicit in the *Mappe-monde papistique* and the *Christian Knight Map,* but the complexity and density of decoration associated with baroque globes and world maps generally reflect the emblematic mode of reading graphic images. Thus Joseph Moxon's *Tutor to Astronomy and Geog-*

raphy, of 1654, fills the space between adjacent hemispheres with the sub-equatorial extensions of Africa and America, avoiding the need for speculative mapping of a southern continent while framing the world with engravings of the Genesis creation narrative and the seven eras of human history (Fig. 6.4).[64] The creating power of the Tetragrammaton emerges from a source of light, while sacred history culminates in St. John's vision of the celestial city.

The globe, "most capable, most simple, [which] doth bend in all parts towards it selfe, sustains it selfe, includes and containing it selfe, wanting no joyning together, nor having any end or beginning in any of its parts,"[65] encouraged seventeenth-century cartographers to treat its representation as a total artwork, in the same manner as the new art of opera or the baroque Mass. If the globe was a divine *Wunderkammer*, its representation was a *Gesamtkunstwerk*. This idea underlay the lavishly elaborated Dutch and French globes and atlases of the seventeenth century. Mercator's 1585 elaboration of Ortelius's *Theatrum* initiated a continuous evolution in the stylistic presentation of the world and its parts. One hundred seven of Mercator's *Historia mundi* maps were completed by 1595, and the work had been translated into English by 1635. The text addresses itself directly to a gentry who "in these tempestuous times" cannot attempt travel with any safety and are obliged to pursue their journeys from their studies. Geographical maps, like emblems, "presenteth to our sight the Globe of Earth as it were a Mirrour or Looking-Glasse, And doth show the beauty and ornaments of the whole Fabricke of the Worlde . . . to omit the neare affinitie which this noble science hath with Astronomie, which mounting above the Earth doth contemplate the Heavens."[66] Mercator's idea of vicarious travel was by now a commonplace, but his intention was to assist the soul as much as to feed the mind and please the eye. In *The Ship of Fools* Alexander Bartlay had addressed the geographer's obligation to circumscribe the soul rather than the earth, and Mercator's five-part atlas, opening with the days of Creation and extending through astronomy, geography, genealogy, and chorography, prefaces a more pious end, the spiritual contemplation of the heavens: "The glory of this thy habitation granted unto thee only for a time, who doth not compare it with the heavens, that he may therefore lift up those minds which are drowned in these earthly and transitory things, and shew them the way to more high and eternal matters."[67]

Such ideas had a particular appeal in the Netherlands, where "visual culture was central to the life of society . . . the eye was a central means of self representation and visual culture a central mode of self-consciousness," and

6.4. Joseph Moxon's world map and narrative of universal Christian history (1654). By permission of the British Library.

where the emblem book achieved especially wide popularity.[68] Jan Vermeer, picturing the quiet simplicity of everyday life in commonplace settings, illustrates profound moral propositions and frequently decorates his interiors with both emblems and maps, inviting the viewer to contrast them as ways of understanding the world.[69] Vermeer's only male subjects were mapmakers, portrayed in *The Astronomer* and *The Geographer*. The former, compass in hand, globe and texts above him and bathed in light from his window, is the very figure of contemplative piety, deploying secular knowledge to raise his mind to higher thoughts. In his *Allegory of Faith* Vermeer explicitly links the purity of the suspended crystal sphere with the geographical globe via an image of the crucified Christ. The Leiden painter Gerrit Dou connects globe and emblem even more directly. A young man is seated next to a table on which a globe supports an open book of emblems. The room is bathed in light from a window whose glazing bars resemble the cartographic grid. At the center of the composition is a violin, symbolizing Pythagorean harmony.[70]

In Protestant Europe the textual authority of the Bible dominated religious belief and practice. Calvinist iconoclasm questioned biblical illustration but accepted the instructional value of maps. Maps of the Holy Land, Jerusalem, or the Mediterranean journeys of St. Paul assisted the reader's grasp of scriptural geography, while others illustrated biblical exegesis. Cosmographic schemes, maps of the postdiluvian repopulation or of Daniel's four-kingdoms dream, and chorographies of Eden all served to "proclaim the Protestant view of the primacy of scripture over theological doctrine, and emphasized both the historical reality and the eschatological promise of scripture by demonstrating its geographical setting."[71] A Haarlem Bible of about 1598 placed Ortelius's *Typus orbis terrarum* at the opening pages of Genesis. Cosmography, world history, atlas, and scripture met in seventeenth-century's sacred geography, as Mercator said, to "lift up those minds which are drowned in these earthly and transitory things, and shew them the way to more high and eternal matters."

Totalizing Order in Dutch Mapping

Seventeenth-century secular cartographic projects also attained a heroic, "global" scale. The Protestants Willem and Joan Blaeu and the Venetian Catholic Vincenzo Coronelli illustrate the scope and complexity of baroque world images and the dimensions of European spatiality—secular and commercial, idealist and imperial, empirical and transcendental. In independent

Holland, globe and atlas making developed dynasties; the Blaeus, Henry and Jodocus Hondius, and Abraham Goos were commercial rivals, plagiarizing and reengraving one another's plates until the Blaeu workshop, with all its documentation and plates, was destroyed by fire in 1672, signaling the decline of Amsterdam's intense cartographic culture.[72] The swelling body of information produced by Dutch exploration and commercial expansion provided their knowledge base; indeed, Joan Blaeu worked directly for the Dutch East India Company. Discovery is celebrated in legends on the globes and maps, but cosmographic rhetoric competes with empirical accuracy. Henry Hondius and Jan Jansson's preface to their five-volume atlas of 1646 extends an implicitly European and male lordship over creation: "All visible creatures made by God are comprised by these two here, Man and the World. The former has been made lord of the Universe, the latter the seat of his empire. The former is the guest and inhabitant of the world, the latter the most magnificent and spacious house for such a great guest. In Man we recognize the image of the excellent Artisan who created him, and in the world, the image of Man"[73] (see Fig. 6.1).

Global images record and celebrate the contest for this imperial domain between Holland, England, Spain, and France in a rhetoric that continues to draw upon the classical model of empire. Thus, in his "Annus Mirabilis: The Year of Wonder, 1666" (1667) the English poet John Dryden describes Britain's wrestling control of the world's riches from Holland in the second Anglo-Dutch War, of 1665:

> Thither the wealth of all the world did go,
> And seem'd but shipwrack'd on so base a Coast.
>
> For them alone the heav'ns had kindly heat
> In Eastern Quarries ripening precious Dew:
> For them the *Idumaean* Balm did sweat,
> And in hot *Ceilon* Spicy Forrests grew.
>
> The sun but seem'd the Lab'ror of their Year;
> Each wexing Moon suppli'd her watry store,
> To swell those Tides, which from the Line did bear
> Their brim-full Vessels to the Bel'an shore.
>
> Thus mighty in her Ships stood *Carthage* long,
> And swept the riches of the world from far:
> Yet stoop'd to *Rome,* less wealthy but more strong:
> And this may prove our second Punic War.[74]

Joan Blaeu's cosmographic *Atlas maior,* illustrating earth, air, waters, and the
fiery celestial bodies (uranography), exemplifies the baroque globe as total
artwork, synthesizing, illuminating, and celebrating an imperial mastery of
creation. Inevitably, such a scheme was never fully realized, but eleven volumes were produced and translated into different languages as official gifts
of the Estates General in the economy of diplomatic exchange between European sovereigns.[75] Opening the 1664 Dutch edition, *Grooten Atlas oft Werelt-
beschrijving,* the terrestrial globe appears as a green-robed female crowned
with the city, marking civilized human dominion, and holding key and trumpet, signifying the knowledge that the atlas unlocks and triumphs. Her chariot is drawn by Ripan figures of the four continents. The text proper opens
with the geocentric cosmos and a double-hemisphere world map, its cities
marked by gilded points. The chariots of planetary God pursue their cosmic
circuits around an earth supported by cosmographic figures holding an armillary sphere and terrestrial globe.[76]

A more concentrated cosmographic expression is Blaeu's multisheet wall
map of 1648, *Nova totius terrarum orbis tabula.*[77] The map is emblematic in
structure with a summary title above a central set of images and a detailed
text below. The terrestrial hemispheres are illuminated by light radiating from
the solar center of a Copernican cosmos. Celestial rays disperse billowing
clouds of dark ignorance to display knowledge of creation. The image is
composed by an array of ocular spaces revealing terrestrial hemispheres,
northern and southern constellations, Ptolemaic and Brachian hypotheses of
planetary motion, together with salamander, whale, eagle, and mole representing, respectively, fire, water, air, and earth. The Ptolemaic ecumene is
superimposed on a cylindrical graticule showing the Aristotelian climates
and zones and revealing the limited scale of ancient knowledge. Ships and
convoys fill the oceans in an inventory of maritime empire that has extended
the classical land imperium *ad termini orbis terrarum.* The whole work acts as
a totalizing emblem of knowledge, illumination, and global acquisition. The
text below explicates the meanings of the map that lie beyond the representational capacity of pictorial images. In Latin and French versions it outlines the principles of cosmography and describes both earth and heavens,
whose countless stars are known individually to God. The reference to Psalm
147 would have alerted a contemporary reader to the lines that follow: "He
sendeth forth his commandment upon the earth . . . he causeth his wind to
blow, and the waters flow," and, given the map's production in the year in
which the Thirty Years' War ended in the Peace of Westphalia, "He maketh
peace in thy borders."

The Peace of Westphalia ended Europe's bloodiest early modern war and established a system of territorial states that consigned to spiritual and representational space alone any remaining dreams that a universal Christian empire might revive that of ancient Rome. Global empire became a secular prize to be contested between monarchal states. Dreams of spiritual universality had fueled cosmographic utopias in the early years of the century—Campanella's *City of the Sun*, Michael Maier's emblematic Rosicrucian republic, and Bacon's *New Atlantis*—and such yearnings may have underlain Enlightenment itself. Stephen Toulmin suggests that the Cartesian philosophical project was born less from a rational rejection of speculation in favor of empiricism than from a desperate desire to sustain the ancient cosmological dream of universal order and harmony in the face of the anarchy unleashed in religious warfare.[78] The self-confident, acquisitive, imperial gilding of the baroque world map hides the more contemplative and pietistic emblem of a disappearing order.

Papal Globe and Jesuit Empire

Unsurprisingly, it was in Catholic Europe that the globe continued to signify universal ecumenism. Louis XIV's Versailles and papal Rome competed as *axes mundi* of Catholic empire, a competition worked through the iconography of ancient empire into their landscapes under the patronage of the Sun King and Popes Urban VIII, Innocent X, and Alexander VII. In Rome, Gianlorenzo Bernini and Francesco Borromini constructed in stone and stucco a *theatrum mundi* for a globe evangelized by militant missionaries such as the Jesuits, who radiated from the city across its expanding surface. Baroque Rome is a series of urban stage sets focused on ancient Egyptian obelisks, reerected as gnomons measuring the Tridentine Church's global centrality. At Piazza di Spagna, the ship of Faith navigates the spaces below the obelisk of Trinità del Monte. At St. Peter's, Bernini's towering double columns embrace an imperial forum designed to accommodate the world's pilgrims. At its center the obelisk rises on the supposed spot of St. Peter's martyrdom. At Piazza Navona, Bernini's Fountain of the Four Rivers raises its obelisk over the four world rivers—the Nile, the Danube, the Ganges, and La Plata—flowing like the streams from Eden across the globe, their sculpted personifications signifying the human variety united in a single faith. "A whole organic world [is] alive with light, water, and air and the forms of animals and plants: in effect a grotto of the original source, turned inside out. Even the force of the wind is present, blowing through the palm tree."[79] The vision of Blaeu's wall map is here realized in a travertine marble confection.

The obelisk was not only a cosmographic instrument whose shadow located time and latitude; its very form denoted a beam of light shafting through the clouds. Illumination was a governing theme of the Catholic geographic iconography: the declared mission of the baroque Church was to enlighten with faith the spaces of European discovery, rendering the globe itself a symbol of belief. Thus on Alexander VII's tomb in St. Peter's, an overwhelming construction of black marble, Faith places her foot on a terrestrial globe engraved with the Ptolemaic ecumene.

Catholic missionary imperialism, cosmographic science, and metaphysical speculation framed the particular goals of the Jesuit order. Founded in 1540 by Ignatius Loyola as an order of Counter Reformation teachers to evangelize at the geographical borders of Christendom, Jesuits located themselves just as earnestly at the frontiers of learning. Their colleges and seminaries housed magnificent libraries, especially in Rome, the center of Jesuit learning and calculation, with which a global network of correspondents communicated. Jesuits such as Christopher Scheiner and Giovanni Battista Riccioli, who illustrated six possible world systems, were major contributors to astronomical science. Global evangelism induced admiration for linguistic, philosophical, and intellectual traditions far beyond those of the Latin West—Byzantine, Coptic, Islamic, Jewish, Chinese. The Jesuit cosmographic and geographic vision also relied heavily upon the power of visual images.[80] Matteo Ricci, following Francis Xavier's mission along Portugal's sea lanes to Asia, penetrated the imperial court of Beijing and produced an Ortelian world map with Chinese phonetic equivalents in 1584. Later updated, the map played a globalizing role in its own right by persuading some Chinese scholars to accommodate their own vision of global space to that of the West.[81]

A century after Ricci, Andrea Pozzo decorated the ceiling of the Jesuit collegiate church of St. Ignatius in Rome with a spectacular global representation (Fig. 6.5). Drawing on a tradition of cosmographic ceiling decoration originating in Byzantium, Pozzo produced a single illusionistic image, an "infinite" perspective that carries the eye along a vertiginous axis past soaring columns and capitals and through billowing clouds to a heavenly source of light high in the vault of heaven.[82] St. Ignatius receives a single ray of divine illumination from the Trinity; it fractures from the saint's breast into a fan of beams illuminating representative figures of mankind, saved by the Ignatian mission. A secondary ray reflects from a great concave mirror held by an angel to shine on the four continents, sceptered Europe resting on the globe of Faith. The textual key to this spectacular drama of universal illumination is Ignatius's claim "Ignem veni mittere in terram,"[83] and Pozzo's

6.5. Global redemption reaches Africa through the light of faith shining from infinite celestial space, in Andrea Pozzo's fresco on the ceiling of St. Ignatius, Rome, 1694.

illusory space is a dramatically appropriate setting for the theater of the baroque Mass, where the soul rises in light and music from gross material ignorance toward the beatific vision via the mystical body of Christ. Christ-Apollo's central position within Pozzo's composition, his body directing the light of faith to the corners of the earth, suggests an iconographic heliocentrism within an order whose scientists were among the most learned of the Church's opponents to Copernicus's cosmic decentering of the globe.[84]

Pozzo's celestial mirror reflecting the light of redemption across the globe was a common feature in Jesuit iconography. It appears on the title page of Athanasius Kircher's *Ars magna lucis,* of 1671.[85] Here too the angelic lens directs a ray of divine illumination toward the earthly sphere while opening sacred celestial space to speculative vision. Jesuits took a keen interest in the issues raised by the extended scales of nature opened to vision by telescopic and microscopic lenses.[86] Kircher (1602–80), a professor of mathematics at their college in Rome, ranged over Ptolemaic science, from his Platonic celestial journey to a geography of China and a chorography of Latium. His display of objects at the Jesuit College, gathered from across global time and space, was one of the spectacles of baroque Rome, effectively one of the first public museums in Europe. No random amassing of curiosities, it operated as the empirical foundation for Kircher's scientific project: to develop a single moral, religious, and philosophical framework for a diverse globe, below whose bewildering variety Kircher sought a primordial logic via syncretic theories combining symbolic correspondence, philological comparison of ancient and modern languages, and Platonic-Pythagorean combinatorial science.[87] His studies commanded respect even among critics such as Galileo Galilei and London's Royal Society. Dutifully criticizing heliocentrism, Kircher does not fit the stereotype of Catholic bigot, and in favoring experimentation he embraced much of the new philosophy. "Yet he diverged from it, as practised by Galileo, Descartes, and many members of the Royal Society, in one important respect: his insistence that wonder was a category of analysis rather than simply a tool to lead men to the contemplation of higher truths."[88] Wonder or marvel underpins emblematic cosmography in Jesuit thinking, and fundamental to Kircher's epistemology was the identification of signs. This drew him to Egyptian hieroglyphics, interpreting those on the obelisk in the Piazza Navona in the hope that they would reveal the wisdom given to mankind directly from the mouth of God.[89] If Egypt represented the origins of human culture and language, it was connected also to the origins of geographic space. Egyptian land and Egyptian civilization were the creation of the Nile, whose flood remained the most enduring of geographic mysteries, although Kircher claimed that the Jesuit Pedro Pais had actually visited its source in 1618.

In his mystical text *Itinerarium exstaticum* (1656) Kircher, entranced by sacred music, ascends through the spheres of a Tychonian cosmos to contemplate each planetary surface, including that of the Sun (whose surface pattern of volcanic fires he maps), before returning toward Earth, whose globe he sees in full (Fig. 6.6).[90] The text follows the model of the Ciceronian *somnium*, or enigmatic dream, familiar from Macrobius's *Commentary* and revived as a scientific-literary form by Kepler, whose lunar dream had appeared in 1634.[91] Among the terrestrial phenomena that attract Kircher's attention are volcanoes (subject of a later book, *Mundus subterraneus*, 1665), tides, and water flows, including Scylla and Charybdis, the whirlpool and tide race in the Straits of Messina, supposedly caused by underground seawater passing below the fires of Mount Etna. The logic behind Kircher's interest in natural mysteries is summarized in his Lullian text, *Ars magna sciendi* (1669), which draws upon astronomy, astrology, and optics to explore

6.6. Athanasius Kircher's cosmographic journey: frontispiece to *Iter exstaticum* (1671). By permission of the British Library.

6.7. The cordiform map of earth provides the altar cloth below the sacred host that brings the light of redemption across the globe's continents in Stephen Eggestein's Jesuit emblem of 1664. Engraving by Bartholomew Killian. Private collection.

"the relations between solid bodies and all forms of simulacra, many of them produced by manipulating rays of light." The work is structured around Kircher's universal instrument of observation, his *horolabiorum,* a form of astrolabe, and offers instructions on the manufacture of celestial and terrestrial globes. The sphere is the recurrent motif of his elaborate illustrations. Two principles underlie this "great art": combination, an attempt to rediscover the initial unity from diverse phenomena; and analogy, through which each individual phenomenon is reflected in every other. "In Kircher's world . . . symbolic codes disclosed the underlying harmonies that connected what would otherwise appear to be mere collections of unrelated things—suns, moons, animals, plants, gods"; that is, it was global.[92]

The imperial thrust of global illumination is summarized in the image of Kircher's Universal Jesuit Horoscope. An olive tree emerges from the haloed head of St. Ignatius, who kneels on the meridian line of Rome as ships depart for new worlds. Its branches hold the names of Jesuit provincial houses, and its leaves indicate the regions within the society's provinces, "diffused throughout the whole terrestrial orb." For every location, a calendar indicates the hours of sunrise and sunset, together forming the Christian device, the letters IHS. Illuminated by candles, Kircher's original in Rome revealed the shadow of the holy name moving over the globe. At the four corners (the conventional map spaces occupied by the sons of Noah or trumpeting angels) the parts of the world are allocated to Jesuit missionary saints, while the text "From East to West, praiseworthy is the name of the Lord" is inscribed in thirty-four languages. The Society of Jesus is the light of the world, saying Mass, administering the sacraments of faith, and enunciating the mystical name of the Christ-Apollo across the planetary surface each minute of every day, illuminating the turning globe doctrinally as the sun lights it physically.

Kircher's emblematics reflected a broader Jesuit commitment to the power of visual images, frequently connected to global themes. In 1640, for example, the Flemish province published a collection of emblems to commemorate the society's centenary. *Imago primi saeculi Societatis Iesu* contains numerous images of celestial and terrestrial orbs: angels turning the machine of the world, holding the arrows of faith over the two hemispheres and the motto Ruscelli had used for Francis II, *Unus non suffi it orbis.* In 1664 Stephen Eggestein's Jesuit apologia used the cordiform world map as altarpiece in an image whose iconography of light and redemption across four continents is precisely that employed by Andrea Pozzo at St. Ignatius in Rome (Fig. 6.7).[93]

Sun King, Empire, and the Emblematic Globe

If the Jesuits used the globe to symbolize the empire of faith, Catholic monarchs harnessed its symbolic authority to more secular ends. Girolamo Ruscelli's emblems for French and Spanish monarchs are based on the globe, and
a century later another Venetian extended this to unprecedented extremes
to flatter the imperial pretensions of the Sun King, Louis XIV. Between 1681
and 1683 Vincenzo Maria Coronelli (1650–1718) conceived and built in Paris
a pair of giant celestial and terrestrial globes, 3.9 meters in diameter. Commissioned by the French cardinal in Rome, César d'Estrées, they were completed in the year of Louis's revocation of the Edict of Nantes, legal guarantor of French religious tolerance. Revocation, promoted by d'Estrées himself,
signified France's replacement of Spain as Catholicism's secular defender and
paralleled a French policy of commercial and territorial imperialism, symbolized by the appointment of d'Estrée's brother as admiral of France. The
great globes were initially destined to be placed in the Petite Orangerie at
Versailles, a public space designed more to flatter and celebrate royal authority than for strategic deliberation or policy formulation. Their dual claim—
to empirical accuracy and symbolic expression of divinely authorized sovereignty—determined their design, their content, and their location.

The dedicatory inscription on Coronelli's celestial globe proclaims it an
image of the heavens "in which all the stars of the firmament and the planets are placed in the very locations where they were to be found at the birth
of this glorious monarch in order to conserve for all eternity a fixed image
of that hour and disposition under which France received the greatest gift
that the heavens had ever offered to the earth."[94] The emblematic rhetoric
of the terrestrial globe is equally explicit. Picturing the earth's surface known
to Europeans in 1683, it "renders continual homage to his glory and heroic
virtues, showing the countries where a thousand great actions have been
executed both personally and through his command, to the astonishment of
the nations, which he could have subordinated to his empire had not his
moderation arrested the progress of his conquests and prescribed the limits
of his valor."[95]

Emblematically, the globes' contents simultaneously illuminate and obscure the world. Their decorative content is astonishing in range and complexity as well as in design and color; many of the legends and inscriptions
that occupy "unknown" spaces are detailed geographic and ethnographic
dissertations. The scale that allowed such detail also meant that the globes'
contents were made apparent to a viewer only by making Apollonian cir-

cuits or navigators' ocular journeys across their surface, and even then with such difficulty that the king had opera glasses specially commissioned to read their content. In 1710 the courtier François le Large transcribed into two notebooks six hundred texts with detailed iconographic and allegorical commentaries, inverting Coronelli's project. As Christian Jacob put it, "The Italian cosmographer, starting with a vast documentation, correspondence, oral information and bibliographical study, encoded knowledge in graphic form using all the resources of the image: allegory, emblems, metonymy and metaphor . . . , representing a single episode to suggest an entire history," while le Large sought to decode the visual evidence of the globes and turn it back into a coherent text.[96] Jacob remarks that the logic of vision and the logic of the text are here brought into sharp contrast, revealed in the high level of redundancy and cross-referencing le Large was forced to adopt.

Coronelli's globes are vast emblems, stores of empirical and allegorical information that serve equally Louis XIV's geopolitical and symbolic purposes. The self-proclaimed Apollonian monarch is positioned by his nativity on the celestial globe. Jean-Domenique Cassini's grand meridian of 1678, surveyed from Dunkirk to the Mediterranean through Paris, connected celestial space directly to the royal capital. From Versailles, after 1682 the center of French territorial space, Louis could encompass the earthly orb, his lenses and le Large's text enabling him to circle the mechanical globe as the actual globe turned around him. Jean-Baptiste Colbert, who oversaw LeVau's construction of Versailles, was also Louis's principal adviser on navigation and commerce, encouraging the king's patronage of Cassini's astronomical work on the meridian and France's rise to prominence in map and atlas making. It was also Colbert who amused the king with model naval spectaculars on the waters of the Petite Venise and established the Petite Academie, whose responsibilities included designing emblems and medals. On its recommendation, the decorative organization of Versailles was to be determined according to "relations among the influences and qualities attributed by mathematicians to the seven planets."[97] Its apartments were planned on a heliocentric planetary model wherein the various subsidiary rooms revolved around those of the Sun King himself. Le Brun's decorative scheme derives from Ovid's cosmogonic description in the *Metamorphoses*.

Coronelli's globes follow this emblematic convention as a visible commentary on the symbolic spatialities of Louis XIV's person, palace, court, and empire. Versailles's courtly rituals revolved around the Apollonian conceit of royal authority illuminating political and intellectual darkness just as the sovereign vision unifies global space. This baroque princely conceit had been anticipated at Louis's birth on 5 September 1638, the date commemorated by the arrangement of the heavens on Coronelli's celestial globe. During his 1639 lecture visit to Paris the Italian utopianist Tommaso Campanella, heliocentrist and hermetic cosmographer, had cast Louis's horoscope, and his theocratic utopia, *The City of the Sun,* almost anticipates the project for Versailles. Campanella's imaginary city is "divided into seven large circular areas named after the seven planets, and the way from each circle to the next is along four roads and through four gates which face the cardinal points of the compass."[98] The city's central temple is decorated by celestial and terrestrial globes, from which metaphysicians, furnished with astrolabes and telescopes, study the heavens, while geographic and ethnographic information is furnished by explorers and ambassadors.[99] The supreme authority in Campanella's utopia was a priest-philosopher-king, like the philosopher-scientists of Francis Bacon's *New Atlantis,* both enlightened and enlightening.[100]

At Versailles, Apollonian illumination radiates from the palace along the axes of Le Nôtre's designed landscape (Fig. 6.8).[101] Marly, where Coronelli's completed globes were actually placed, in square pavilions decorated with astronomical and geographical symbols, was the mechanical center of Le Nôtre's scheme. The pavilions accommodated machines that allowed the globes to be moved and seen and that produced the light and shadow effects of eclipses. Marly also housed the enormous hydraulic machines that controlled the waterworks and fountains of Versailles. Le Nôtre directed the royal gaze west along the gardens' principal axis to the great fountain of Apollo, where a gilded statue of the sun god sinks into Ocean. His system of canals represented the rivers of France, while the great basin, like the oceans of Coronelli's globe, was often filled with ships from across the world.[102]

Versailles was thus a cosmographic conception:

> The four parts of the world were omnipresent, in garden statuary, in the pictures decorating the Ambassadors' staircase, and those of the royal apartments, which also represented the planets. The interest accorded by the King to the knowledge of the world was underlined especially in the rooms of Mercury, protector of the arts and sciences. In the King's rooms, one saw Ptolemy in conversation with the scientists in his Library, and Alexander receiving the animals of the whole earth, permitting Aristotle to write his natural history.[103]

Following its iconographic logic, Coronelli's globes emblematized Louis's moralized cosmos, "symbols destined for the palace of Apollo, where the ter-

6.8. Postcard showing an aerial view of the palace and gardens of Versailles, looking west, c. 1990.

restrial globe would bear witness to a reality too distant for a king whose attention was focused on the immediate eastern and northern frontiers, while the celestial globe is a 'relief figure of the Prince's nativity,' the prince charged by Campanella to realize the project of religious unity and universal theocracy described in *The City of the Sun.*"[104] From the Hall of Mirrors (where at one stage Coronelli proposed locating his globes) the kingly eye could pass from the ceiling decoration of Apollo crossing the Rhine and overawing the narrowly commercial Dutch, to the princely reflection in the endless glass mirrors, through the gallery windows to follow the line of the *grand allée* to the western horizon. Coronelli's globes extended the gaze empirically and symbolically over terrestrial and celestial space to infinity.

Under Colbert's authority France explored and exploited a North Atlantic empire, and through the work of La Hire, the Cassinis, and French mapmakers such as Sanson it challenged Dutch cartographic hegemony. The *Neptune François* atlas of 1693 mapped oceanic space at a scale and lavishness rivaling the Blaeu atlases. By the late seventeenth century heliocentric images, globes, and maps were being manufactured everywhere in Europe as its monarchs, merchants, and philosophers, their rhetoric often exceeding their imperial grasp, competed for the symbolic authority of the globe. Adam Oleario produced globes three meters in diameter for Duke Frederick of Holstein, and Gerardo Weigel produced a ten-meter armillary sphere, his *Pancosmo,* at Nuremberg in 1699, a powered mechanical wonder demonstrating universal time and space together with the earth's physical and political geography. The earth was supported jointly by Hercules and Athene, its internal fires—and the congealing gems coveted by Dryden—demonstrated by erupting volcanoes blowing smoke across the device.[105] But none matched Coronelli's emblematic globes in conception or decorative elaboration.

Coronelli's relentless synthesizing included an alphabetic encyclopedia, the *Biblioteca universale sacra-profana* (of which only seven of a planned forty volumes, covering the letters A to C, were published), and the thirteen-volume *Atlante veneto,* a summary of cosmography illustrated with more than eleven hundred images, more than two hundred of them Coronelli's own maps based on his vast correspondence and copies of maps by earlier Venetian mapmakers. The extended title of the *Atlante* proclaims it "a geographical, historical, sacred, profane and political description of the empires, kingdoms, and states of the Universe, their divisions and boundaries, with the addition of all the newly discovered countries, augmented with many geographical maps, never before published." It contains much of the material on the Marly globes, a Francocentric record of geographical science and discovery highlighting, for example, Cavelier de La Salle's exploration of Louisiana. The *Atlante* opens with a summary of cosmography, listing ancient and modern contributors to an understanding of the globe. Cosmography opens with the divine fiat, moving through "the order of Creation" to individual cities *(iconografia),* palaces *(scenografia),* and rivers *(potomografia),* a conception graphically captured in the opening image, *Idea dell'Universo,* where concentric spheres expand from the lowest house of hell, at the center of the earth, through purgatory and limbo, immediately below the terrestrial surface, to the four elements, the seven heavens, the fixed stars, and the *primum mobile*. Smaller globes illustrate cosmographic relations between zodiacal spirits and earthly metals, eclipses and the flux and reflux of the tides, and the planetary influences under which continents and countries lie. Volumes 2 and 3 are Coronelli's *Isolario* (1696), while volume 10, the *Libro dei globi* (1697), is a practical coursebook in globe making containing printed gores for Coronelli's celestial and terrestrial globes. Volumes 6–8 constitute a *Teatro della città* (1696–97), inspired by Braun and Hogenberg's work: 310 plates of the world's cities.

Coronelli's publications represent the culmination of emblematic cosmography. The grandeur of their scale and the incompleteness of their execution signify the difficulties of sustaining a singular global vision in the face of excess in observational knowledge and a weakening metaphysics. Thus, while Coronelli's globes echo the speculative geography of Europe's early confrontation with a global earth, they illustrate more frequently rational and critical surveys of observed evidence. On Coronelli's terrestrial globe an African cartouche shows a venerable geographer drawing aside a veil to reveal the sources of the Nile (Fig. 6.9), believed by Coronelli to start twelve degrees south of the equator with a possible underground link to the Niger at Lake Chad and a lagunar source in Lake Zaire. Pyramid, obelisk, and column signify the Nile as the source of human knowledge, language, and architecture, recalling Kircher's beliefs and the debates among Bembo, Ramusio, and their circle. Indeed, Coronelli's pictogram of Mount Etna and the Straits of Messina in the *Atlante veneto* explaining Scylla and Charybdis is attributed to Kircher and recalls Bembo's *De Aetna,* written in the opening years of the new cosmography.

The Nile cartouche is one of six hundred such inscriptions, a visual encoding of "mnemonic images, bibliographic references, recent discoveries, theories, natural phenomena, technical processes, accounts of navigation and exploration, tales of conquest and war,"[106] whose location responds as much to a design logic (placing texts in the blank spaces of knowledge) as to a geographical one (pinpointing the location of a battle). This is an emblematic principle that le Large's manuscript analysis seeks to interpret, referencing each image or text to its coordinate position before explaining it.[107] Thus, on longitude 305 between latitudes 50° and 75° south a series of images describe Patagonia: a cabin constructed of branches and two canoes, a picture of "two savages . . . hunting ostriches with arrows," and an elaborate cartouche with figures of Mercury, a flightless bird, and decorative rocks, shells, sea serpents, seaweed, and bales. It is inscribed "primus me circum dedisti." Le Large points out that these images illustrate the habitat and habits

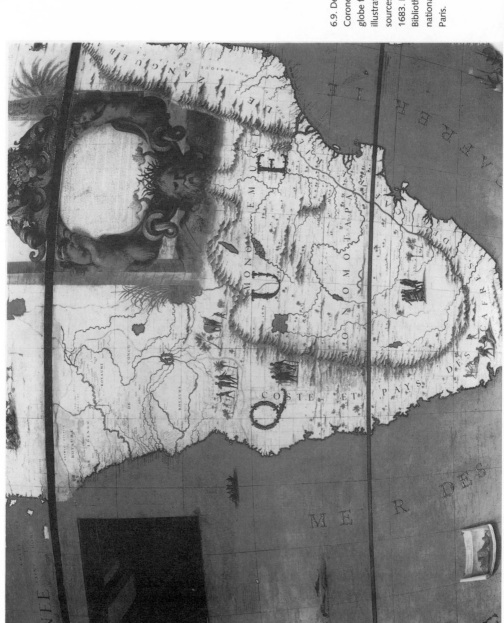

6.9. Detail of Coronelli's terrestrial globe for Louis XIV, illustrating the sources of the Nile, 1683. Photograph by Bibliothèque nationale de France, Paris.

of Patagonians, and he refers the reader elsewhere in the text for a fuller discussion of the canoe and hunting practices. The Latin inscription recording
Magellan's circumnavigation launches a long description of Magellan's circumnavigation and the dispute over the Moluccas between Portugal and
Spain that produced the voyage. Magellan is supported by Mercury to indicate the value of his discovery to Commerce, indicated also by the bales; the
flightless birds are curiosities of the region, killed by the sailors for supplies;
the other decorative elements signify the sea.

Beneath Coronelli's emblematic celebration of Louis's Apollonian ma-
jesty, le Large's iconographic analysis signals a secondary global discourse of
commerce and geopolitics. Coronelli's oceans are filled with ships of all
nations, mainly European but also Turkish, Arab, Indian, and Chinese, competing for the world's trade. There are also naval fleets in battle for control
of the seas. Le Large is explicit about the commercial purpose of globes:
commerce is "the outcome of geography and navigation," whose principal
advantage is that it brings the peoples of the world into mutual communication for the common good of nations and the glory of their sovereigns.[108]
Commerce allows items found in a single location to be made available in
all places, it promotes the perfection of science through the exchange of
ideas, it brings the word of God to all peoples, it brings home the riches of
the sea, and it allows overcrowded peoples to rise from their poverty through
colonization of new lands. The language of universal Christendom is giving
way to that of universal exchange of material goods and ideas. This is not
free trade, however, but trade governed by a geopolitical imperative. In uniting the crowns of Spain and Portugal, le Large claims that Phillip II was the
first sovereign to attain the possibility of controlling global trade and thereby
achieving global empire. But the riches of the Americas had rendered Spain
lazy and vulnerable. Annotating Coronelli's globe in 1720, le Large saw only
three European powers capable of mastering universal Commerce: France,
Spain, and Turkey. Which, he asks, is most likely to succeed? Unsurprisingly,
the answer is France, not for teleological reasons but overwhelmingly for
practical reasons that lie within the technical competence of well-run states.
To be sure, the most important is preordained: it is a large, well-populated
country conveniently located for navigation. But beyond this are matters of
low excise, control of one's own ships, manufacturing, and common internal pricing policies.

In 1680 Coronelli, as "Cosmographer of the Venetian Republic," established a geographical society, the Accademia degli Argonauti, with 261 members drawn from across the continent and including key figures of Catholic

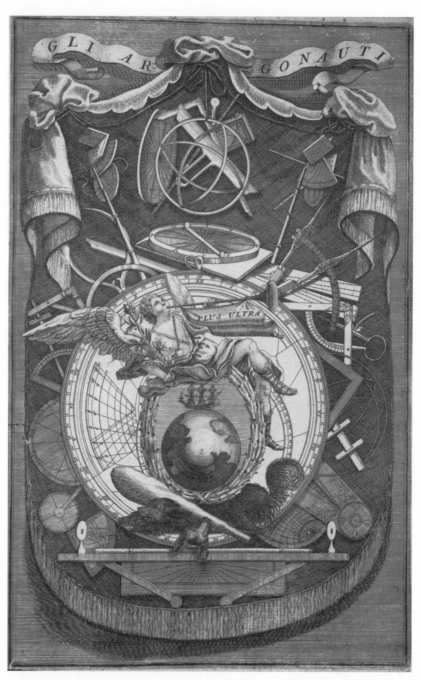

6.10. Vincenzo Coronelli's emblem for the Accademia degli Argonauti with a navigated globe and the imperial motto *Plus Ultra*. Harry Ransom Humanities Research Center, The University of Texas at Austin.

Europe's scientific and political elite, as well as the 30 Russian cosmographers sent to Venice by Peter the Great as part of his imperial project for a Europeanized Russia. Coronelli designed an emblem for the academy as an allegory of its intellectual ambition. A trumpeting angel unfurls a banner with the predictable motto *Plus Ultra,* while geographical and navigational instruments decorate the device like the triumphs on a Roman arch. A ship surmounts the globe, and Hercules' club and bearskin suggest that the Moderns have superseded the greatest of the ancient heroes in reaching to the ends of the earth (Fig. 6.10).[109] In the humanist tradition of Ramusio, Gastaldi, and Kircher, Coronelli brings the ends of the earth within a single representation in order to illuminate the unity hidden beneath the globe's apparent variety. But the vision allegorized in this image was fading; le Large's discourse on Commerce suggests that the ship at its center sailed more for reasons of state than for reasons of heroic *telos.*

In the forty years between the manufacture of Coronelli's globes at Marly and le Large's commentary in the decade of Louis XIV's death a fundamental change is apparent in the way Europeans viewed and imagined the globe. The divine light that Kircher spread with the name of Jesus across global space was fading before the source of illumination deified as Reason.

SEVEN *Enlightened Globe*

We speak of the right of all humans to ask foreigners to enter their
society, a right based on the common ownership of the Earth's surface, its spherical form forces them to support one another, since they
would not be able to disperse themselves to infinity, and originally no
one has any more rights than anyone else to a country.

—Kant, *Principles of Lawful Politics*

In the middle years of the eighteenth century the painter Pietro Longhi
(1702–85) gently satirized the pretensions of bourgeois Venetian families who
were educating their daughters in the refinements appropriate for a young
lady. In a series of small images innocent youth is sullied by the lascivious
intent of older teachers.[1] In *The Geography Lesson* (1750–52) (Fig. 7.1) a fashionably dressed young woman is seated next to a celestial globe, to which
she holds a pair of compasses in the classic act of measurement. The light
that illuminates her figure also falls across an open atlas at her feet, while in
the dim recesses of the room dark-bound tomes bear the weight of geographical scholarship. Four other figures are present, their dress and de-
meanor, like the room's furnishings, indicating an environment of middling
wealth. Two are gentlemen tutors, engaged in the young woman's geographical instruction, while behind her left shoulder can be seen a female
servant bringing a tray of refreshments, with a second servant in the shadows. The image plays on the word *globe*. The studied globe shares the compositional center with the young woman's uplifted breasts. Other globes are
figured in the bulging stomach and ruddy cheek of the portly, seated tutor,
balancing his geographical maps. The standing instructor raises his spectacles
to his eye the better to examine a globe. But his gaze is directed, not at the
globe upon which he discourses, but at the more fleshly example emphasized by the girl's tight bodice. The tutor's surreptitious glance is spotted by
the maid's disapproving eye, so that the picture space is constructed through

a geometry of sightlines connecting the three pairs of visual globes, none of them actually focused on the terrestrial orb.

Longhi's painting serves well to introduce an Enlightenment globalism. There are traces here of the emblematic globe placed in a rich domestic setting as a *vanitas,* signifying the distractions of temporal riches from more spiritual concerns, symbolized by light entering the room. But the atmosphere of spirituality and illumination found in Dürer's *Melancholia,* Vermeer's *Geographer,* or Gerrit Dou's *Young Musician* is absent. The globe has become an instrument of secular learning, the focus of an objective, disinterested gaze, while the moral point is made by a knowing satire on fleshly desire.[2] Vision and knowledge contribute to secular advantage rather than spiritual advance, in this case the young woman's increased value in the marriage market. The globe is an object of property and the planetary consciousness it signifies is, as François le Large recognized, commercial.[3] The intense, contemplative gaze into metaphysical worlds found in comparable seventeenth-century images gives way to the outward-looking, even vacant glance of a rather bored young woman. Seemingly unaware of her tutors' voyeurism, she turns her own eyes away from the globe and the dully lit volumes toward the viewer. The entire group seems disconnected from the subject matter of the lesson. Such changes do not exclude other continuities in representing and imagining the globe; the Enlightenment was by no means a simple cultural shift.[4]

The Orrery and the Celestial Globe

An image contemporary with Longhi's also pictures a small domestic group gathered around a global image for the purposes of learning. *A philosopher lecturing on the Orrery* (1766), by Joseph Wright of Derby, illustrates the spread of scientific learning beyond the confines of court, college, and aristocratic academy into a more professional and provincial world. Wright's group attends an exposition and performance of Newtonian planetary theory, demonstrated with the aid of a mechanical model of the heliocentric cosmos in which a candle takes the place of the solar light source.[5] Mechanical orreries took their name from the device designed for Charles Boyle, fourth earl of Orrery, by John Rowley in the first decade of the eighteenth century to illustrate the Copernican system, and they became popular for visual instruction in the new science. By the time of Joseph Wright's painting these sophisticated systems of cogs and gears had effectively replaced the armillary sphere, rendered obsolete by Isaac Newton's decisive demonstration of

7.1. A play of globes: Pietro Longhi's *The Geography Lesson*, 1750–52. Fondazione Querini Stampalia, Venezia.

heliocentrism in *Philosophiae Naturalis Principia Mathematica* of 1687. The armillary's Aristotelian summary of cosmic harmony corresponded neither to
the theory of universal gravitation nor to observation of the heavens and
belief in infinite space. Newton's impact on the Enlightenment imagination
is captured in two global images: Etienne-Louis Boullée's 1784 design for an
architectural monument to the English scientist in revolutionary Paris (Fig.
7.2) and William Blake's 1795 image of a seated figure bending his compass
to a geometric diagram, an imitation of Blake's cosmographic image of the
Creator compassing the sphere (1784). In both, pure geometry and inorganic space signify Newton's work. Boullée's vast sphere had a planned diameter of seventy meters, and it was to be illuminated internally only by the
light penetrating the heavenly bodies inscribed into its surface. The visitor
passes briefly across its base, dwarfed by the immensity of empty spherical
space and the infinite points of light in its firmament.[6] Comparing Boullée's
dome with the Roman Pantheon's single shaft of light illuminating a circle
of divine images during a diurnal round signals the magnitude of the conceptual shift. For Blake, Newton's discovery offers creative power to the individual, inscribing human will onto the spaces of earthly nature.

Like his predecessors Galileo and Kepler, Newton himself did not sub-
scribe to the godless mechanical universe attributed to him by later commentators: "It seems probable to me," he wrote, "that God in the beginning
form'd matter in solid, massy, hard, impenetrable, moveable particles, of such
sizes and figures, and with such other properties, and in such proportion to
space, as most conduced to the end for which he form'd them."[7] The spherical form still denoted divine *telos*. The *Principia Mathematica* was grounded
in a theory of universal attraction quite foreign to Descartes's mechanical
philosophy. Universal gravitation actually bore a distinct resemblance to Neoplatonic conceptions of divine love as a cosmic power of attraction and repulsion.[8] Nevertheless, the Creator's disengagement from an active cosmic
presence was implicit in the new cosmology and had profound implications
for global images and meanings. Unlike the armillary, the orrery's meaning
lies in motion: inert matter is driven by forces that once set in motion continue to operate independently as the variously sized spheres revolve at diver-
gent speeds.

Newton's planets and constellations are contingent matter, subject to external forces rather than a living cosmic presence. By 1754 the use of the
telescope on navigations to ever higher southern latitudes had added twenty-
six new constellations to Ptolemy's forty-eight. Their very naming reflects
the disenchantment of cosmography; Antlia, the Air Pump, for example,

7.2. The globe of Reason: Etienne-Louis Boullée's 1784 design for a monument to Sir Isaac Newton in revolutionary Paris. Photograph from Bibliothèque nationale de France, Paris.

recalls one of the most popular demonstrations of the new science, itself one of Joseph Wright's subjects.[9] The seventeenth-century debate over Ptolemaic, Tychonian, and Copernican world systems betrayed a sustained attempt to confine the celestial globe within the bounds of religious conformity. Andreas Cellarius's astonishing *Atlas coelestis seu harmonica macrocosmica* of 1661, a celestial equivalent to Blaeu's lavish terrestrial atlases, illustrates the competing systems dispassionately, while representing Ptolemy blindfolded and ignorant. Cellarius also devotes two lovingly wrought plates to Julius Schiller's 1627 Christianizing of the classical constellations as biblical personalities. His *Caelum stellatum Christianum* populated the northern and southern skies with New and Old Testament figures, respectively, with the zodiac as the twelve Apostles.

Comparing these celestial maps or the moralized planispheres in Joseph Moxon's *Tutor to Astronomy and Geography* (see Fig. 6.4) with a characteristic cosmographic image from less than a century later is profoundly revealing. Gabriel Dopplemayr and Johann Baptist Homann's celestial hemispheres for a 1742 Nuremberg *Atlas coelestis* take their authority from "the most recent astronomical observations," returning the skies to elegant classical personifications surrounded by tables of mathematical calculation and illustrations of astronomical observatories in Hven (Denmark), Paris, Greenwich, Nuremberg, and Berlin, mapping a geography of celestial knowledge across

European centers of calculation, power, and commerce. The gathering and transfer into Europe of astronomical knowledge from navigations across the globe reflect eighteenth-century competition among European states within a political economy of knowledge that was truly global. Europe's historic encounters with the Pacific were initiated by astronomical concerns, notably observation of the transit of Venus from scattered points across the globe to determine the accuracy of its spherical measure and the commercially critical plot of longitude and latitude.

The disenchantment of the globe figured by the orrery and changing iconography can be overstated. The new natural philosophy and the wonders performed by manipulating natural laws and properties held their own enchantment. No recourse to metaphysical speculation was necessary to enhance the wonder in Wright's faces observing the orrery or air-pump experiments. Indeed, negotiating the line between entertainment and education, theatrical performance and scientific demonstration, "quack" and "scientist," was fundamental to constructing Enlightenment geographies of knowledge, faith, and magic.[10] Assessing the cultural meanings and significance of globes and global images in eighteenth-century Europe is thus by no means simple. Longhi's, Wright's, and many similar images, let alone the huge numbers of globes, planispheres, and atlases manufactured in eighteenth-century Europe, indicate that knowledge of the terrestrial globe, its place in the solar system, and its geographical patterns was a prerequisite for educated men and women.[11] Joseph Moxon's late-seventeenth-century pocket globe was a popular geographical toy. In the 1780s, James Furguson was one of a number of craftsmen producing miniature globes, less than seven centimeters in diameter, of plastered wooden spheres with terrestrial gores attached to their surface. These were enclosed within a spherical leather case lined on the inside with celestial gores, so that the earth was contained within an image of the skies. Once the possession of divinely appointed monarchs, the globe was now contents of a merchant's or broker's frock-coat pocket. Coronelli's vast, decorated confection for Louis XIV's Apollonian eyes had, by the year of Louis XVI's revolutionary overthrow, been reduced to the size of a tangerine. Although this indicates the spread of a distinct planetary consciousness across a growing literate class, it is difficult to distinguish the sense of global authority, knowledge, and power these baubles might imply from the sense of wonder and marvel always associated with the image of the globe.

Terrestrial Sphericity and the Longitude

Universal gravitation posited a planetary motion distorted by the differential attraction of other planets, and Cassini's measurements indicated that the pure sphericity of the earth could no longer be assumed. The measurement of tidal variation across the globe, one of the oldest geographical arcana, lent empirical support to Newton's principles. Spherical perfection had been the foundation of Platonic, Pythagorean, and medieval cosmological thinking.[12] But by the end of the seventeenth century a debate had developed between proponents of an oblate earth and those of a prolate one, between those who believed the earth to be flattened at the poles and those who believed that it was compressed at the equator. Careful measurement of degrees of latitude down longitudinal arcs offered an empirical resolution since their changing value would indicate the precise location of deviations from pure sphericity. This required the physical capacity to encompass the globe. Thus, Jean-Domenique Cassini's son's and grandson's continuation of their forbear's science led to French expeditions to Lapland and La Condamine's to Spanish Peru during the 1730s to obtain arc measurements in regions of the earth where distortion would be maximized. The matter could not be resolved by linear measurement and terrestrial mapping since instrumentation was insufficiently accurate to contain the error within 23.4 meters per degree. Pendulum rather than linear measurement had led to Newton's belief in an oblate globe. One significant intellectual consequence of the debate was "the alienation of geography, as earth description, from its very foundation, the measurable earth."[13]

If Newton's explanation of planetary space and the form of the earth constituted one aspect of Enlightenment global consciousness, another was the invention of a simple means to determine longitude at sea and thus ensure the accuracy and consistency of navigation and graphic representations of the globe. While the Ptolemaic graticule had long determined the spatial logic of such images, the precise relationship between the 360 degrees of longitude on the Euclidean sphere and actual terrestrial space was altogether arbitrary. In an absolute sense this has to be so: latitudinal circles are astronomically determined, whereas the relationship between longitudinal lines and the global surface depends upon the selection of a prime meridian, for which no astronomical reference exists.[14] Practically and imaginatively, the compass points East and West have different resonances from North and South. The poles may be regarded as fixed locations, but East and West remain open, fluid geographical conceptions, their location relative and endlessly deferred in navigation, giving rise to practical problems and imaginative possibilities for inscribing meaning on a spherical earth.

Setting a meridian is a primary act of global representation, practical in fixing absolute location and imaginative in signifying origin and ends on a turning sphere. Various meridians have been privileged in European spatial representations; the ancients used Rhodes, Delphi, the Pillars of Hercules, and the Fortunate Isles, whereas Christians have favored Jerusalem and Rome. The papal line fixed at Tordesillas was the first to have material geopolitical effects, intensified in struggles over its extension through the Moluccas.[15] The astronomical observatories established by baroque states needed to determine their own meridians in order to calculate almanac tables and ephemerides. Thus, Toledo, Bologna, and Florence were among the cities that served as originating points for global graticules. For oceanic navigation the matter was more practical than intellectual, and for Atlantic navigators coasting Africa or following the trade winds to the New World, the island of Tenerife, on the western edge of the ancient ecumene, had both practical and mythical appeal. Its peak, soaring more than 3,700 meters out of the Atlantic, was long considered the highest on the globe, while the lunar landscape of its lava fields offered Dantean visions of hell.[16] Determining a prime meridian for the globe controls its representation, especially on the two dimensions of a world map whose margins and center are fixed by meridians. Centering a planisphere on the Atlantic, thereby cutting the Pacific at its left- and righthand margins, may have been a practical way to avoid the question of America's connection with Asia at the Strait of Anian, but its ideological effects in centering Europe on the globe are undeniable. Setting the meridian is an ineluctably ideological act.[17] Thus, in 1634 Louis XIII of France ruled that the Isle de Fer (Hierro) should define the prime meridian for French navigation and forbade "all pilots, hydrographers, designers or engravers of maps and terrestrial globes to innovate or vary from the ancient meridian passing through the most westerly of the Canary Islands."[18] Predictably, the European states and empires that emerged from the 1648 Peace of Westphalia competed in using the meridians of their capital cities—Paris, Madrid, London, Copenhagen, Turin—not only as the point of origin for their territorial triangulation and chorographic mapping but as the prime meridian for their geographic and global images. Jacques Cassini set the meridian of Paris at the French Royal Observatory, and John Flamsteed performed the same operation for England at Greenwich, authorizing the global centrality of their respective capitals.

National meridians were used as baselines for territorial triangulation of

European states. Jacques Cassini achieved this for France between 1739 and 1744,[19] and most other states had systematic triangulations by the late eighteenth century. At sea, however, the problem remained of how to determine longitudinal location on a ship out of sight of a known coast and thus provide a consistent log of the voyage with replicable data on the locations of phenomena observed. The problem was technical rather than theoretical; simple celestial geometry indicates the sun's movement through one degree of longitude every four minutes. An hour's time difference between two places is thus fifteen degrees of longitude. The necessity is for an accurate means of determining time simultaneously at two separate points on the earth's surface. Lunar eclipses traditionally offered a means of such calculation, being simultaneously visible at all points on the globe. The longitude of Tenochtitlan/Mexico in relation to Toledo had been fairly accurately estimated by this means as early as 1584. By 1514 Europeans had also recognized that lunar-zodiacal observations allowed longitude to be fixed at sea, at least in theory, and published ephemerides and almanacs, globes and pendulums, all provided help in making the necessary and complex calculations. But none of these methods guaranteed accuracy, especially given the problems of parallax in celestial observation and the number of simultaneous sextant observations required, often from the deck of a heaving ship. Gross errors were common in both navigational and cartographic recording, so that by the later sixteenth century maritime states such as Venice, the United Provinces, and Portugal were following Spain's 1567 offer of a reward for a more reliable and simple solution to the problem. The invention of the telescope offered the alternative of measuring the regular eclipsing of Jupiter's moons, which occurred much more frequently than terrestrial lunar eclipses. But each method presented enormous practical problems, for example, in obtaining a stable observational platform or of overcast weather, and each demanded complex mathematical calculations, in which the possibilities of error multiplied rapidly. Easy, accurate measurement of longitude at sea remained intractable well into the eighteenth century, principally for want of a timepiece guaranteed to remain constant and accurate during the course of a sustained or rough sea voyage. Le Large devotes twenty-seven pages in his commentary on Coronelli's globes to a detailed outline of the longitude problem, an indication of its centrality in contemporary scientific and technical discourse.[20] Navigators themselves, rather than undertaking demanding and dubiously accurate calculations, often adopted the technique of aiming "two to three hundred miles seaward of the desired landfall, observing the Sun or Pole Star until the latitude required was reached, then [sailing] east or west keeping to the same latitude—running down the latitude with observational checks as often as the weather allowed—until the destination was reached."[21]

Resolving this costly navigational problem was a principal stimulus to state funding of astronomical observatories such as that at Greenwich in 1676. These observatories sought to increase the number of named and location-ally fixed stars on the celestial chart and thus more accurately determine lunar movement and distances. Flamsteed's successor at Greenwich, Edmond Halley, personally navigated the Atlantic for the purpose of verifying lunar observations and compass calculations, initiating the tradition of state-sponsored scientific navigations, which would eventually transform European globalism. In Britain resolution of the problem was given added impetus by the loss of five Royal Navy ships and nearly two thousand men on the Scilly Islands in 1707, apparently because of inaccurate longitudinal readings. In 1714 a huge bursary—up to £20,000—was offered by the British Parliament to anyone who could produce a foolproof method of calculating longitude at sea, and a Board of Longitude was established to test claimants' proposed methods.[22]

This longitude quest dominated the scientific work of eighteenth-century European observatories, producing not only increasingly detailed star charts but significantly closer observation and improved understanding of the lunar globe and its motions. One outcome was the publication from 1766 of *The Nautical Almanac and Astronomical Ephemeris.* The work, subsequently published annually, was based on tables produced by the German astronomer Tobias Mayer, tested at sea and printed by Nevil Maskelyne, English astronomer royal from 1765. Based on the meridian of Greenwich, the almanac gave accurate lunar distance tables for longitude fixing, which were reprinted in the official almanacs of rival maritime nations. John Harrison's invention of the marine chronometer about 1760, its accuracy in keeping shipboard time decisively proved on Cook's 1772–75 voyage, solved the longitude problem and signaled a decisive step toward a universal coordinate system for Western global mapping.

The techniques of longitude calculation meant that while the metaphysical relations between terrestrial and celestial globes had been transformed by Copernican-Newtonian cosmography, use of the navigational almanac reinforced the traditional centrality of the individual observer in relations between Earth, the planets, and the stars, but in secular form:

> The lunar distances predicted in the almanac are geocentric; that is, they assume the observer to be at the centre of the Earth [inevitable if they are to used anywhere on its surface]. The following require-

ments have also to be fulfilled: that the observation is taken some distance above the Earth's surface (dip of the sea horizon), that both
bodies appear too high, the amounts depending upon their respective
altitudes (atmospheric refraction), that it is the limbs of the sun and
moon that are observed, not their centres (semi-diameter), that the
observer is not at the centre of the earth (parallax).[23]

Thus, the complex operations required for practical use of the almanac's
globalized spatial information demanded imaginative centralizing of the individual in global space.

The intellectual and imaginative consequences of success in setting the
longitude were geopolitical as well as commercial. It led to calls for establishing a universal prime meridian. From the time of the publication of
Maskelyne's *Nautical Almanac* this was likely to pass through Greenwich, the
location from which time variation had to be calculated if most published
tables were to be usable. By 1884, when the universal prime meridian was
established at Greenwich, more than two-thirds of the world's shipping used
Greenwich-based charts, giving its meridian line a practical primacy that
would only be challenged in an era of satellite observation and global positioning systems. In the context of rational Enlightenment, a universal prime
meridian represented a further western shift of Europe's imperial *axis mundi,*
previously located in Jerusalem, Rome, and Paris, now justified in terms of
commercial practicality rather than metaphysical speculation or doctrinal
authority.

Eighteenth-century debates over a universal prime meridian offer insight
into associations between Enlightenment globalism and ideals of universal
humanity. As early as 1796 Pierre Simon Laplace argued for a "natural" location to avoid the geopolitical implications of a socially determined one, such
as a capital city. A peak such as Tenerife or Mont Blanc, universally recognized for prominence and permanence, "would introduce into the science
of geography the same uniformity which is already enjoyed in the calendar
and the arithmetic, and, extended to the numerous objects of their mutual
relations, would make of the diverse peoples one family again."[24] For this
French Enlightenment thinker, a geometric determination of global representation would contribute practically to the dream of a common humanity within a single ecumene. The American polymath Benjamin Vaughan
followed a similar line of thought in 1811, proposing a universal meridian
that would be "'immutable and everlasting' at least with respect to 'transitory' and 'fluctuating' human civilization, and would accordingly outlast

political changes."[25] The model of the cosmos as a celestial clock allowed Vaughan to develop the conceit of "nations in the Moon," rather like the inhabitants of Kepler's Levania, counting their days by the passage of the earth's continents and oceans. To Vaughan, such an image of the earth suggested that the continental islands of Eurasia-Africa and the Americas should be separated by the prime meridian. Palma, in the Canaries, "is to be considered as suitable for the *general* meridian of our globe, or at least that portion of it which is most united by the tie of Christianity, science, civilization, and the means of mutual intercourse."[26] Vaughan's words neatly tie practical argument for a "natural" prime meridian to an ecumenical Christianity in a characteristic Enlightenment conflation of "science," "civilization," and commerce ("mutual intercourse"). The longitude is transposed from the physical world to human society as the rational space of global representation becomes a "natural" justification for social organization.

Island and Oceanic Globe: The Changing Discourse of Empire

Work on the longitude was in key respects the foundation of an "imperial science." A declared purpose of James Cook's first Pacific voyage, in 1768, was to observe the transit of Venus and confirm the accuracy of navigation tables, and a goal of his second, in 1772–75, was to test the reliability of Harrison's chronometer. A circle of scientific observation and calculation, navigation and discovery, colonization and further calculation, connecting Europe's "centers of calculation" through global space was completed when Maskelyne's assistant at the Royal Observatory, William Dawes, was appointed surveyor for the Botany Bay colony in 1788. His work combined plans and geometrical plats for colonial settlements with communicating to Greenwich astronomical observations from the only permanent observatory in Australia.[27]

Eighteenth-century navigator-explorers such as William Dampier, George Anson, John Bryon, Louis Antoine de Bougainville, and James Cook concentrated on an oceanic global hemisphere antipodal to Europe. Straits, isthmuses, and peninsulas whose connections to continental landmasses had remained unresolved on the great baroque globes and maps were fixed on navigational charts, while the outline of *terra australis incognita,* the great southern continent, was finally determined. The Strait of Anian, in the northern Pacific, anticipated by Giacomo Gastaldi and subsequently appearing intermittently on globes and planispheres, was proved by Bering to separate Asia and America; California was not the island Coronelli and others had pictured, whereas Siberian Sakhalin was. By the time of Napoleon's attempted resurrection of Rome's territorial empire in Europe, *termini orbis terrarum* had been pushed to the northern and southern poles. Dividing the global surface by continents and oceans rather than by cosmographic climates makes Europe more or less central to a land hemisphere, while the sea hemisphere is dominated by the Pacific Ocean. The colonization experience in the New World had dramatically broken the ancient Aristotelian connection between the latitudinal *klimata* and actual climatic conditions long before Alexander von Humboldt's isoline map of 1817 illustrated the divergence. European discovery in the Enlightenment was concentrated overwhelmingly in the ocean hemisphere, a hemisphere of islands scattered across whole degrees of latitude and longitude. Fixing the coordinates of discrete specks of land in otherwise undifferentiated space had to be precise if they were to be revisited, let alone accurately recorded; "running down the longitude" was simply not a practical option if such space was to be effectively incorporated into a global episteme.

National maritime atlases, pilots, or neptunes, such as Edmond Halley's *Atlas maritimus & commercialis,* of 1728, fixed the changing image of an oceanic world for a scientific readership. Halley, astronomer royal and intellectual associate of Newton, opens his world atlas with a discussion of advances made on the longitude question before describing how his own global projections are based on the method described in the "24th and last chapter of the First Book of [Ptolemy's] Geography."[28] Rhumb lines are indicated by spirals in order to assist taking of bearings. His fifty-seven maps include large folded ocean-centered charts of the world's coasts and islands, local depth charts and drawings of harbors, as well as hemispheric images of the fixed stars with text descriptions of methods for finding bearings using simple means such as thread and ball. Halley's image of the world characterizes the revivified oceanic vision that emerged from the competition of maritime commercial empires.

A number of consequences flowed from the oceanic character of the Enlightenment globe. One was to reemphasize the imaginative significance of ocean and island themselves as geographical forms. In 1750, when Giovanni Battista Tiepolo completed the largest and most dramatic of his paintings of *The Four Continents* on the ceiling of the Residenz, home of the prince bishop of Würzburg, using conventional continental iconography, he unconsciously signaled a conservative and dying genre.[29] Not only was the number of continents about to increase but the global imagination of Europeans was no longer captivated by land. It was dominated by the island, whose self-contained spatialities offered the perfect geographical template for the systematic theories of "nature" debated by rationalist philosophers and naturalists such as Montesquieu, Rousseau, Linnaeus, and Buffon.[30] Australia was figured as an island continent, its interior a blank canvas to Europeans and its natural history more radically different from Europe's than even the Americas' was. It more than satisfied both the enduring fascination with "marvels" and the *esprit du systeme* that sought to incorporate its flora and fauna into global classification systems. Daniel Defoe's 1709 figure of Robinson Crusoe, "natural man" on a deserted island, remains the Enlightenment's most durable literary creation.[31] Despite the technical advances represented by fixing the longitude, the optical uncertainties caused by atmosphere, indeterminate horizon lines, human tiredness, and the disorientation caused by endless months in the liminal spaces of ocean all served both to increase the island's imaginative power and to undermine the accuracy of records and charts.[32]

A second feature of the oceanic emphasis was geopolitical. The last formal reference to the idea of Europe as a *respublica christiana* was in the Treaty of Utrecht of 1713. By this date the political and economic decline of Spain and the commercial wealth of the United Provinces, France, and England had crucially changed the global imperial venture upon which Europe had embarked with Renaissance Atlantic navigation. Outside an increasingly marginalized Rome the evangelical rhetoric of Iberian imperialism had little purchase, increasingly replaced by the mercantilist and colonial arguments of territorial states, although Jesuit domination of education, especially in France, until mid-century ensured the survival of theological interest in cosmographical description as well as more secular measurement.[33] But even le Large's commentary on Coronelli's globes placed evangelism well down on the list of benefits of navigation. A century of religious warfare had served to exacerbate rather than resolve the differences within European Christianity, dividing the continent into self-consciously Catholic and Protestant states. Spain, whose universal monarchy Tommaso Campanella had proclaimed in 1601, was by 1713 quite clearly a spent global force, its population low and stagnant, its economy dependent upon imports of virtually all manufactured goods, its colonial trade fenced by protectionism, and its social order dominated by religious intolerance. The War of the Spanish Succession had lost the Spanish crown its European territories, "and with them . . . the basis of [Spain's] ideological claims to universalism."[34] The European discourse of empire turned increasingly away from a model of continuous territorial expansion justified by "religion and reputation," the mission and

glory of conquest, *ad termini orbis terrarum,* toward a more sober and practical vision, better suited to the experiences and goals of independent merchants and manufacturers than those of aristocrat and princely defenders of
faith:

> The old universal empire based upon conquest would finally be
> transformed into a new political society, similarly universal but based
> now on the more enduring principles of trade and manufacture. . . .
>
> If the three European overseas empires had begun the sixteenth
> and early seventeenth centuries as different kinds of society with
> different structures, objectives and imperial matrixes, by the end of
> the eighteenth century they had converged upon a common set of
> theoretical concerns. These were overwhelmingly concerned with
> undoing the deleterious consequences of the "spirit of conquest" and
> the military ethos of glory, Machiavellian *grandezza* and, its ecclesiasti-
> cal counterpart, evangelization and doctrinal orthodoxy.[35]

Such geopolitical change in the global empire paralleled the altered spatiality of the globe in which continental "Europe," self-positioned at the center of a territorial hemisphere, found its Other in the oceans and islands of
the water hemisphere.[36] Efficient and secure navigation was the only effective way to secure empire through commerce. Control of the seas by a powerful navy was more important than territorial conquest by land armies. "By
the 1760's the recognition that in the modern world power depended on the
maritime commerce had been firmly established."[37] As the Venetians had
discovered in the eastern Mediterranean, the island and the coastal enclave
were the critical strategic spaces for organizing a commercial empire, so that
the Netherlands and Britain offered a more appropriate model than Spain
for imperial extension into the Pacific hemisphere. In this context, the competition between European powers for the technical means of navigating
and representing maritime space—above all the struggle to secure the longitude—coastal charting, and the production of maritime pilots and atlases
are immediately comprehensible.

The ocean island offered more than simply a trope for an altered vision
and practice of commercial empire. Island discovery in the South Seas, especially Cook's of Hawaii and Bougainville's of Tahiti, netted within the Euclidean graticule from the limitless expanses of the open Pacific, offered
European thinkers the opportunity to examine what they took to be living
ethnographic and environmental cabinets of curiosities, microcosms of the
empire of nature previously colonized by medieval encyclopedists, Renais-

sance cosmographers, and baroque collectors such as Francesco Calzolari, Ulisse Aldrovandi, and Athanasius Kircher.[38] Island worlds also suggested the possibility of finally resolving some of the encyclopedic questions about the globe's natural and human order, universal history, and geography that had remained open since Aristotle. Enlightenment rationalists, self-consciously freed from what they took to be the speculative epistemologies and superstitions of earlier ages, believed they could frame these questions with a new objectivity and new evidence gleaned from the oceanic ends of the earth.

Critical to this project were the same means of representation that allowed successful oceanic navigation. The systematic collection and locational recording of information and items collected and returned to Europe, their listing, categorization, and systematic location for scientific taxonomy and display, were all critical to theoretical production of meaning. Spatial and visual rather than narrative and textual classification and presentation is one definition of *mapping,* and a characteristic distinction between European sixteenth-century Atlantic and eighteenth-century Pacific navigation and discovery was the latter's increased attention to accurate and systematic recording of the location and spatial relations of phenomena and artifacts found across the globe. Such information was transported back to Europe with material artifacts for study and display. Universally agreed coordinates allowed the map to become the archive and classificatory grid of the earth's contents, and a cartographically based science of geography could emerge, charged with recording and accounting for the globe's natural diversity. Such demands are reflected in the style of representation on globes and maps themselves.

The elaborately wrought cartouches, decorative flourishes, and emblematic inscriptions that characterize Coronelli's globes and Blaeu's world maps disappear from eighteenth-century images, replaced by sober and severe visual presentation. Title cartouches might still contain allegorical images, but these were restrained, denoting commercial activities or natural history in the areas depicted. An example is the maps engraved by Emmanuel and Thomas Brown for atlases and as book illustrations. Extraneous spatial calculations and navigational or topographic information might surround the map or occupy open oceanic spaces, but they are not allowed to interfere with the accurate delineation of coasts and islands, while unknown or unexplored spaces are acknowledged as such and left blank rather than exploited for iconographic elaboration. Representations of geographical distributions of phenomena across cartographic space are similarly sober, in the style known to historians of cartography as "plain representation." The map's scientific authority as an instrument of rational knowledge demanded the conscious restriction of adornment and decoration, which might seem to distract from its presentation of empirical data. Looking at a map such as *Clark's Chart of the World* (Fig. 7.3), published in London in 1822 and "exhibiting the prevailing religion, the form of government, state of civilization, and the population of each country together with the various missionary stations," we recognize that "plain representation," like James Cook's apparently objective topographical naming of discovered places, is actually no less culturally informed than the most flamboyant baroque artwork.[39] Representational style betrays the intention of Enlightenment makers of globes and maps to construct a geographical archive of global diversity and to array within it the information gathered from the reports of what were commonly state-sponsored scientific expeditions.

Exploration and Mapping the Self

The best-known contributors to the Enlightenment's global archive represented Europe's principal national competitors for overseas empire: the Frenchman Louis Antoine, comte de Bougainville, and the Englishman James Cook. Their fame extended beyond the immediate commercial or scientific value of the geographical information their Pacific navigations returned to Europe. Public opinion, a new social phenomenon, expressed through new media such as newspapers, gave individual navigators the status of national heroes, uniting both a distinctively modern nationalism and a more transhistorical response to the voyager into unknown space. The same public opinion was informed and stimulated by the wide circulation of published materials, including maps and atlases.[40] Cook's achievements, for example, were lauded as patriotic contributions to Britain's prestige and imperial status, and he was rapidly apotheosized after his dramatic death in Hawaii in 1779. Public pageants celebrated him as a fallen hero whose destiny had pushed him to the limits of global space and time, quite literally in charting Britain's global antipodes of New Zealand. The marriage of science and imperialism, "so commonplace as to be conventional,"[41] was celebrated in popular culture. This is apparent in the case of the rather less celebrated antipodean navigator and Cook's successor in antipodean waters, Matthew Flinders (1774–1815).[42]

Flinders's life was dominated by islands. He is recorded as the first European to circumnavigate Australia's island continent and to establish Tasmania's island status. He was imprisoned on the French island of Mauritius

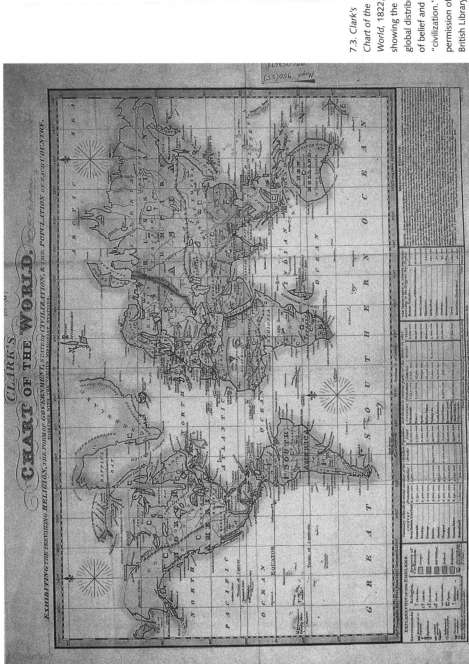

7.3. Clark's Chart of the World, 1822, showing the global distribution of belief and "civilization." By permission of the British Library.

during his return voyage, and he used the experience to dramatize his published scientific account, *A Voyage to Terra Australis.* His narrative characterizes the emerging "scientific" mode of Enlightenment presentation, written in a "plain" style, the textual parallel to contemporary cartographics. It is illustrated by Flinders's own maps and longshore topographic drawings, which are models of the unadorned and topographic delineation taught to all British naval officers at the Royal Mathematical School of Christ's Hospital and later the Hydrographic Office by masters such as Alexander Cozens and Alexander Dalrymple.[43] Flinders's *Voyage* exemplifies a complex interrelationship between graphic and textual modes of representation among "explorer-artist-writers,"[44] who claimed the authority of truthful and objective "natural" vision via "transparent" media of representation, which were actually carefully learned and highly regulated graphic and textual codes through which the exotic and the "tropical" were graphically regulated and normalized within a European knowledge space.

Thus *A Voyage to Terra Australis* reveals more than its claims to naturalism and empiricism intend: it is "the first great Australian work of spatial history."[45] Flinders devotes considerable space to his personal experience and psychological state during the voyage, using his presence as an eyewitness as a means of authenticating his graphic representations. Paul Carter finds coherent expression of this strategy in Flinders's choice of names for the South Australia coastal features and islands that his observational surveys and maps seek to delineate with such painstaking scientific accuracy. These betray a more localized biography mapped into the imperial act of naming locations discovered and appropriated for the European globe. In Flinders's case it is the towns and villages of his native English county of Lincolnshire, carefully selected and transposed onto antipodean space according to a spatial logic derived from the position of the towns and villages on the English county map. In the transfer of names and spaces, as in the remembering and forgetting through which tropical coastal regions became fixed on Admiralty charts,[46] a characteristic feature of Enlightenment epistemology is revealed:

> The country Flinders discovered did not in any naive way reflect his own personality or personal history. Yet it was, in some sense, a product of his mind: what he saw, where he went, and how he ordered the results, all of which was an intellectual activity. And, if the organisation of this complex experience of exploring, and the discoveries it brought, were not to be arbitrary, it had to be grounded in some theory of knowledge. Without some grasp of how the mind worked, it was impossible to know anything about the world.[47]

The arbitrariness and hybridity of representing discovery and exploration relates to a globe over whose coordinated sphere the human hand of discovery rather than the divine hand of creation took responsibility to inscribe names and meaning. Examining Australia's looming coast from the deck of a ship, the human rather than the divine mind was given responsibility for giving order and coherence to alien space. Precisely because of their instrumentation and the scientific nature of their projects, Enlightenment navigators apprehended the spatial scale more immediately than their Renaissance predecessors but lacked the cosmological certainties of da Gama or Columbus. Imaginative authority could only be exercised if such spaces were reduced to a manageable scale, a scale that was phenomenological as much as cartographic. Flinders chose the landscape of his youth as the source for cosmogonic naming of blank spaces on the globe, planting the psychological and ideological center of self onto its furthest margins.

Universal Cartographics

But the Western self was also challenged by the scientific information gathered and returned from across an expanded global surface, a surface inhabited by a greater diversity of peoples and customs than Europeans had previously imagined, named, or classified. Secular rationalism, central to the assumptions of the European philosophes who constructed the idea of enlightenment, offered no place for ideas of a single humanity constructed through Christian redemption. The Jesuit order, for example, whose dominance over French education remained complete until 1762, was suppressed across Catholic Europe in a mere twelve years, between 1759 and 1771, forfeiting the authority of both its theocratic global geography and its universal missionary vision. The moral and philosophical questions of human difference and unity did not disappear; they were reformulated within philosophical debates over universal reason and human rights whose debt to the Judeo-Christian tradition is greater than the Jesuit-educated Montesquieu and Voltaire or Thomas Paine would have acknowledged. Working in part from the evidence of navigation, Enlightenment philosophers erected theories of universal history and global geography that recentered Europe in universal space and time and endowed its citizens with a "civilizing mission" caught in complex dialogue with the rational abstraction of "natural man."

In his statement quoted at the opening of this chapter Immanuel Kant uses the unitary geometry of global form to derive a conception of natural and equal rights as the common inheritance of a common humanity. Like Plato, Kant emphasizes the unitary form of the globe rather than the earth's geographical diversity, recalling the spherical geometry with which Boullée, Kant's contemporary, had designed his vast commemorative sphere to Isaac Newton. As one who supported his philosophical work by lectures on geography, Kant knew well that the pure sphere was hardly an apt description of the earth's human geography. He states a cosmopolitan ideal rather than an empirical truth, positing a boundless social world corresponding to global geometry. The rights he allocates to humans seem neither to be conferred by individual birth nor to be grounded in a specific soil but to derive socially from common occupancy of the earth's surface and the human capacity to reason forms of society.

But the matter of spatial and anthropological differentiation of the earth's surface is implied twice by Kant: in the references to "foreigners" entering a "society" and to original rights to a country. A century earlier John Locke's claim that "in the beginning all the world was America" left ambiguously open its signification of space available for *de novo* construction of society: by native inhabitants or colonizers? Enlightenment reasoning suggested that philosophy may inform about "natural" human rights only by considering space and time. Humanity may be defined by possession of a rational mind, but geography locates embodied minds, while the movement of history produces "society." Both remove actual humans from pure intellect and an ideal or original state of nature. The uneven consequences of geography and history for a global humanity were being revealed empirically by global exploration. Images of widely and wildly varying environments and human customs circulated, no longer as travelers' tales, but as measured, recorded, located, and thus authoritative observations. For rationalists, human differentiation had to be a product of secular history rather than of Creation, Fall, and Redemption.[48] Precisely how universal social evolution might relate to the geographical differentiation of a global ecumene was the central question of the European human sciences for nearly two centuries, between about 1750 and the triumph of genetics in the second half of the twentieth century.

In humanist cosmography, the Christian paradigm of universal history unfolding teleologically through seven eras, from the creation of Adam to the Second Coming and the celestial city, and across geocentric space, illustrated, for example, in Moxon's 1674 image, mapped the spatialities of the classical polis across gender, age, and social status. Crowning Europe with a city in Renaissance iconography extended this conception of nature, culture, and civilization across the expanding ecumene of early modern discovery. Describing such thinking in terms of "culture" and "civilization," we draw in fact on an eighteenth-century terminology explicitly intended to refigure Christian ideas of empire and redemption in favor of a rational geography and anthropology and a universal history of human progress authorized, like Flinders's maps, by the human mind. All else in creation is subject to the natural laws that mathematics and experimental science could demonstrate. Locke's 1690 *Essay on Human Understanding* had proclaimed the limitless mind as the characteristic feature shared by all humanity, the blank space upon which global geography and history were inscribed. Yet if humans shared a single nature, how were the differences in human culture revealed by exploration to be explained? Whereas earlier writers had taken the cultivation of the earth as the signifier of human society, it was the cultivation of mind as the foundation of "culture" that Enlightenment philosophers grasped to explain the anthropological evidence revealed by global navigation. In so doing, they reconstructed the sociohistorical narrative through which human societies were imagined to advance through world-historical stages of economic and environmental organization, from hunting the forests and wilderness, through herding and cultivating the meadows and fields, to commerce, city building, and the arts. What was decisively new in the universal schemes proposed by Enlightenment writers such as Charles Louis Montesquieu (1689–1755) and Anne-Robert-Jacques Turgot (1727–81) was their attachment of a secular historical process to a scientifically mapped global geography.

Universal history gave rise to a new and distinct mode of spatial representation. The 1747 *Atlas complet des révolutions que le globe de la terre a éprouvées depuis le commencement du monde jusqu'à présent* was the first atlas to "break the Ortelian mould" by using the same base map to present sixty-six views of world history as it unfolded across the classical ecumene.[49] Thirty-two maps are devoted to antiquity, thirty-one to the medieval period, and three to the years following 1500. Although Ortelius had included historical maps in his *Parergon,* the *Atlas complet* illustrated to an entirely secular world history based on the historiographic principle of a spreading "civilization" that had brought mankind from an originating state of nature to a perfected state of enlightened reason and ordered society. History progressed through social stages rather than epiphanies (Eden, the Deluge, Christ's Incarnation, the Second Coming). These stages occurred at different moments over space and time, but their order was consistent, spreading globally from a European center. The atlas was followed in 1763 by *Les révolutions de l'univers,* illustrating the geographical evolution of the Roman Empire and France's inheritance of its civilization. German examples of universal historical atlases predictably

illustrated an alternative inheritance via the barbarian invasion and rejuve-
nation of Rome.

Alternative explanations were offered for the global geographical pattern
of culture and civilization revealed in world-historical atlases. Montesquieu
and Georges Louis Buffon (1707–88) emphasized variations in the natural
environment, arguing that the supposed European superiority in the hierarchy of human evolution derived from the geographical advantages offered
by its climate and soils, apparent also in the size and variety of its vegetation
and fauna compared, for example, with those of the Americas. This approach
tended toward historical stasis, theoretically limiting the capacity of humans
to alter their condition. Turgot, by contrast, drew more explicitly on John
Locke's philosophy of mind to emphasize the human capacity to develop
and overcome natural limitations. His 1751 *Plan d'un ouvrage sur la géographie
politique* contained notes on seven world maps illustrating periods of world
history, in which he interrogated the ethnographic information derived from
French exploration for evidence of unconstrained human progress through
universal stages—hunting, pastoralism, agriculture, and commerce.[50] Human
history in Turgot's thesis moves toward the perfectibility that European society has most nearly achieved, and evidence of previous stages is progressively
apparent as one moves away from Europe to remoter parts of the globe. At
the ends of the earth, viewed from a center fixed by the cartographic coordinates of Paris or Greenwich, are islands of prehistory, of which Australia
was the most distant and most imaginatively dramatic. In its elaboration over
succeeding decades, Turgot's idea became the defining feature of Enlightenment global thinking about humanity: "The philosophical traveler, sailing
to the ends of the earth, is in fact travelling in time; he is exploring the past;
every step he takes is the passage of an age. Those unknown islands that he
reaches are for him the cradle of human society."[51]

The idea of global history unfolding from a Judeo-Christian center is
dramatically illustrated in Edward Quin's *Historical Atlas,* first published in
1830, whose maps, "constructed at a uniform scale, and coloured according
to the political changes of each period; accompanied by a narrative of the
leading events exhibited in the maps: form . . . together a general view of
universal history from the Creation to AD 1828," rework the ancient conceit of raising the reader above the earth's surface, "like a watchman on some
beacon-tower . . . view[ing] the hills and peopled valleys around him, always
the same in situation and form, but under every changing aspect of the hours
and the seasons. Now basking in the meridian sunshine, then slinking into
gloom of even, and again emerging into the light of returning day" (Fig.

7.4). The outward spread of "history" from its Mediterranean core is suggested by "the appearance of a cloud over the skirts of every map, exhibiting at each period only the known parts of the globe, and lifting up or drawing off this cloud as the limits of the known world gradually extended."[52] Such was the dominant reading of Pacific island societies such as Tahiti or Hawaii. In the words of Edmund Burke, "The great map of mankind is unroll'd at once; there is no state or Gradation of barbarism, and no mode of refinement which we have not in the same moment under our view."[53] Across Kant's boundless globe a common humanity is differentiated by the stage it has reached in an inevitable and progressive social evolution. The implied moral imperative—to bring all humanity to the levels of civilization enjoyed by those furthest advanced—became by the early nineteenth century both a powerful justification for European empire and an equally powerful basis for condemning slavery in colonial societies as a denial of natural liberty and the benefits of progress to those "children" of history positioned on the outer edges of Europe's global geography.

In all these discussions an insistent question posed itself: "If science could not peer into the heart of things, could it at least construct a picture of the external world which would be coherent and orderly and law governed— a picture which might not be deeply 'true,' but could at least be self-consistent?"[54] Orreries, globes, and maps, their accuracy verified by ever more refined instrumentation and subjected to the interrogation of Wright's observers, sought to answer this question in the affirmative. The Enlightenment navigator, accompanied as the eighteenth century drew to its close by an increasingly large and specialized team of observers, draftsmen, and technicians, acted as Europe's eyewitness to the desired rational order of global geography and to human societies at different stages of development within universal history. Authenticating and validating explorers' claims at inconceivable distances from Europe were vitally important to a discourse that claimed to replace faith with observation, experimentation, and rational proof as the basis of secure knowledge. The claims of natural philosophy had to be demonstrated in appropriate spaces and according to ever more strictly defined rules of measurement and experimentation if they were to be awarded the status of truth.[55] Locating exploration information within the frame of the measured globe or the surveyed geographic map, accurately coordinated and liberated from rhetorical flourish or fanciful illumination, came to guarantee the integrity of geographical knowledge for those unable to observe for themselves, the paper equivalent of the ordered and restrained space of the Enlightenment laboratory.

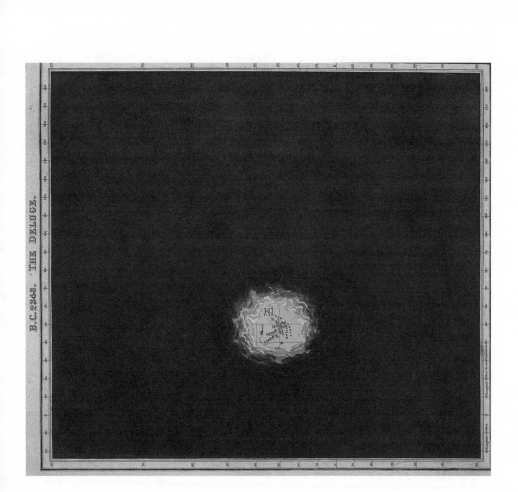

7.4. Edward Quin's images of clouds of unknowing retreating across the globe before the expansion of empire, in his *Historical Atlas* (1830). By permission of the British Library.

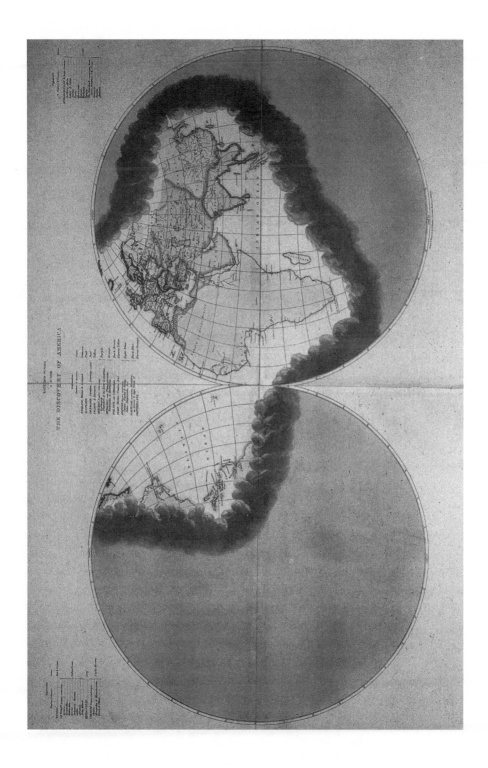

THE DISCOVERY OF AMERICA

Once published on the globe or map, new knowledge left the hand of the explorer to circulate in the public domain. As Longhi's salon testifies, "by 1800 the map had been thoroughly naturalized within modern culture so that map use warranted little comment among members of the general public who were neither cartographers nor cartographic boosters."[56] In France more than anywhere the culture of the map became the foundation for the active construction of an orderly, secular society and human geography. For example, the triangulated map of France at a scale of 1:86,400, begun by César Cassini in 1744, provided the basis for the practical work of the École des Ponts et Chaussées, whose rational curriculum of arithmetic, survey, hydraulics, drafting, stonecutting, architecture, physics, and chemistry was complemented by field practice, with the aim of creating across France a smooth, rational space over which *circulation* would be uninterrupted by "accidents" of nature.[57] This cartographic imperative, whereby the map precedes the space it claims to represent, is similarly apparent in Thomas Jefferson's French-inspired scheme to allocate the future spaces of western North America to colonists, "dispersed to infinity." And the smooth space of cartography could also enlist the past in the re-creation of universal society.

Edme François Jomard (1777–1862), a member of the first graduating class of the École Polytechnique, established in revolutionary France to train the rational leaders of the new civilization, undertook the mapping of Egyptian antiquities during the Napoleonic Survey, later editing the twenty-two-volume *Description d'Egypte,* through which Napoleonic France sought historical legitimation for refounded empire in a study of civilization's origins. A founding member of both the geographical and ethnographical societies of Paris, Jomard's "understanding of geography started, culminated and ended with maps."[58] He used precise cartographic survey and analysis and a critical reading of ancient geographers in an attempt to break the code of hieroglyphics, believing ancient Egyptians to have been "inherently geometrical," thereby reworking the assumed founding connection between language and number in terms of France's new, "rational" metric system of measurement. As the director of French exploration in Africa, Jomard set out the methodology of Enlightenment universalism as an extension of the global mapping exercise:

> In order to know the earth, it was necessary to start by determining the position of particular places: to establish their real distances and their respective situations, their relative and absolute elevation, to study finally their natural productions. In other words, it was neces-

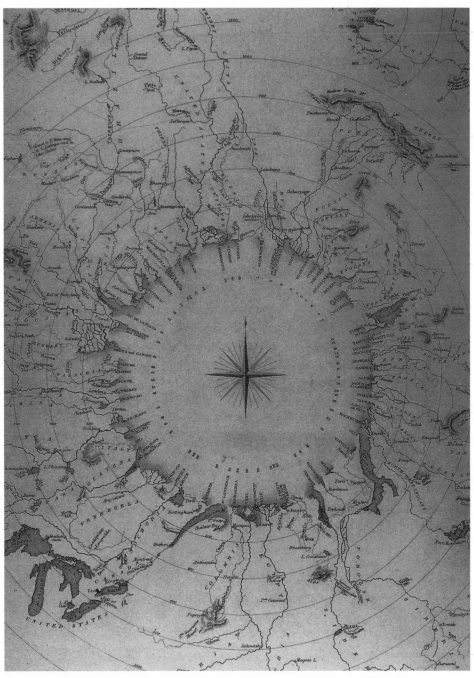

7.5. A measured globe: lengths of the earth's principal rivers mapped by the Society for the Diffusion of Useful Knowledge, 1844. By permission of the British Library.

sary to start by "geography" and physical geography. Today the map
of the world can neither satisfy our avid curiosity nor the progress
of contemporary knowledge. It is, in any case, sufficiently advanced to
allow us to turn our efforts to another, more important, direction.
Here I am speaking of the distinction of the human races and the
universal knowledge of their idioms, of the character of their phys-
iognomy, and their social state. This is what almost all of the nations
of Europe are starting to do. . . . After all shouldn't this be the final
aim of the description of the inhabitable earth?[59]

Ancient knowledge and historical authority are approached with the same
critical and disinterested eye as the contemporary measurements of navigators and land explorers. Thus Jomard's last work was *Monuments de la géographie,* a facsimile atlas of pre-Ortelian maps and globes of the earth. One of
the first histories of global mapping, it aimed to aid public understanding of
human progress through a history of graphic images themselves rather than
written texts. As mapping was brought within the orbit of applied rational
science, so a history of "cartography" became a distinct field of study, offering its own testimony to progress. Alexander von Humboldt would use
Jomard's and others' facsimile maps to understand the link between imagination and navigation in the history of geographical knowledge.

Jomard's projects synthesize the various concerns connecting global thinking and global representation in the European Enlightenment. In the academic study and the laboratory, even in the drawing room of Longhi's young
geography student, the globe and map became active in shaping the spaces
they claimed to represent with scientific accuracy (Fig. 7.5). Subjected to the
disembodied gaze of the rational eye, classified and illuminated by the light
of sovereign European reason, global space became evidence of universal historical progress. The totalizing images of global space indicate the moment
of modernity.

EIGHT *Modern Globe*

Modernity begins when the real space of the world becomes a stage, and when this stage, mastered by a stage manager, turns inside out like a glove or simple geometrical line drawing to the eye and collapses into the utopia of a knowing, interior, intimate subject. This black hole swallows up the whole world.

—Michel Serres, "Gnomon"

In 1793, nine years after designing his cenotaph to Isaac Newton, Etienne-Louis Boullée again used a vast empty globe to design an imaginary architectural project, in this case to represent universal space. Boullée's *Temple to Nature and Reason* pressed still further the relationship between the planetary globe and the virtual space of the human mind. The design married a smooth and featureless upper hemisphere of pure Reason, to a lower hemisphere of Nature, roughly quarried out of the earth and riven at its depths with a great black fissure descending to infinity. This enormous "world" stage was to be viewed internally from a colonnade encompassing the equator line, where conceptual and material hemispheres meet. Neither hemisphere could be physically accessed; they were to be gazed at. The cosmic unity of Boullée's globe is a creation of the eye and the intellect of an individual observer poised at the midpoint between the striated and accidented globe of nature and the smooth, undifferentiated sphere of the mind. Although never constructed, Boullée's grandiose staging of the globe anticipates modernity's characteristic reworking of the idea of seeing the globe: "Without a separation of the self from a picture, moreover, it becomes impossible to grasp 'the whole.' The experience of the world as a picture set up before a subject is linked to the unusual conception of the world as an enframed totality, something that forms a structure or system."[1]

The globe is a paradigmatic form within the modern experience and modern image making. In 1783, as Boullée was designing his conceptual globes, two other great spheres, inflated with hot gases, rose into the air over Paris; these first balloons were the earliest realization of the Ciceronian dream of flying over the earth's surface. Thenceforth human flight advanced erratically, powered by the engines of scientific curiosity and youthful adventurism.[2] The vision offered by physical flight was anticipated in the nineteenth-century taste for drawn or painted panoramas and dioramas, whose subject matter was dominated, significantly, by imperial images of open continental interiors such as the American West and the European metropolis. Panoramic images of London in 1792 and Paris in 1802 by Henry Barker and Thomas Horner's celebrated 1821 view from the cupola of St. Paul's using Cornelius Varley's graphic telescope, anticipated the actual views of London offered to the public from the Regent's Park Colosseum from 1829. Public georamas were constructed in Paris in 1823 and 1844, while from 1851, complementing the Great Exhibition, which brought the peoples and products of the world to London, Wyld's Great Globe in Leicester Square, at the heart of the city's entertainment district, displayed plaster casts of the world's continents and oceans.[3] At the center of Europe's first global city the earth was offered to a mass public as a theatrical novelty. Since the mid-nineteenth century the image of the globe has emerged as the icon for the interrelated processes of connection, communication, and control that characterize modernity; it is an image that rests precisely on the idea of the globe's *visibility*.

Enlightenment assumption of transparent correspondence between actual form and patterns and their scientific representation in models and maps underlay nineteenth-century surveys of global space, while visual clarity in representation underpinned the scientific and technical authority of globes and geographical maps themselves.[4] Since John Harrison's chronometer and Jesse Ramsden's sextants, theodolites, and barometers, instrumentation for measuring and recording the globe and its geography have been continuously refined, up to today's orbiting instrumental satellites. Instrumentation defined what counted as exploration knowledge as much as it responded to already formulated questions. The map's visual language, embodied in mathematical survey, graphic rules, and design codes, offered a scientific alternative to the inability of words and language to convey the sheer *difference* of discovery. Thus, Thomas Jefferson's instructions for Meriwether Lewis and William Clark's 1804 expedition into Louisiana Territory emphasized precise recordkeeping and surveying and the need to restrict "imagination." The explorers' journals indicate a slow fragmentation and eventual collapse of language in the face of overwhelming difference in the West. Sentence structure, syntax, and descriptive language all broke down as the expedition penetrated spaces that had been blank on the European map.[5] Such failure of language placed ever greater emphasis on nonlinguistic representations— numerical tables, graphs and statistics, topographic drawing and painting— as accurate registers of vision.[6] Spatial movement across unknown territory or around a revolving diorama emphasized the limitations of the static eye. Experimentation in ways of seeing and representing space, especially Charles Wheatstone's simulation of stereoscopic vision in 1832, was accelerated by the appearance of photographic images in the final years of the same decade and underpinned the photograph's powerful claims to mimetic truth and scientific naturalism.[7] It was a trajectory that "would calibrate the shift from the transcendental to the material, from the numinous unity of the cosmos either to the ecological unity of the organic and inorganic or to an absorp- tion with the individuality of discrete phenomena—the selfsame shifts that gave impetus to scientific theorizing."[8] However, some of those who were most committed to the new graphic language also saw in the mapping of material space a way to achieve more humanist ends, as the late-nineteenth- century internationalism that produced universal exhibitions, universal time, and plans for a universal language also created the International Geograph- ical Union and proposals for an international map of the world.

In this context, the West's completion of its long, self-conscious expan- sion of the Greek ecumene to the ends of the earth was simultaneously an act of imaginative and practical vision and their expression in mapping. I focus my discussion of this modern "closure" of the global image around three strategic spatial metaphors: interior *penetration* of continental and tel- luric space; axial *advance* along the meridian to the polar ends of the earth; and *encirclement* of the latitudinal arc, appropriating global space by bringing East and West together within a single imaginative realm. A military lan- guage of space, also used in planning, seems appropriate for a set of processes that served to align scientific and commercial with geopolitical concerns.

Penetrating the Land Globe

Geopolitical and cartographic *penetration* of the continental interiors of the Americas, Asia, Africa, and Australia coincided with the arc of the nine- teenth century. In 1800 Alexander von Humboldt was a year into his four- year exploration of the Andean and tropical interior of South America. His dramatic presentation of the journey to an audience in Philadelphia was a stimulus to Lewis and Clark's two-year passage through the trans-Mississippi West. Their exploration opened a federally funded "great reconnaissance" between 1867 and 1875 under soldiers subsequently promoted as heroes of the new nation—Clarence King, Ferdinand Hayden, John Wesley Powell, George Wheeler.[9] Lavishly illustrated multivolume scientific reports recorded geology, climate, vegetation, and ethnography as well as topography over vast areas. They were published at federal expense for wide distribution to the states and, in the tradition of the baroque atlas, as official gifts for visiting dignitaries. In these reports, American science signified and legitimated the manifest destiny of American empire. Australia and Asia were similarly surveyed by transcontinental expeditions, while Portuguese and British explorers competed to cross equatorial Africa in both the cardinal directions and resolve Europe's most ancient geographical question by fixing the source of the Nile on the global graticule. The image of the river steamer with pith-helmeted passengers penetrating to the deepest interiors of equatorial "jungles" has become the symbol of imperialist high culture, captured in Joseph Conrad's *Heart of Darkness.*[10] The image of European technology charting the humid spaces of the torrid zone was paralleled on the globe's temperate plains and savannas by the steam locomotive advancing over transcontinental iron rails.

In organization and discourse continental exploration projects further refined the scientific model of Enlightenment oceanic expeditions. The culture of exploration emphasized both the physical discipline and the endurance conventionally associated with masculine adventure and the legitimating discourse of scientific and technical precision in survey, observation, and recordkeeping. Modernity thus reworked the requirements that had characterized expansive exploration since antiquity—youthful heroism and the deployment of hard-won, specialized skill in uncovering arcane knowledge. If popular press reports and autobiographical accounts of explorations emphasized heroic confrontation with raw nature, ignoring local knowledge in the form of guides and porters to focus on the lone individual, state agencies and scientific societies emphasized instrumentation, accurate measurement, and rigorous procedures for gathering, recording, and transferring knowledge.

Geodetic survey and topographic mapping, together with the naming and spatial classification of observed phenomena, became measures of exploratory success and legitimated both imperial claims to authority over the spaces surveyed and national prestige in contributing to the advance of universal "geographical science." A U.S. senator thus praised the congressional Pacific Railroad Reports of 1853, whose publication cost twice the amount paid for four army reconnaissance surveys. "Every unusual swell of the land, every unexpected or unanticipated gorge in the mountains has been displayed in a beautiful picture. Every bird that flies in the air over that immense region, and every beast that travels the plains and mountains, every fish that swims in its lakes and rivers, every reptile that crawls, every insect that buzzes in the summer breeze, has been displayed in the highest style of art, and in the most brilliant colors."[11] The survey is more than a linear tracing of lines of surface pattern. It produces a comprehensive, accurate *picture* of the terrestrial surface, a scaled reproduction of the real. Whole sections of global space are "framed" according to universal criteria of observation and representation in which the aesthetic is folded into a notion of scientific accuracy.

Lieutenant Joseph Christmas Ives's *Report upon the Colorado River of the West,* published in 1861, exemplifies the demands on cartographic representation that "global" understanding entailed. Ives's 1857–58 expedition had traveled by steamboat up the Colorado River from the Gulf of Mexico to the limits of navigation in the Grand Canyon before continuing overland to the Rio Grande and western Texas. The expedition report, composed by Ives himself and illustrated by the German artist and topographer Frederick W. von Egloffstein, was the first attempt to represent the American continent's most dramatic telluric feature—the Grand Canyon—a physical realization of the fissure at the depths of Boullée's *Temple to Nature and Reason.*[12] Egloffstein's problems in seeking to represent the mile-deep gorge, the limitations of his geological accuracy, and his divergence from Ives's written text have attracted critical comment since the publication of the report. The billowing gloom of the picturesque style he adopted to illustrate Ives's own plainly drawn geological sections failed to capture the horizontal stratification that gives the Grand Canyon its drama, hinting at the limitations of a naturalistic vision. Confronted with a mile-deep ravine where the physical distance and visual detachment required by the picturesque gaze are simply impossible, the eye is forced back into a body dwarfed much more completely by nature's spaces than in Boullée's cenotaph. The combination of scale and lithology in the Grand Canyon was beyond anything a European-trained painter could possibly have experienced before, and Egloffstein lacked the pictorial language to convey such space, a common problem faced in picturing nineteenth-century exploration spaces.[13] By contrast, his panoramas and the area map Egloffstein produced of this huge region are visually convincing as a representation of how its spaces might appear to a disembodied eye located a mile above the earth (Fig. 8.1). The maps gain authority from the naturalizing powers of relief shading taken from the panoramas, while the panoramas themselves were coordinated and given locational names

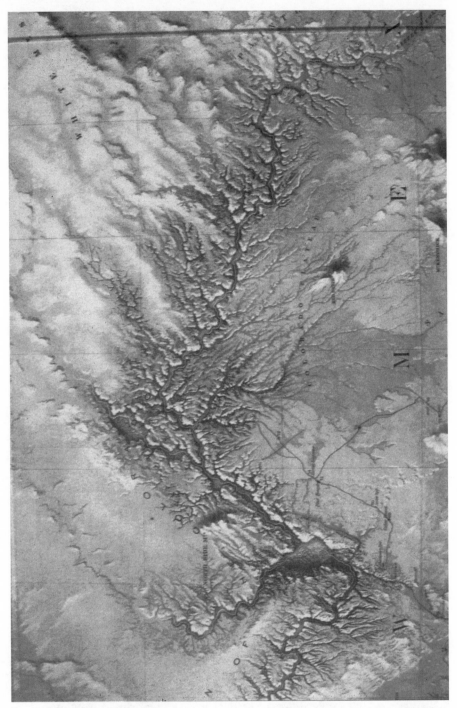

8.1. Frederick W. von Egloffstein's imagined aerial view of the Colorado River, *Rio Colorado of the West* (1858), from J. Ives, *Report upon the Colorado River of the West* (Washington, D.C.: Government Printing Office, 1861).

derived from the map: "The maps also privilege 'actual appearance to the eye' but extend beyond the situated view of the panoramas to a planimetric framework . . . it is this transformation and compilation into planimetric form that lends authority to the numerous observations of the survey."[14]

Egloffstein's work was quite specifically inspired by the cartographic innovations of his older compatriot Alexander von Humboldt developed during the latter's own continental exploration. For Humboldt, the mapped image could itself test an experimental hypothesis, revealing aspects of the world invisible to the eye. Although it was not his invention, Humboldt was responsible for exploiting the isoline to illuminate global phenomena, such as climatic and vegetational zones, that can be known only through their representation. Thus, his 1817 world isotherm map at 5°C intervals, *Des lignes isothermes et la distribution de la chaleur sur le globe,* conclusively unshackled global surface temperatures from latitudinal belts, replacing the relationship between climate and the Aristotelian *klimata* that had structured the globe's physical geography since antiquity.[15] Based on his technical experience of mapping mining geology, Humboldt recognized that cartography offered powers beyond description. The thematic map could actively generate hypotheses that could assist in the achievement of the longstanding cosmographic goal of revealing hidden order below an apparently accidented globe:

> Humboldt was seeking a more analytical spatial language which would allow the almost intuitive transfer of understanding from one graphic genre to another and from one specialist body of knowledge to another. Humboldt was trying to find a language capable of expressing his vision of the unity of nature with the newly found rigour of the systematic sciences. To that end, he experimented with isolines, distribution maps, flow maps, a map of error, proportional squares, something he called "pasigraphy," and a multi-dimensional pictorial graph. These eloquent graphic arguments about dynamic relationships in space, distributions and interactions which often revealed patterns not visible to the naked eye and created a new systematic, rather than geographic time/space, seemed to Humboldt to suggest that his intuition about the unity of nature was sound.[16]

The map, like the scientific laboratory, could generate as well as illustrate new knowledge, creating worlds that could then be explored empirically. Map *projection* here takes on a secondary, epistemological meaning. Both Humboldt and Egloffstein had initially mapped geological phenomena. Not only was the nineteenth-century geological science instrumental as a model for inferential mapping of unseen stratigraphy but by illustrating "deep time" it empirically undermined the biblical account of universal time conventionally associated with global mapping. Relating geological processes to continental and oceanic mapping restimulated the search for global symmetry, never completely abandoned with the discrediting of Aristotelian patterns. Ortelius had commented in his *Thesaurus geographicus* on similarities in the Atlantic coasts of Africa and South America, and Francis Bacon had declared of global symmetry that "similar instances are not to be neglected, in the greater portion of the world's conformation; such as Africa and the Peruvian continent, which reaches to the Straits of Magellan, both of which possess a similar isthmus and similar capes, a circumstance not to be attributed to mere accident . . . the new and the old world are both of them broad and expanded towards the north and narrow and pointed toward the south."[17]

The geologist Charles Lyell developed a theory of antipodal asymmetry, while the late-nineteenth-century Russian scientist A. P. Karpinsky traced geological harmonies in which continental mountain chains make up a global tree branching across the sphere from roots in Antarctica. Geological "deep time" eventually permitted the reestablishment of pattern logic within the continents and oceans through the 1920s theory of continental drift and later widespread acceptance of plate tectonics and the conception of a mobile global surface.

Global social patterns too could be revealed by mapping selected distributions and indexes—language, religious belief, agricultural practice, diet, custom—and related in turn to global physical geography. Ptolemaic chorography mutates intellectually into the geographical "science" of "chorology" by way of the ecological fallacy of theorizing causal relationships between sets of phenomena through observation of their cartographic correspondence. The "geographical experiment" that dominated the Western interpretation of global social order and the geopolitics of empire was cartographically based.[18]

Humboldt's own cartographic innovations had a more conventional philosophical dimension. The first two volumes of *Cosmos: A Sketch of a Physical Description of the Universe,* completed in 1844 and translated into English by 1848, constitute a philosophical and historical argument for a graphically based, five-volume cosmography "uniting, under one point of view, both the phenomena of our own globe and those presented in the regions of space."[19] Humboldt was profoundly committed to Enlightenment principles of rational thought and empirical observation, drawing on the most recent scientific discoveries, but the intention of *Cosmos* is to demonstrate an aesthetic if not a directly metaphysical harmony "to show the simultaneous action and the connecting links of the forces which pervade the universe."[20] Describing the structure of his work, Humboldt reverts to the time-honored conceit of imaginative flight and descent, in which the earth reveals ever greater detail to the Apollonian eye:

> Beginning with the depths of space and the regions of the remotest nebulae, we will gradually descend through the starry zone to which our solar system belongs, to our own terrestrial spheroid, circled by air and ocean, there to direct our attention to its form, temperature and magnetic tension, and to consider the fulness of organic life unfolding itself upon its surface beneath the vivifying influence of light. In this manner a picture of the world may, with a few strokes, be made to include the realms of infinity no less than the minute microscopic animal and vegetable organisms, which exist in standing waters, and on the weather-beaten surface of our rocks.[21]

This intuition of unity, visible from above, is as much imaginative as empirical, so that *penetration* shifts from Humboldt's earlier physical exploration of equatorial forests to an image held in the mind's eye:

> A considerable portion of the qualitative properties of matter . . . is doubtless still unknown to us; and the attempt perfectly to represent unity in diversity must necessarily therefore prove unsuccessful. Thus besides the pleasure derived from acquired knowledge there lurks in the mind of man, and tinged with a shade of sadness, an unsatisfied longing for something beyond the present—a striving towards regions yet unknown and unopened. Such a sense of longing binds still faster the links which in accordance with the supreme laws of our being connect the material with the ideal world, and animates the mysterious relation existing between that which the mind receives from without, and that which it reflects from its own depths to the external world.[22]

The cosmic perspective on the globe produced a unitary vision that Humboldt's romanticism attributed to the human spirit, drawn by "a certain secret analogy" shared with cosmic being. His debt to German idealism, and specifically to Goethe, is readily apparent. Goethe's own "total art work," *Faust,* had reworked Seneca's cosmographic vision of the great theater of the world.[23] But as the epigraph from Serres at the beginning of this chapter suggests, modernity's Faustian bargain offers the human subject power and control precisely by disenchanting nature and cosmos.

Disenchantment and the physical limitations of vision were more acutely apparent among nineteenth-century explorers themselves, although their conversation with the geographical texts and reconstructed maps of the ancient world indicates a self-conscious attachment to a lineage of Western globalism. Thus, Richard Burton's account of his search for the Nile's source with John Hanning Speke devotes an entire chapter to the river's mapping back to Greek antiquity, and their journey is dramatically narrated as a personal and physical quest to both physical and spatial limits, one of the men ultimately incapable of walking and the other of seeing.[24] Equally dramatic was the 1870s leveling of the Saharan chotts under the direction of Colonel François Élie Roudaire, prompted by a desire to survey the ancient Sea of Triton south of the Atlas Mountains, through which Jason and his Argonauts supposedly sailed. "Temperatures often approached 30°C during the day and usually fell below freezing at night. Surveying was hampered by the heat haze and the blinding reflection from the white salt crystals. . . . Each man covered at least 20 kilometers on foot each day and was permitted only four hours sleep a night."[25] Little wonder that the surveyors suffered from hallucinations and that some eventually went blind. The ultimate witness to cartographic authenticity could be the destruction of the bodily organ upon which vision depends.

Roudaire's survey was undertaken with the aim of constructing a canal to connect the Libyan Sea at Gabès with the chotts of Tunisia and Algeria, a string of saline depressions below sea level. His aim, shared with the French engineer of the Suez Canal, Ferdinand de Lesseps, was to flood the sea once more and thus reverse the environmental decline believed to have set in since the end of Roman imperial rule. As heir to Rome, imperial France would arrest the economic and social decay into which Islamic rule had supposedly thrust the southern regions of the Middle Sea. Such "Orientalist" gendering, framing an Islamic "East" as passive subject to the moral and physical authority of an active West, structured the moral geopolitics of European modernity, offering metaphysical justification for often ruthlessly commercial geopolitical global projects such as the Suez and Panama Canals, presented in St. Simonian terms by Ferdinand de Lesseps as global engineering projects to "trace across this very globe the *sign of peace* and to truly forge a link between two parts of the Old World, the Orient and the Occident."[26]

The search for empirical justification for speculative stadial and evolutionary theories of world history and human distinction also drove continental exploration.[27] Global mapping of climate and physical environments and of biologically defined human groups underpinned geographical theories of race. Such theories inevitably raised questions about the origin and diffusion of human groups, including self-constituted Europeans themselves. As a self-declared new nation occupying a "new" hemisphere yet profoundly attached to cultural and ethnic origins in northwestern Europe, and with race-based identity an inevitable corollary of slavery, such questions were particularly rehearsed in the United States. Thomas Jefferson himself instructed Lewis and Clark to look out for traces of ancient life in interior America—dinosaur bones and the lost tribe of Israel—wishing for evidence of natural and cultural origins of American distinctiveness in its unmapped spaces. By the later nineteenth century American science was generously funded by private as well as state sources, for example, by the Vanderbilt and Carnegie fortunes, derived from the conquest of continental space.

By 1900 a significant proportion of Andrew Carnegie's Washington Institute's $100,000 annual grant program was devoted to funding geographical expeditions to study such longstanding global questions as terrestrial magnetism. The institute promoted both the heroic individualism of expeditionary exploration and the scientific concern with human origins and their relations with global patterns of environment. Carnegie's personal stipulation was that funds be directed toward "the exceptional man," above all the heroic young explorer. One such explorer was Raphael Pumpelly, who led an expedition to Turkestan, in the heart of the Eurasian landmass, in 1903–4. Its purpose was to determine "the physical basis of human history."[28] Pumpelly's personal aim was to locate the cradle of "Indo-European civilisation." Two members of his expedition, the geologist W. M. Davis and the climatologist Ellsworth Huntington, would draw upon its findings to construct global geographical hypotheses of causal relations, between climate and landforms in Davis's case and between climate and civilization in Huntington's.[29] Led by a heroic individual into the landlocked center of the continents, exploration "science" married the most sophisticated technical instrumentation, careful recordkeeping, and measurement with global theories of origins and individual destiny that bear the enduring hallmarks of the mythological imagination.[30]

Heroes of Modernity: Advancing to the Poles

This modernist marriage of science and unacknowledged myth in the construction of meaning for the globe was most dramatically apparent at its poles, in absolute landscapes of illumination and shadow. The poles had long prompted eschatological musings. In antiquity, Arctikos, the frigid region of the bear, had been the home of "happy" Hyperboreans. Navigational problems caused by ice and fog and by magnetic variation in high latitudes continued to frustrate the European search for symmetrical northern passages around both old and new worlds.[31] Cook had crossed both polar great circles; his was the first European ship to enter antarctic seas. But the islands, oceans, and ice shelves of spaces at the longitudinal ends of the earth were genuine theaters of modern discovery, in which the distribution of lands and seas remained imprecise into the final quarter of the twentieth century. Greenland's island status was determined only in 1892; Canada's northern archipelago was accurately outlined by air reconnaissance during World War II, while Antarctica's coastlines are traced by question marks on a 1966 atlas map.[32]

Movement in high latitudes was always beset with unique difficulties and uncertainties, the annual patterns of summer light and winter night and the risks of foundering on floating ice or being crushed by monstrous bergs being only the most obvious of these. The compass becomes increasingly unreliable as the magnetic pole is neared and as longitude lines gather; atmospheric and light effects make distance measurement uncertain, with superior mirages bringing coastlines, islands, and mountains hundreds of kilometers nearer to the observer than their true location. The certainty of vision, the authorizing basis of modern science, is thus profoundly compromised, so that observational claims may come to seem like deranged fantasy.[33]

These problems were compounded by storm and cold, making accurate instrumentation almost impossible, and sea mists and fogs that appeared and disappeared with startling unpredictability. On land and on ice the vision and directional reckoning of both the walking and the sledding observer are equally threatened by the blinding whiteness of blizzards, the undifferentiated expanses of ice or snowfields, and the obscured dangers of crevasses, sastrugi, and pressure ridges. Such conditions, along with the emptiness, silence, and light effects of constant night or daylight and of the aurora served to heighten these regions' grip on the imagination.[34] Indeed, the ultimate goals of polar exploration were themselves imaginary. The poles exist only in their scientific and cartographic representations, and they are multiple—geographic, magnetic, geomagnetic, each indeterminate and mobile.[35] Both the Amundsen and Scott expeditions of 1911 spent entire days at the South Pole making sorties and circular marches to ensure that they had covered the actual location of the polar point.

The poles thus represent the final ends of the earth, global destinations of ultimate inaccessibility. Their "conquest" offered individuals and nations a competitive sense of global mastery comparable only to circumnavigation by sea or air or the ascent of high mountains.[36] Modernist narratives in the form of news reports and popular written accounts negotiate the contradictions between heroic individualism tested to its physical and psychological limits and assumed scientific and engineering superiority in the conquest of nature's toughest obstacles. Thus, in accounts of the 1911 race to the South Pole Amundsen's folk knowledge of high latitudes gained among Suomi peoples of northern Norway and his reliance on dogs is often contrasted with Scott's pride in technology and his importing an automobile into Antarctica. The competition emphasizes the significance of polar regions as theaters of nineteenth-century competitive heroism, a competition not unconnected to racial theories of human differentiation and hierarchy. "Newer" nations sought to establish prestige at the ends of the earth and claim polar territories in competition with established imperial states such as Britain, France, and Russia. In Nordic Europe and in the colonial nations of Australia, New Zealand, Chile, Argentina, Canada, and the United States, public opinion regarded participation in polar exploration and claims to polar territory as indexes of national heroism, scientific acumen, and progress. Like the summits of the world's great mountain ranges, the "purity" of white, empty polar regions acted as an imaginative opposite to equatorial "hearts of darkness." Devoid of disturbing human presence, they were silent stages for the performance of white manhood.

The drama ended as often in tragedy as in triumph. The Franklin and Scott expeditions are exemplary. The former, equipped with the latest scientific instruments and supposedly manned by the doughtiest naval officers, left Britain in the spring of 1845 to navigate the Northwest Passage. Only after three years' silence was a search mounted for the missing expedition. Initially, the complete absence of any evidence of its fate captured the national imagination, as ultimately did the evidence of cannibalism among its dying survivors. An ancient fear of bestial behavior at the limits of space coincided with the publication in these very years of Charles Darwin's *Origin of Species*. The lines mapping the borders between humanity and Tennyson's raw-red nature, and between "civilized" Europeans and *anthropophagi* at the ends of the earth, were suddenly and decisively smudged. The story of the Scott expedition of 1911–12, defeated by Norwegian explorers and expiring gruesomely but politely on the return journey, offered a redeeming narrative of cultivated self-sacrifice while signaling a challenge to technological modernity as disturbing as the loss in the same months of the ocean liner *Titanic,* sunk by an iceberg in the North Atlantic.

The scramble for Antarctica at the turn of the twentieth century paralleled that for Africa a quarter-century earlier. The southern continent, projected by ancient desires for global symmetry, diminishing in size from Ortelius's Magellanica, and demonstrably uninhabitable since Cook's 1774 voyage, was trodden for the first time only in 1895. Within decades the overlapping claims of Western states had sliced the last island continent into the wedges of a cartographic pie with purely representational boundaries whose statistical significance in competitive calculation of total global territory far outweighed their practical value. At the opposite pole, Norway realized the dream of northern circumnavigation in 1893–96, while Robert Peary claimed controversially to have reached the North Pole in 1909 and demonstrated the absence there of both land and open sea. The territories of the late twentieth century's opposing global power blocs enclosed all but ten degrees of longitude around the North Pole, giving the top of the world a powerful practical and symbolic role in modern geopolitical strategy.

During the 1930s, arctic navigation and exploration were vigorously promoted as demonstrative evidence of the revolutionary Soviet order. In 1932 the icebreaker *Sibiriakov* successfully navigated the entire Northeast Passage. Two years later more than one hundred people were rescued by Soviet pilots after being stranded for two months on ice floes following the sinking of their research vessel *Cheliuskin.* And in 1937, two years after the American millionaire Ellsworth had overflown Antarctica, a Soviet plane landed at the North Pole, establishing a Soviet scientific base, which was later overflown on a transpolar flight between Moscow and the United States. A Promethean struggle against an imperious nature, cast at the continental scale of the Soviet empire, was a key feature of Georgii V. Plekhanov's influential fusion of socialism and modernism, which constructed a Soviet ideology out of Marxism and geographical determinism. The struggle was most apparent in the arctic empire, where human individuals and machines shared the honor of socialist progress and conquest at the ends of the earth. "The vast empty expanses of the Soviet Arctic served as a perfect blank slate, a discursive *tabula rasa* on which the Soviets could inscribe their visions of the new world they were building."[37]

For forty years the American and Soviet empires confronted each other across the North Pole in the language of technological high modernism. Intercontinental ballistic missiles (ICBMs) routed through the polar skies, nuclear submarines submerged below the icecap, and electronic early-warning systems encircled the tundra deserts. This *cold* war was aptly named. At its height came the International Geophysical Year (IGY), 1957–58, a defining moment for twentieth-century globalism, initiated by American atmospheric geophysicists. The militarization of the North Pole redirected competitive research toward Antarctica, including studies of the aurora as a source of information on terrestrial magnetism and of sunspot activity and American detonation of hydrogen bombs to examine auroral effects. There is a direct line of descent from this work to the discovery of the Antarctic ozone deficit and fears of global climatic catastrophe.

Modernist and materialist visions consistently incorporate echoes of mystical, if not metaphysical, enchantment with the polar ends of the earth, a "boreal exoticism."[38] Coleridge's *Rime of the Ancient Mariner,* published in 1799, introduced tropes for a polar imagination that endured through more than a century of polar exploration and representation: isolation in empty, oceanic space, the floating mysteries of ice and fog, and the saintly albatross, the only creature capable of rising over polar space and seeing across the very top of the world. Even Soviet writers spoke uncharacteristically of the "call" of the polar region and the mysterious hold of "bewitching," elemental spaces where "the cold of space smote the unprotected tip of the planet," and dreamt of gaining the aerial view of the great bird. In a passage that echoes Seneca's rhetorical response to the aerial vision of the globe, "How puny are the doings of men," the radio operator of the *Cheliuskin,* Ernst Krenkel, wrote of arctic space in decidedly un-Soviet and unmodern terms: "All movement ceases, the sky clears, the heavens are brass; the slightest whisper seems sacrilege, and man becomes timid, affrighted at the sound of his own voice. Sole speck of life journeying across the ghostly wastes of a dead world, he trembles at his audacity, realises that his is a maggot's life, nothing more. Strange thoughts arise unsummoned, and the mystery of all things strives for utterance."[39]

It was powered flight, perfected in the same years as Westerners reached the poles, that offered the vision formerly reserved to the albatross. The "airman's view," examined more fully in chapter 9, was critical for Antarctica's incorporation into the modern image of the globe. A 1950s project, the Falkland Islands Dependencies Survey Expedition (FIDASE), is exemplary. A series of mapping sorties backed by ground-based terrain work was designed to produce detailed topographical records of Graham Land and the Antarctic Peninsula as part of Britain's attempt to sustain its imperial claims in the face of challenges from Argentina.[40] Like maritime navigation, this work was severely hampered by the problems of vision encountered in polar environ-

ments, so that the accurate mapping of Antarctica had to await satellite sensing technology, developed only in the closing years of the twentieth century. Even today, with azimuthal maps of polar regions a standard feature of
any atlas and with the white expanses of frozen ocean and continental Antarctica inescapable features of space photographs of the earth, there seems to
be an unwillingness to contemplate these global regions without anxiety;
they remain eschatological ends of the earth, whence ozone depletion or
ice-sheet meltdown threatens life across the globe. The Antarctic Treaty of
1959 placed in abeyance territorial claims to the continent, declaring it a
global wilderness—an empire of nature—thus marking both a final boundary between the zones of culture and nature and, in theory at least, the deter-
ritorialization of the Apollonian perspective, through which it is largely
known. However, the research undertaken in Antarctica is not disinterested:
"intellectual claims are interlaced with other assertions of ownership," and
imperial geopolitics still map the ends of the earth.[41]

Encircling Space and Time

High European imperialism at the turn of the twentieth century also witnessed the achievement of the Enlightenment vision of global encirclement,
attributed by Henri St. Jean de Crèvecoeur in 1780 to the destiny of the
"new" American, who "will complete the great circle." Crèvecoeur's prediction projected the American republic's imperial destiny across the continental spaces of the New World by reworking the Western myth of civilization and human progress following Apollo's chariot. A "manifest destiny"
in the West was the key theme in American political discourse and popular
iconography as the republic carved a continental empire toward the Pacific;
Asher Durand's painting *Progress* and Thomas Cole's five-part *Course of Empire* both figure an imperial destiny in the iconography of classical Rome.[42]
Circumnavigation has remained an act of enormous imaginative significance
since Magellan. A number of scientific, political, and technological achievements in the closing decades of the nineteenth century brought the image,
idea, and implications of an encircled globe strongly into focus in the Western consciousness. Specifically, the idea of a *closed,* or circled, globe generated new hopes and fears about the ends of space and time.[43]

In 1884 twenty-six nations attending the International Meridian Conference in Washington, D.C., agreed to recognize the Greenwich longitude
as the prime meridian of the globe. East and West, like North and South,
would henceforth be fixed hemispheric spaces on a quartered globe cen-

tered upon London. In 1911 a Paris conference divided the earth into twenty-four time zones of 15 degrees of latitude, with an International Date Line fixed 180 degrees from Greenwich, with deviations taking into account the eastward extension of Siberia, the westward scatter of the American Aleutians, and the pattern of islands in the South Pacific. As we have seen, the choice of Greenwich was preordained as much by maritime practice as by ideological intent, but it inscribed Eurocentric assumptions into a hegemonic global image.[44] Regularizing the graticule was critical in an era of global navigation, trade, commerce, and finance, dramatically emphasized by the profound modification of global sea routes by continental engineering of the Suez and Panama Canals. Over the course of more than a century since fixing the meridian a suite of technological inventions have brought globally distant regions into regular contact over ever smaller units of time. Laying telegraphic cables across the ocean floors, the Atlantic cable in 1866 and the Pacific by 1902, and railroads across every continent in the same years was rapidly followed by international postal services, telephones, and radio communications. The competing capitalisms of Western industrial powers effectively annexed global space, leaving no continent free from colonial rule and the competitive reach of commercial, industrial, and finance capital.[45]

These competitive processes, driven by market economy and nationalist territoriality, accentuated anxiety about the "closure" of global space, which was expressed culturally in various ways by intellectuals and political leaders. A pervasive sense that spatial limits had been reached dominated the geographical imagination of many Westerners in the last years of the nineteenth century.[46] Completing their "great circle," Americans, tutored by Frederick Jackson Turner's 1894 "frontier hypothesis," reflected on what would follow the "consummation of empire."[47] Pointing out that the end of the process of European westward advance across North America, which had begun in the early seventeenth century, had been officially announced in the 1890 federal census, Turner reflected anxiously on the necessity of open space for the continued survival of American "civilization."[48] Such anxiety was not confined to the United States; it became cause and consequence of pioneering European settlement in Australasia, southern Africa, and temperate South America. Global theories relating race and climate sought both to explain existing population distributions and to justify the allocation of land and labor among different groups within a colonial world.[49] Within Europe itself competitive nationalism developed similar theories suggesting that dynamic and healthy "races" inevitably engaged in competition for "living space."[50] The longstanding connection between global represen-

tations and universal history was reworked in the context of modern biological and ecological science by writers such as Ellen Churchill Semple,
Arnold Toynbee, and Ellsworth Huntington.

These concerns generated an explicitly global geopolitics among Western states, a discourse that sought to base political and military strategy on
what were regarded as enduring patterns of lands, seas, climates, and resources
across the terrestrial sphere. In 1904 the British academic Halford John Mackinder promoted such a theory in response to the fears of closed space circulating in the dominant imperial power. "The Geographical Pivot of History," delivered to the Royal Geographical Society, became a foundational
document for a geopolitics based upon global images and maps that was variously elaborated by later political theorists.[51] Seizing upon the insight of
Alfred Mahan, an American admiral who had pointed out the military significance of mechanized naval power capable of ranging across an oceanic
globe, Mackinder mapped the pattern of continents and oceans around the
"world island" of Eurasia and Africa in an unintentional reframing of the
medieval *mappae mundi* surrounded by Ocean and an outer circle of islands
and peninsulas, including the new worlds of the Americas and Australia.
Upon opening his lecture Mackinder pointed to the polar journeys of the
Norwegian Fridtjof Nansen, among others, claiming that these represented
the final acts of a four-century "Columbian epoch," after which "we shall
again have to deal with a closed political system, and none the less that it
will be one of world wide scope."[52] The spatial and political imperatives of
the Christian globe had indeed returned to haunt Mackinder's Europe. Global imperatives of environment and cultural history generated, in his opinion, a secular clash between universal forces of "civilization" and "barbarism,"
the former represented by a necklace of maritime powers, then dominated
by Great Britain, whose alliance was necessary to contain the tendency of a
single land pivot, the Russian empire, possessed of the strategic advantages
of land mobility and retreat in depth, conventionally by means of horses and
in the modern world by rail: "The oversetting of the balance of power in
favour of the pivot state, resulting in its expansion over the marginal lands
of Euro-Asia, would permit the use of vast continental resources for fleet
building, and the empire of the world would be in sight."[53]

Mackinder's analysis reflects the contemporary concerns of an imperial
Britain whose commercial empire had constructed and depended upon the
spatialities he was describing: an oceanic lifeline marked by naval stations
linking northwestern Europe, the Mediterranean, the Suez Canal, India,
Singapore, Hong Kong, and the China treaty ports. Britain's imperial encir-

cling of the "world island" was most immediately threatened by Russia, conceived as an "Eastern" power, in the Himalayan passes of its North-West
Frontier Province of India. The geopolitical vision relates a narrative of universal history to the spatialities of a global geography, with explicit reference
to the evolution of Western cartographic images,[54] and betrays a profound
anxiety about the end of history as a defeat of human civilization by a barbarism exploding from the very spaces in which Western culture itself was
believed to have emerged. The Pumpelly expedition, discussed above, was
currently searching for the origins of the Aryan "race" in the same Trans-
caucasian steppe regions that caught Mackinder's imagination as the realm
of the horseman, striking out of "vacant space" at vulnerable commercial
civilizations on the oceanic edge of the world island. Mackinder's fear in-
verts the conventional spatialities of culture and nature. Whereas civilization
had been the attribute of the spatial center, with nature and barbarism progressively evident toward the periphery of the habitable earth, Mackinder's
post-Columbian map, by giving strategic dominance to the former, achieves
precisely what Serres claims of modernity: the globe becomes a stage upon
which a universal drama is enacted, driven by the abstract spatialities of the
map. In one form or another this Manichean spatiality has since dominated
Western strategic thinking, especially from the creation of the Soviet Union
in 1917. It determined the strategic thinking of the American-led North
Atlantic Treaty Organization (NATO) during the cold war and was figured
as a world-historical struggle between Christ and Antichrist by the Roman
papacy, especially under Pius XII. For all its apparent modernity, global geopolitics in the twentieth century betrayed distinct echoes of much more enduring themes of Western ecumene on an ocean-bound earth, civilization,
and fear of an Asiatic Other.

Geopolitics was overwhelmingly a discourse of mapped images, generating concepts and theories of space relations out of cartographic representations. Although regional-scale geopolitical theories were developed,[55] the
insistent appeal of the discourse has been the global scale, generating concepts such as "pan regions," which divided the globe into a limited number
of zones or spheres of influence, each under the political, military, and economic hegemony of individual Western states. Geopolitics, as Mackinder
clearly recognized, is about empire to the ends of the earth. Both its imaginative strength and its practical limitations are clearly demonstrated on an
Italian map of 1941, published in the journal *Geopolitica*, proclaiming the pan
region of Africa to be destined by location as the continental reserve for an
imperial Italy, already in control of Libya (Fig. 8.2). Africa's rail system and

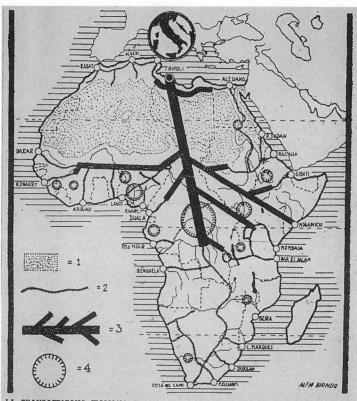

LA TRANSAFRICANA ITALIANA SECONDO L'IDEAZIONE [ED IL PROGETTO DELL'AUTORE
1) la barriera desertica   2) le ferrovie africane esistenti   3) il tracciato della Transafricana italiana
e sue diramazioni verso ferrovie esistenti di sbocco agli Oceani (Atlantico, Indiano)   4) Possibilità
di sfruttamento idroelettrico in Africa

8.2. "La Transafricana Italiana," a geopolitical vision of Fascist Italy's imperial destiny, tapping Africa's hydroelectric potential. From A. Biondo, "La Transafricana Italiana," *Geopolitica* 3 (1945): 571.

hydroelectric potential are graphically tapped by the "Transafricana," flowing north across the latitudes to Tripoli, Italy, and Europe like sap rising through a global tree.[56]

The Globe as a Sucked Orange

Globalization is the social expression of modern "time-space compression," driven by the exigencies of capital accumulation and circulation, in which returns on invested capital are increased through geographical extension into new spaces of exploitation and reduction of the time period during which capital remains unproductively fixed in any one location.[57] Speculatively, finance capital exploits any available time advantage to secure information affecting investment outcomes. There is an overwhelming imperative to reduce the time-frictional effects of distance, answered by technological advances in communications. The imaginative consequences for representing the globe are as profound as the practical. Jules Verne was the most successful writer to exploit the public fascination with globalism in the closing years of the nineteenth century. In novels such as *A Journey to the Centre of the Earth* and *Around the World in Eighty Days* precise attention to scientific and geographical detail combines with a cosmic vision acute to the poetic appeal of an elemental earth.[58] A more formal indication of the globe's centrality within nineteenth-century culture is its role in public education, both in formal state schooling and in the more recreational role played by globes and global images in mass culture.

Recognition of the growing public demand for geographical knowledge informed the work of the philanthropic Society for the Diffusion of Useful Knowledge, formed in England in 1826. In the decades immediately preceding compulsory state education, the society aimed to provide general knowledge at affordable prices, financing much of its publishing operation by producing globes, such as Malby's eighteen-inch standing globe of 1858, and world maps, among them highly novel representations of global phenomena, such as that showing the courses and relative lengths of the world's rivers flowing into a central, circular sea (see Fig. 7.5). Mass-produced globes, wall maps, and atlases such as Blackie's *Imperial Atlas of Modern Geography* (1860) became standard features of European and American schoolrooms during the first century of public education. German commercial publishers such as Humboldt's map publisher, Heinrich Berghaus, exploiting new techniques of steel engraving and especially lithography, which allowed cheap and rapid reproduction of colored sheets, dominated European mapmaking until the 1890s, when French and Scottish firms began to compete. France's *Atlas General Lablache* appeared in 1894, and the *Times Atlas,* actually an English-language version of the German *Allgemeine Hand Atlas,* was first published in 1895. The Edinburgh firm of John Bartholomew redesigned the atlas immediately after World War I, and the Edinburgh Geographical Institute, formed by the union of Bartholomew with the London firm of Thomas Nelson, dominated the production of wall maps and school atlases in Britain and its colonies during much of the succeeding half-century. The title of their most popular school atlas, *The English Imperial Atlas and Gazetteer of the World,* first published in 1892, clearly expresses the principal educational message conveyed by its images. An address by L. W. Lyde, a professor of geography at London University, to a conference of teachers in 1908 makes clear how these educational aids were to be used.

The goal of geographical education for children ("boys" throughout the lecture) is to train the imagination in forming images of "the sphere of space" that cannot directly be seen. In a formulation for which we can recognize a formidable historical precedent, he requires that this proceed from "the known" to the "unknown":

> The known is, obviously, the Homeland—its forms of land and water and its climatic forces, observed till the boy can make distinct mental pictures of them, and the distinctness of the pictures tested by systematic re-production and re-presentation; he must model, and he must draw maps.
>
> What is the unknown? The Globe and nothing else. Here, too, there is no substitute, no alternative. The Globe is the simplest unit of imagination, and the boy must make one mental jump—and only one—from the tiny seen and known, measured and modelled, forms of the homeland to the largest unseen and unknown forms that he will ever come in contact with. Then, whatever area he may study subsequently, it has already been grasped implicitly in that one complete step from the known to the unknown.[59]

Lyde's idea of geographical education is primarily visual, based on globes and maps. Its goals are political ("to picture truly the condition of foreign peoples in distant lands"), moral ("to picture truly the condition of one's fellow countrymen at home"), and above all commercial ("to anticipate . . . needs, and so . . . find new customers and new markets"). Imperialist images of the globe, emphasizing through projection, color, and naming the universal reach of Europe's hegemony, are unambiguous representations for young citizens of their own and other peoples' geographical and social place in a world order.

Lyde's emphasis is on the capacity to visualize, to "see" the spaces of both one's homeland and the globe as the foundation for conceptualizing political, moral, and commercial conditions. An ability to distinguish visually the observer from the object seen (the boy from the globe or map) and the seen object from the reality it seeks to represent (the globe or map from the earth itself) has been fundamental to a modern Western geographical imagination. It is apparent well beyond formal pedagogy, dominating what Timothy Mitchell has referred to as "the cognitive habit of world as exhibition," which has found its most characteristic expression since the late nineteenth century in the global exhibition or world's fair, a cultural phenomenon dating from the Great Exhibition of 1851 at the Crystal Palace in London and peaking in universal exhibitions held in virtually every Western capital during the decades leading up to World War I.[60] Exhibitions were invariably held on specially designated and designed sites within the metropolitan cities of Europe and the Americas. Part public instruction, part commercial entertainment, part trade fair, the exhibitions were global in various ways. They sought to concentrate and display places, peoples, cultures, and products from across the world at its metropolitan center, as spectacles in imperial Rome featuring exoticized peoples and animals had done in antiquity. They attracted spectators from across the world, even from those otherwise positioned for display, to witness the spectacle. Explicitly modern constructional materials, industrial and transport technologies, and above all communications systems underpinned the conception and achievement of these spectacles, so that technological modernity has been an insistent theme.

Closely connected to the world exhibition, and often sharing the same urban space, have been other globalized events: the Olympic Games, for example, initiated in 1896, along with world unions, congresses, and conferences. All share the characteristics of high visibility through the choice of striking urban and architectural settings, rigorous ordering and categorization of exhibits often linked to the celebration of science and technology, statistical measurement and mathematical control in a rhetoric of peace and progress, world citizenship and shared interest among peoples across the diversity of human cultures. Almost invariably, their iconic motif has been the globe itself. Many of the largest globes ever constructed have been designed as centerpieces for world's fairs and exhibitions. In the London of the Great Exhibition Wyld's globe vied for public attention with Burford's diorama of the *Investigator*, a British polar-exploration ship snowbound for the winter.[61] A globe with a circumference of forty meters was erected for the first Paris Exposition in 1889, and a still larger sphere with an external tramway was proposed by Elisée Reclus for the second in 1900. A monumental globe formed the center of the formal gardens at Chicago's Columbian Exhibition of 1892, and the "Unisphere" was the centerpiece of the New York World's Fair of 1964–65, both connecting gigantic or technologically sophisticated global images to progress and modernity.

Basic to the conception and commercial success of world exhibitions was their capacity to attract a mass spectatorship. Only world cities could hope to mount them successfully. By definition, therefore, they have both responded to and driven modern mass culture, as the debate over the dome located at the Greenwich meridian in London for the year 2000 clearly demonstrates. Placing enormous models of the globe in public spaces is not a consequence of increasing scale of the models themselves;[62] rather, the increased scale of the model reflects command over a shrunken real earth. What had been presented by Blaeu or Coronelli for the eyes of the sovereign in his palace was now offered to a mass public in the imperial cities of Europe and America. Coronelli's own globes have been displayed in Paris only twice during the past three centuries, in 1875 as the centerpiece of the exhibition of cartographic history at the Bibliothèque Nationale connected with the founding of the International Geographical Union and at the opening of the Pompidou Centre in 1981, the key architectural symbol of French modernity in the postwar years. Media, communications, and transport organizations are the private bodies attracted to the iconography of the globe. Newspapers, for example, especially in the United States, have commonly used the words *world* and *globe* in their titles, illustrating the graticule in their titles and constructing model globes on their office buildings. In the 1930s a two-ton aluminium globe measuring twelve feet in diameter marked the *New York News* building in Manhattan, one of the landmarks of the paradigm Modernist skyline.

Columbia Studios in Hollywood has used the globe as its logo in the opening sequence of all of its movies, and the association of global image and modernist culture was given the ultimate imprimatur when the Museum of Modern Art in New York, cultural arbiter of modernism, erected a hollow globe thirty feet in diameter from within which the public could view the world from a central stage—a contemporary reworking of Boullée's 1793 conception.[63] The collapse of distinctions between high and mass culture that has characterized modernist art movements is captured in the ubiquity of the globe's use as image and symbol in the modern world: the same form erected at MOMA is the centerpiece of Disney theme parks in California and Florida, where the EPCOT Center and the seventeen-story silver geosphere, Spaceship Earth, continue the tradition of the globe as centerpiece of universal spectacle, as does the 1980s Globus, marking Stockholm's claim to the status of world city (Fig. 8.3).

The globalism of universal exhibitions and world's fairs reaches beyond simple display of monumental models. Model globes are emblems of an *idea* that such spectacles share with other globalized exhibition spaces of the modern metropolis, such as the zoo and the botanical garden, which gather the diversity of the earth at a designated center, using the public display of knowledge to signal authority and possession of an empire of knowledge.[64] From their inception universal exhibitions have placed a premium on the

8.3. Globus, Stockholm's postmodern emblematic claim to global status. From Alan Pred, *Recognizing European Modernities: A Montage of the Present* (London: Routledge, 1995).

scientific accuracy of their reproductions of exotic environments and artifacts, drawing on the authority of science and public education but connecting it to an undiminished public appetite for marvels located at the edges
of time and space. The implications of reducing the earth's complex diversity to a set of visual metonyms was recognized as early as 1901, when visitors to the Buffalo Pan-American Exposition were offered trips to the Moon,
the terrestrial globe supposedly having jaded the geographical imagination:

> The prodigal modern Midway is fairly using up the earth. A few
> more Expositions and we shall have nothing left that is wonderfully
> wonderful, nothing superlatively strange; and the delicious word "for-
> eign" will have dropped out of the language. Where shall we go to
> get a new sensation? Not to the heart of the Dark Continent; Darkest
> Africa is at the Pan American. Not to the frozen North; we have met
> the merry little slant-eyed Eskimos behind their papier-mache glacier
> at Buffalo. Not to the far islands of the Pacific; Hawaiians, and little
> brown Filipinos are old friends on the new Midway; not to Japan; tea
> garden geisha girls, and trotting, jin-riksha men have rubbed the
> bloom off that experience. Not Mexico, not Hindoostan, not Ceylon,
> not the Arabian Desert, can afford us a thrill of thorough-going sur-

> prise. . . . The airship Luna leaves in three minutes for a Trip to the
> Moon . . . not satisfied with exhausting the earth, they have already
> begun upon the universe. Behold the world is a sucked orange.[65]

Martin Heidegger characterized modernity as "the age of the world picture." Drawing on non-European visitors' responses to the Paris Exposition
of 1889, Timothy Mitchell has argued that such exhibiting is part of a much
broader modern Western attitude of disengaged visual examination wherein
a colonized world is ordered up as an endless exhibition, literally "de-mon-
strated" as the global QED of modern science.[66] The exhibition offers the
world as a stage, a performance, not unlike Joseph Wright's orrery, set before
a curious subject and indicative of a much more profound epistemological
stance that non-European outsiders such as Egyptian visitors to the Paris
Exposition remarked everywhere in European culture. It was a stance that
privileged classification and organization of the world according to visual
criteria, staging reality as a dramatic performance within the frame of a proscenium arch. In Europe "one was continually pressed into service as a spectator by a world ordered so as to represent." Three features characterize this
experience:

> *First,* it has a remarkable claim to certainty or truth: the apparent cer-
> tainty with which everything seems ordered and organized, calculated
> and rendered unambiguous—ultimately, what seems its political
> decidedness. *Second,* there is a paradoxical nature to this decidedness:
> its certainty exists as the seemingly determined correspondence
> between mere representations and reality, yet the real world, like the
> world outside the exhibition, despite everything the exhibition
> promises, turns out only to consist of further representations of this
> "reality." *Third,* there is what might be called its "colonial nature":
> the age of the exhibition was necessarily the colonial age, the age of
> world economy and global power in which we live, since what was to
> be made available on exhibit was reality—the world itself.[67]

The relationship between this represented globe, with its visual order and
classificatory lack of ambiguity, and the experienced earth, known always
locally and partially, is inevitably tense and unstable, opening space for anxiety, contestation, and challenge.[68]

Modernist Cosmodrama

Some in the twentieth century sought to mobilize the association of global vision, progressive science, and education exploited commercially in the universal exhibition to the goals of universal social utopia, deploying the image of the globe to re-enchant what they took to be the cold materialism of nineteenth-century science. They ranged from bizarre schemes with limited impact to theories of considerable influence in shaping the twentieth-century world. An example of the former is Cyrus Teed's extraordinary Koreshan Unity in Estero, Florida. Fascinated, like many late-nineteenth-century people, by electromagnetism, Teed developed an entire cosmography based on the idea of a concave earth. In his *Cellular Cosmogony,* of 1898, he declared on the basis of hydrostatic studies and observational experiments that "earth is a concave sphere, the ratio of curvation being eight inches to the mile, thus giving a diameter of eight thousand, and a corresponding circumference of about twenty-five thousand miles. . . . The alchemical-organic (physical) world or universe is a shell composed of seven metallic, five mineral, and five geologic strata, with an inner habitable surface of land and water. This inner surface . . . is concave."[69] He developed a syncretic social religion that drew upon alchemy, the Bible, socialism, and "science" to attract followers to an experimental community located on Florida's east coast. Here, experiments in community life, horticulture, and observational science were conducted in the early years of the twentieth century. One of many utopian schemes in America, Teed's is fascinating in its specific foundation on a model of the globe and the cosmos.

The influence of the Scottish sociologist Patrick Geddes was much greater. Drawing on strands of French Enlightenment thought that run from Turgot through St. Simon, Frederick Le Play, and Elisée Reclus, Geddes explored ways of bringing together science and religion and of educating a modern public to see a global coherence across history and geography, hierarchically structured from the immediate locality to the whole earth. Once the principles of coherence were grasped, he believed, *planning* space and society would realize the goals of universal human progress. An early illustration of his idea was Geddes' camera obscura, mounted atop a tower in the heart of Edinburgh, a public exhibition in which an observer passed through the spatial hierarchy from the globe to the city by ascending the tower, where he or she would find the directly projected image of the city.

The self-consciously modernizing Sociological Society of Britain, standard-bearer for Geddes' progressive beliefs concerning rational social and spatial planning according to ecological principles founded on a scientific understanding of the physical and cultural worlds, mounted an exhibition in 1922 at London's Imperial Institute. The exhibition illustrated the themes of a conference on living religions within the empire, whose goal was to demonstrate "the essentials of vision," which, the society claimed, religion had traditionally expressed but which the analytic tendency of a modern scientific gaze tended to ignore. A description of the exhibition points to the significance of its location at the Imperial Institute: it is at the heart of a district of London devoted precisely to museums. "Albertopolis," built with the profits from the Great Exhibition, contains the Victoria and Albert Museum, devoted to human art and design, and the Natural History and Science Museums, displaying the universe of nature.[70] Nearby are two great Christian buildings, an Anglican church and a Roman Catholic oratory. Like the museums, these churches "express an ordered unity of Man and Nature and the Ideal." But the description claims that the impression left by these museum displays is of a chaotically disordered world of artifacts plucked at random from across the globe for display, of an unsystematic *Wunderkammer.* The empirical truth of the exhibits is unimpeachable, the Sociological Society argues, but they give no hint of the world's unity and coherence. Its own exhibition, by contrast, offers a humanist and scientific "vision" for a modern and scientific culture that can stand alongside "Ancient Faith." It achieves this through a series of graphic images that depend on the theme of the globe as a stage for the historical dramas of mankind's social life.

The opening image of the exhibition is a relief map of the world, "that marvel of condensed and shrunken landscape." Alongside are celestial charts and diagrams of planetary relations between the globe and the heavens emphasizing diurnal and seasonal illumination:

> Are not all these maps and charts, pictures and drawings, charged with an emotion intense and arousing? Yet they are also scientific documents, accurate in fact and where necessary drawn to scale. They *exhibit our world-without* in its largest being and becoming. They are also evocative symbols. They stir the imagination. At their call our world-within is kindled with the sense of a spectacular and mysterious Cosmodrama, and in no vague way but luminous with clear ideas and vivid imagery.[71]

The "world-without"—the geographic globe—is thus established as the primary stimulus to human vision, a graphically conceived imagination. This initial global image is succeeded by a succession of others—a "Biodrama"

8.4. The fallacy of global knowledge: "Erudition fails when it repeats the recurrent error of religion by violating the sacred unity of Folk, Work, Place." From P. Abercrombie et al., *The Coal Crisis and the Future: A Study of Social Disorders and Their Treatment* (London: Williams & Norgate for the Sociological Society, 1926).

showing seasonal change in natural landscapes, a "Technodrama" of the connections between the Geddesian triad of "Place, Work, Folk," an anthropological "Ethnodrama" of the world's peoples, and finally a "Chronodrama," the cycle of civilization, deemed to have begun in biblical Jerusalem and classical Athens, moving through imperial and papal Rome to modern London. Each of these dramas is pictorially illustrated, realistically or emblematically. The Technodrama uses the passage from mountain to sea to illustrate stages in the progressive evolution of human society, from "primary" activities such as hunting to more advanced modes of livelihood such as manufacturing industry. Among the exhibition's final images is that of a skeleton garbed in academic robes in a room full of volumes, a vulture on its arm, standing astride a globe (Fig. 8.4). The globe lacks any surface features, and the text comments that verbal analysis would be superfluous. "Erudition fails when it repeats the recurrent error of religion by violating the sacred unity of Folk, Work, Place."[72]

The exhibition's message is resolutely modernist and secular, yet the transcendental is not dismissed. It offers the power of "The Plan" as an almost metaphysical solution to human life on earth. Modernist cosmography consciously seizes the visionary imperatives of traditional religious cosmographies, the Society urges, arguing that while religion tends to distract the mind from the actions necessary to change and transform the material globe, so modern science risks losing harmony by neglecting the inner truth of the world, which is available only to "vision." Iconographic and thematic echoes of Renaissance and baroque cosmography are present throughout the exhibition, but there is a conscious attempt to bring together two types of vision: the analytic gaze of modern science and the contemplative insight of more traditional thought. Dominating all is the imperative to plan, to engage actively in the world's transformation, to engineer the globe.

Given its futuristic intent, there are surprising gaps in this modernist cosmodrama. Its graphic images take the form of paintings, engravings, and maps. There is no reference to the use of photographs. And no mention is made of powered flight, the specific technology that more than any other has transformed the capacity of humans actually to *see* the earth's surface and that would give the twentieth century the first and only eyewitness visions of the terrestrial globe. The marriage of these two technologies has created a virtual globe, the subject of the final chapter.

NINE *Virtual Globe*

Do not deprive me of the joy of thinking I could rise in flight and
see in twenty-four hours the earth revolve below me, and I would see
so many different faces pass by, white, black, yellow, olive, with caps or
turbans, and cities with spires now pointed, now round, with the
Cross and the Crescent, and cities with porcelain towers and lands of
bells, and the Oroquois preparing to eat alive a prisoner of war, and
the women of the land of Tesso busy painting their lips blue to please
the ugliest men of the planet, and those of Camul, whose husbands
pass them to the first newcomer, as Messer Milione tells us in his
book.

—Umberto Eco, *The Island of the Day Before*

In the past century the many facets of the Apollonian dream—arcing passage over the turning sphere, grasp of the geometric perfection and brilliant
diversity of its terraqueous surface, imperial reach toward the ends of the
earth, together with spiritual reflection on human insignificance against the
measure of its vastness—have been experienced physically as well as imaginatively. Associations historically attached to seeing the globe remain potent, but physically viewing the earth has generated the anxious paradox of
a humanity at once isolated on a fragile solar satellite and lacking any special distinction from other life on that sphere. Rhetorics of world empire,
geopolitics, the airman's vision, universal brotherhood, one world, whole
earth, and globalism shaped planetary social discourse in the twentieth century as the inheritance of the Enlightenment and modernity, while as a new
millennium opens, the poetics and politics of the globe emphasize fracture,
difference, and locality, individuating human dignity and rights of embodied
men and women. Near the dawn of an air age Friedrich Nietzsche recognized that if each individual views the world through a personal camera obscura, a single, transcendental world picture is impossible. With photographic

images of the globe now a banal element of our everyday visual diet and the camera itself a disposable consumer item, the contradictions of Apollonian vision have intensified.

Twentieth-century global images and imaginings may be considered primarily in terms of the converging technologies of flight and photography. Their rhetorical coupling with modernity and progress should not obscure their reworking of longstanding Apollonian themes of global unity and perfection, imperial authority to "the ends of the earth," heroic individual *telos,* and a redeemed humanity. Western politics and the Western practice of empire, virtually *ad termini orbis terrarum,* produced military struggles that were themselves truly global. At the millennium, the dream of territorial empire seems dead, buried practically and morally. But imperializing global discourses resist, recast in terms of an altered spatiality of globalization, as connection and communication, networks of infinite individual points linked across invisible channels over a frictionless surface, generating and transforming a virtual globe.

Powered Flight and Aerial Vision

In *Spirale tricolore su Roma,* a 1923 painting by Roberto Marcello Baldessari, the emblematic red, white, and green of Italy's national flag trace the dramatic, descending curve of a fighter pilot over the Eternal City (Fig. 9.1). The Roman landscape is emblematized in a combination of imperial and modern structures—the Colosseum, the Arch of Septimus Severis, the Palatine's imperial palaces, the monument to Vittore-Emmanuele II and a united Italy, smoking factories, and the radio mast whose broadcasts reunite the nation. Rome is not "mapped" in a conventional sense; rather, iconic fragments are selected, constructed, and composed across multiple perspectives. Baldessari's compositional geometry derives from experiments in pictorial representation in the years leading up to World War I, partly in response to changing ideas of vision and representation generated by photography. His image captures some of the cultural resonance of early powered flight and the new perspectives it offered. It links past and future, earthly attachment and aerial detachment, patriotic nationalism and abstract, universal space, capturing something of Italy's enthusiastic response to powered flight. A territorial state and European "power" only since 1870, yet guardian of Roman patrimony in the *fin de siècle* years of self-conscious imperial historicizing, Italy's push toward modern colonial empire coincided with the years when Wilbur Wright was popularizing his Flyer in the capitals of Europe.

9.1. Aerial vision connects ancient and modern visions of empire over Rome: *Spirale tricolore su Roma,* 1923, aeropainting by Roberto Marcello Baldessari. Private collection, Rovereto.

In his 1909 "Futurist Manifesto" Filippo Tommaso Marinetti linked artistic innovation to technological progress, to speed and to powered flight: "We want to hymn the man who holds the control stick, lancing his spirit across the Earth, along the circle of its orbit."[1] Together with the writer Gabriele d'Annunzio, Marinetti joined crowds spellbound by the performances of early aviators over Verona and Turin. Aviation displays vied with great exhibitions as the most popular mass spectacles in Europe's great cities in the years immediately preceding 1914. The futurists' fascination with flight was shared by the journalist Benito Mussolini, whose postwar association with Marinetti introduced futurist themes of speed and vision as features of a Fascist modernity parallel to those of renewed imperialism and ancient Rome.

Baldessari's is an early aeropainting *(aeropittura),* a genre theorized by Marinetti in a 1929 manifesto and given official legitimacy in a state-sponsored exhibition in 1932. More than merely the machine, it was the human pilot, at one with it, that fired the popular passion for aviation. The "intoxication of flight" was bound up with the heroic personae of pilots, who were styled as "young gods." Their "sensation of escape from the constraints of earth" and exhilaration at the visual and kinetic sensations and new panoramas offered by powered flight were never dissociated from the Icarian risk of sudden descent and death.[2] Vision, freedom, and mastery over space accompanied a sense of breaking beyond the anxious closure of mapped space in the late nineteenth century. In the words of a French pilot: "Freedom, this leap that detaches you from the ground and opens you to the sky. Freedom, this road without limit that can cross all roads at any altitude and in any direction. Freedom, this infinite conquest of trees, of towns, of plains and mountains."[3] Coming at the moment of supreme European confidence in its global centrality and destiny, the conquest of the air seemed a confirmation of Western dominion, while the courage of individual airmen assuaged fears of degeneracy and imperial decline.

These same themes were celebrated in D'Annunzio's writing, which took to violent extremes the elitism, militarism, and misogyny inherent in much of the early response to flight. In the 1902 poem *La conquête des étoiles* Marinetti connected the aspiration to flight with a masculine will to power that escaped the contingent and the local, broke the chains of gravity and soared toward the stars. In 1908, within five years of the American Wright brothers' first powered flights, Marinetti published *L'aeroplano del papa,* a free-verse flight down the length of the Italian peninsula, across the classical and papal landscapes that Ortelius and Hoefnagel had explored on foot and Egnazio Danti had captured in paint three centuries earlier. But futurism was not about the past; space was to be made anew. Marinetti's 1912 "Technical Manifesto of a Futurist Literature" forecast the dissolution of all forms of representation in response to the new spatiality: "In an airplane, seated on a cylinder of gasoline, my belly warmed by the aviator's head, I felt the ridiculous absurdity of the old syntax inherited from Homer. A furious need to liberate words, liberating them from the prison of the old Latin sentence! . . . This is what the propellor told me as I flew at two hundred meters over the powerful smoke stacks of Milan."[4] In his own experiments with words Marinetti sought to realize the unexpected sense of fragmentation that came with the aerial view. Concrete poetry combined the pattern of words on the page and the sounds they created with new arrangements of letters, looping the loop or diving down the page. In a novel of 1909, *Mafarka le futuriste,* Marinetti describes an African king, simultaneously ancient and modern, who fathers a mechanical son, the winged Gazurmah, freed from all constraints of space and time. In reality, Africa was invaded two years later by Italian aviators under the command of Captain Piazza, taking Libya from the Ottomans and establishing Italy's claim to a reborn imperial mission in Africa.

Marinetti's dreams of aerial freedom and those of the artists and intellectuals who followed him were reinforced by the experience of World War I in 1914–18. For an entire generation of Europeans this was an experience of chthonic horror from which powered flight offered the only possible form of physical or imaginative escape. On every battle front across the continent, four years of mechanized carnage produced nothing more than stasis as armies sporadically advanced and retreated across trenched strips of shell-churned mud scarcely miles wide. Nothing could be further removed from prewar Nietzschian visions of escape and freedom. Only the small and elite bands of youthful pilots whose flimsy craft circled and scrapped over the trenches attained a more distant horizon. Their aerial perspective was the antithesis of the vision of the infantryman, imprisoned in an elemental mire of trench, mud, and wire. The air ace thus came to be figured as a crusading knight, mastering his craft high over the battlefield in aerial jousts with machine gun and white scarf, a model of individual chivalry liberated from the landscape of anonymous, mechanized death.

Seizing upon the poetic and ideological power of this image, D'Annunzio, who had written an early aviation novel, *Forse che si Forse che no,* persuaded a group of young pilots in Verona during the closing months of the war to make a dramatic collective flight over the Alps and release futurist leaflets over the imperial capital of Vienna. This grandly rhetorical gesture, in which the group took a blood oath to press on regardless of danger or loss in the mountain flight, is celebrated in a 1933 aeropainting that maps the spaces of imperial Vienna in the outline of a plane, seen from a cockpit as fragmenting into multiple perspectives across the curving horizon. Alfredo Ambrosi's painting closely resembles the composite aerial photographs that were the most practical contribution of powered flight to strategy and tactics in World War I. Invented in 1915 by the German Oskar Messter, the airborne automatic camera allowed pilots to film a 60-by-2.4-kilometer strip of land surface in a sequence of frames at the scale of conventional topographic maps. A new mode of geographical representation was created: "a flattened and cubist map of the earth," which demanded new skills to relate the image to the ground. With increasing altitude, the airborne camera came to replace the human eye in military reconnaissance, and aerial photography evolved into "a new way of seeing in which . . . the earth became a target as far removed from the personal experience of the observer . . . as a distant planet."[5] This new mode of photographic representation would eventually yield the century's most potent images of the earthly globe.

Italian aeropainting was popular in the 1920s and 1930s, when aviation feats and competitions rapidly extended the reach of the new technology. Although some women artists, notably Benedetta Cappa, whom Marinetti married, produced lyrical aeropaintings of rural landscape, the genre, like aviation itself, was dominated by male artists and by urban visions. Its intentions complemented, and its images often pictured, the cult of male youth elaborated in Fascism's ideology of the "new man," who embodied in both physique and character Italy's imperial destiny, at once ancient and modern. The paradigmatic figure was the air ace Italo Balbo, who led the Italian *argonauti,* a group of young military pilots, in a series of record-breaking, long-distance intercontinental and transoceanic flights during the late 1920s and 1930s.[6] Like Coronelli's use of the classical Argonauts' name for his geographical academy, Balbo's reference to Europe's foundational imperial narrative associated a contemporary project with a long pedigree and celebrated the adventure of closely bonded males heroically testing the limits of courage and destiny. While women, from Raymonde de Laroche and Eugenie Shakovskaya to Amy Johnson, were significant pioneers of long-distance aviation, its cultural reception was overwhelmingly masculinist, casting females principally "as admirers of [male] pilots, their prize for having accomplished a daring exploit, or an obstacle that stood between them and the fulfillment of their destiny in the sky."[7] Long-distance aviation became a matter of competitive nationalism in interwar Europe, with those who opened new routes or broke distance records fêted as national heroes. The Argonauts' epic journeys culminated in highly choreographed, spectacular landings at Ostia, symbolically significant as Aeneas's landfall in Italy and the port city of imperial Rome, commanding *mare nostrum,* the Mediterranean at the heart of the ancient ecumene. A seaplane terminal whose decorative iconography of aeropaintings employed the theme of global spatiality through the combined geometry of compass and revolving propeller was constructed there in 1930.

This interwar connection between futurist long-distance flight and historically constituted imperial narratives was not confined to Italy. In Portugal, Admiral Couthino made the first transoceanic flight from Lisbon to the Brazilian coast in a craft named *Santa Cruz,* emblazoned with the red crusader's cross that had decorated Vasco da Gama's sails, using the same Atlantic island steppingstones as earlier navigators.[8] Appropriately, Couthino was a critical commentator on Camões's *Lusiads*. Like Couthino's, the routes favored by other European aviators reflected the patterns of their respective national empires, for obvious practical reasons of having landing points. Their flights thus echoed the imperial narrative. Meanwhile, as the radio mast in Baldessari's painting signifies, radio was also linking scattered imperial territories through global broadcasting networks radiating from the imperial center. The virtual spaces of the imperial powers described by their interwar networks of air adventure and radio broadcast are traced today in the route maps of European national airlines—KLM, Air France, Alitalia, Iberia, TAP, Sabena. And as noted in chapter 8, the reach of long-distance flight could extend the Western imperial domain to those few parts of the globe that had remained beyond its maritime grasp. Aerial observation and mapping of Antarctica, begun in 1926 by Admiral Richard E. Byrd, who discovered and mapped new land at a rate of 4,000 square miles per day, continued into the 1950s with the British Falkland Islands Dependency Survey, explicitly designed to reinforce British imperial claims to the Antarctic.[9]

As the global spatialities of European imperialism were extended by the aviator's reach and vision, so the imperial center was itself reconfigured by the aerial view.[10] Aeropainting's recurrent images are urban—Verona, Turin, Milan, above all Rome. As Fascist Italy became explicitly imperial in the 1930s, Mussolini redefined and reconstructed Rome, recovering, relocating, and reconstructing ancient Augustan monuments within a carapace of modernist urbanism.[11] This program complemented and celebrated Italian imperial expansion into Abyssinia and Somalia, using aerial bombardment as a primary military weapon. Futurist aeropainters responded with dramatic images of fractured and subordinated landscapes, urban and African. Their vision reached a predictable apotheosis with the invention of the parachute in the mid-1930s. The parachute at once reduced the most deadly risk of aviation while producing a radical extension of freedom and vision on the part of the airman. It freed the pilot from the machine, to be held, Daedalus-like, as a slowly descending, free-floating body at one with the air, and offered unimpeded vision of a turning earth below. The parachutist is sometimes pictured as an embalmed Christ figure, released from the shroud in an act of resurrection high above the earth.

The Movie Camera and the Air View

If Messter's airborne camera produced a new military cartography in the form of the composite reconnaissance photograph, its development extended historical time, revealing in shadows and tones visible only from the air archaeological sites, ancient fields, and lost settlements. Aided by the camera, aerial vision could offer new perspectives on time as well as space, disclosing distant human origins as well as destinies. Composite photographic images demanded a different way of looking than the still photograph did. The eye moves over the virtual space of the image as across a map, parodying in some measure the kinetic vision of the flyer and enhancing that experience of vicarious travel that Ortelius and others had recognized as a characteristic of looking at globes and maps. Over time the aerial photograph and, more recently, remote-sensed images have become codependent with the map. "Ground truthing" by direct survey is as essential for coordinating aerial images with real space as aerial survey is for maintaining the accuracy of detail in conventionally surveyed maps.

But the kinetic aspects of the aerial view are most intensely captured by the movie camera. The quasi-simultaneous invention and development of powered flight and cinematic photography in the early twentieth century has had a profound impact on global images. Promoting his flying machine in Italy, Wilbur Wright invited a cinematographer to join him on a low-altitude flight over the Roman Campagna, and in 1912 the Italian film director Elvira Notari actually made Italian air pilots' role in the Libyan colonial war a part of the earliest aviation movie, *The Heroism of an Aviator in Tripoli*. The film interweaves a story of male heroism against a primitive enemy with that of a passionate love affair. The genre rapidly became a cinematic staple, complete with spectacular aerial filming, offering a vicarious experience of flight and sight in the virtual space of the cinema screen. It is difficult today to grasp the imaginative power of cinematic flight in the decades before mass air transportation, but it is menacingly deployed in the opening frames of Leni Riefenstahl's 1936 Nazi propaganda film, *Triumph of the Will,* in which the airborne camera follows the Führer's aircraft through the clouds over southern Germany, allowing the viewer both to share and to marvel at Hitler's mastering gaze over the moving landscape below.

This kinetic vision of a turning earth, first witnessed through the powerful realism of film, would become a commonplace in the postwar world of passenger jetliners, and thus its association with the heroic vision of the air ace would be radically diminished. In the late 1930s such democracy of aerial vision could only dimly be imagined, and the totalizing, global implications of what became known during World War II as the airman's vision evolved into a powerful trope not only for military strategy in a war of fighter planes, massed bombers, and parachute invasion but also for political shaping of the postwar global order.[12] The idea that *vision* in the form of a mastering view across space and time was uniquely available to an aviator disengaged from the limited perspective of earthbound mortals became a recurrent feature of geopolitical discourse at mid-century.

Geopolitics, the Airman's Vision, and a Reconstructed Globe

The sense of closed global space that had disturbed the strategic thinking of so many Westerners at the beginning of the twentieth century found expression in a global geopolitics of competitive imperial struggle as the frontiers of Western empires converged at the ends of the earth. Mapping the globe's climates and physiography into "natural" regions could naturalize patterns of control and strategy among the imperial powers. Geopolitics suggested that unalterable geographic "realities"—the distribution of lands and seas, of landforms, natural resources, or "races"—had to be exploited if a state was to survive, compete, and prosper. Such "realities" were illustrated by means of powerful cartographics; indeed geopolitical arguments were image-driven, as the "geopolitical syntheses" carried by the Italian journal *Geopolitica* demonstrate. Geopolitical realities supposedly operated at all spatial scales, most dramatically at the global scale of "pan-regions" (see Fig. 8.2). Ecological metaphors representing the territorial European state as an organism passing from youth to degeneracy, its native peoples competing globally for living space to fulfil a national destiny, found some expression in most European countries.[13] Countering the imperialist and nationalist jingoism that fed on such ideas, and in a direct line of descent from Enlightenment declarations of the "Rights of Man" and antislavery campaigns, were universalist theories attracting liberal and radical opinion. Marxism, especially in the Leninist critique of imperialism, was the most radical of these; the more liberal found expression in universalist, often anticolonial movements promoting world government, a universal language, and universal human rights, all Western in their cultural assumptions. In 1887, for example, the Jewish liberal humanist Ledger Ludwig Zamenhof had created Esperanto, a "universal" language designed to promote a global brotherhood, *homaranism*.[14] The Irishman James Cousins sought to fuse Hindu pantheism with ideas of brotherhood within an anti-imperialist, quasi-Geddesian "geocentrism," rooted, like Teed's Florida community, in turn-of-the-century theosophy. Cousins stressed "universal oneness" across the geographical differences in creed and color that geopolitics sought to exploit.[15]

But in practice, if not in name, it was geopolitics, the calculated use of geographical science for state strategy, that prospered in the academies and chancelleries of interwar Europe. Its formulation in Fascist academic journals and its apparent influence on Axis strategy generated a powerful reaction among American foreign-policy advisers.[16] From Woodrow Wilson's presidential attempts to frame a postwar global order until America's entry into the European conflict in 1941 a debate over "isolationism" dominated U.S. foreign policy. Was its destiny to be realized as a New World, separated by geography and history from the tricontinental Old World island as President Monroe had proposed in the early years of the republic, or as a global power directly involved in initiatives for universal harmony such as the League of Nations? These questions intensified as European powers collapsed again into war in 1939 and the imperial ambitions of Japan impinged from a new compass point upon American strategic integrity. The issue went far beyond sending troops and materiel to the Old World; it turned on whether and how the United States might inherit the implicitly imperial Western claim to military power and political influence *ad termini orbis terrarum,* and indeed American engagement did result in its global economic, political, and cultural dominance in the second half of the twentieth century. The inheritance of an imperial globe crossed the Atlantic as territorial empire slipped from the grasp of European capitals after 1945. Global images have played a crucial role in shaping America's academic and popular-culture response. Although strongly promoted as humanity's common property, it is an American image of the globe that has come to dominate late-twentieth-century Western culture.

In 1942, in the early days of American involvement in the war, a symposium was sponsored by the influential journal *Atlantic Monthly.* Its subject was the political shape of the world at the end of the global war, and specifically how the United States, clearly destined to emerge as the most powerful nation on earth, should avoid the imperial imperatives that had brought about the conflict. An Anglo-American vision of global economic order based on international free-market trading had been outlined in the Atlantic Charter of 1941, and America's inheritance of Enlightenment humanism, with its liberal, democratic institutions, now had to be reconciled with political and economic centrality on a globalized earth. The conference proceedings were published in a collection edited by Hans Weigert and Vilhjalmur Stefansson in 1944. Weigert was a German political scientist who had made an academic career in the United States, and Stefansson was a freelance writer of Scandinavian-Canadian origin who had written popular books recording his participation in Arctic exploration, *My Life with the Eskimo*

(1913) and *The Northward Course of Empire* (1922). He had also produced manuals and guides on the polar region for the U.S. Army and Navy and had been appointed adviser on northern operations to Pan American Airlines in 1935.

The editors declare their aim to generate a "new vision," shaping American politico-geographical thinking and planning for a shrunken postwar world. It was an explicitly liberal vision, rooted in the principles of peaceful cooperation among nations, for which they believed the United States was fighting:

> We have seen the dangerous beginnings of an American geopolitics,
> with blueprints for an American imperialism riding the waves of the
> future. It favors a disillusioned balance-of-power solution on the basis
> of regional groupings, in preparation for what the sponsors of such
> "realistic" plans consider inevitable: the Third World War. The editors
> and writers of this book . . . agree that acceptance of the ideology
> and creed of geopolitics would be a dangerous step towards interna-
> tional Fascism.[17]

The twenty-eight contributors constituted a roll call of Anglophone intellectuals concerned with the global role of geographical and cartographical science in resolving political, demographic, and cultural questions. Isaiah Bowman, director of the American Geographical Society of New York, whose work as an adviser to the Wilson administration's delegation at the 1919 peace conferences was summarized in his 1921 text *The New World,* sought here to distinguish "geography" from the false "science" of geopolitics.[18] Sir Halford Mackinder's final geopolitical essay, "The Round World and the Winning of the Peace," which had appeared earlier that year in *Foreign Affairs,* was revised for inclusion in the book. The aging strategist sought to adjust his heartland concept, initially focused on a tricontinental "world island," to the pattern of the "round world" created by human flight. Other authors included Ellsworth Huntington, a veteran of the 1903 Pumpelly Expedition and an influential writer on relations between global climate, eugenics, and civilization, who restated his thesis on climatic determinism, and Griffith Taylor, a veteran of British Antarctic explorations in 1910–13 who wrote on environmental limitations to white settlement within the British empire and dominions. The China specialist Owen Lattimore compared Canada's and China's strategic positions on the globe. The maps that liberally illustrate the collection were prepared by the cartographer Richard Edes Harrison, whose widely reproduced and revolutionary polar projections had "sensitized the public to geography in the 1940s, [torn] out of magazines and snatched . . . off shelves." His "One World, One War" map for the August 1941 issue of *Fortune* became a standard wall decoration in American homes (Fig. 9.2). Significantly, Harrison had had no formal cartographic training; he was a designer whose images were based on tracing photographs taken of a large physical globe from a chosen angle and "owed more to the persuasive look of advertising than to cartography." Resembling an actual photograph of the earth, they "brought home the world's sphericity by moving the viewer out to a fixed point above the earth."[19]

The introductory chapter to *Compass of the World* was written by neither a geographer nor a political scientist but by Archibald MacLeish, poet, literary essayist, and Librarian of Congress. "The Image of Victory" is a direct appeal to the Western humanist tradition, which for MacLeish, like Harrison, supported the principles of human liberty upon which America's national destiny was founded and for which America had entered the conflict.[20] Given their determination to overcome Fascism and nationalism, MacLeish asks what victory should mean to Americans, given the political failure, depression, and war that had succeeded their victory in 1918: "Land? Islands? . . . Empire?" Of Americans he says, "It is difficult to talk these days of empires. They think of victory in the future: they think of empires in the past. They have no patience with those who talk of empires or of islands now. They wish to know how they are to imagine their victory in terms they can believe in and understand."[21] For his answer MacLeish turns to the airman, whose tenure is a new vision of mankind's destiny. Whoever wins the current conflict, MacLeish claims, will win the future of the world, "its geography, its actual shape and meaning in men's minds," and victory is going to depend upon air power.

But air power alone is insufficient; it is the airman's ability to *see* and conceive of the earth as a single globe that is most significant: "Never in all their history have men been able truly to conceive the world as one: a single sphere, a globe having the qualities of a globe, a round earth in which all the directions eventually meet, in which there is no center because every point, or none, is center—an equal earth which all men occupy as equals. The airman's earth, if free men make it, will be truly round: a globe in practice, not in theory."[22] For MacLeish, only the airman is capable of rising above the surface of the earth, disentangling himself from provincialism, the poison of nationalism, and the narrow attachments of everyday life. A young Apollo, the airman could see things "as they truly are." It is the perspective Harrison's maps explicitly sought to replicate, turning "the viewer into a pilot

9.2. "One World, One War," Richard Edes Harrison's image of a global future for Americans, *Fortune*, August 1941.

floating above the horizon,"[23] and one to be powerfully restated a quarter-century later in the context of Americans landing on the surface of the Moon.

The book's second section is entitled "New Directions and Skyways." It consists of five essays that address MacLeish's theme in terms not only of military air power but also of the role commercial airlines will play in shaping a postwar global vision. The editors explain that like radio transmission, long-distance air transportation constructs entirely new spatial relations, promoting new ways of imagining and picturing the globe. These will constitute the most significant feature of the emerging global vision. Air transportation renders the distinction between continent and ocean meaningless; Europe is decentered on the global map as routes across the poles link together cities and global regions formerly connected via its ports.[24] And "because the conception of the air is inextricable from the conception of a more unified world, many writers are convinced that air power makes total internationalism inevitable."[25] Although made within a discussion of postwar regulation of airline routes and competition, these claims signal the close connection between globalism and internationalism in mid-century American thought. It would give rise to a suite of postwar globalizing institutions—the United Nations Organization, the Food and Agriculture Organization, UNESCO, the World Bank, and the International Monetary Fund—created in the immediate postwar years of reconstruction in the context of President Harry Truman's commitment of the United States to the global defense of democracy.

"The Home of Mankind": Images of Global Unity and Diversity in Mid-Century Popular Culture

While MacLeish and the contributors to *Compass of the World* were writing for the relatively sophisticated readership of *Fortune* and *Atlantic Monthly*, their ideas were by no means confined to a narrow American elite. Twenty-five thousand copies of the first edition of Harrison's 1944 *Look at the World* had been sold before it reached the stores, and the U.S. Army ordered eighteen thousand copies of his maps of the North Pole for its Air Corps.[26] The Dutch-American writer Hendrik van Loon (1882–1944) broadcast weekly on U.S. network radio in the months between the fall of France and Pearl Harbor, seeking to persuade isolationist Americans that their interests were no longer confined within continental limits and arguing that America's constitutional and cultural inheritance of Western humanism condemned it to a global destiny. Van Loon's use of radio was an extension of his lifelong commitment to introducing the European humanist tradition into American mass culture. On its appearance in 1932, *Van Loon's Geography*, better known through its many subsequent editions as *The Home of Mankind*, had topped the American bestseller list for six months, with more than a hundred thousand copies sold. It remained in print until 1960, selling more than a million copies over twelve reprintings and becoming a standard text in American homes and a favored prize for successful school students. It was translated into most European languages. Its companion volume, *The Story of Mankind* (1921), a history of Western culture, was also hugely popular. Translated into twenty languages, including Braille and Esperanto, it is still in print today. Van Loon had been educated in Holland and Germany, writing a thesis on Erasmus, whose ideas he celebrated as the model of a universally valid, tolerant, civilized humanism.[27] Erasmian humanism structures both *The Story of Mankind* and *The Home of Mankind* and was the foundation of his appeal to an isolationist America in 1940.

The Home of Mankind is a modern, secular version of the universal history and geography traceable from Pliny and the medieval encyclopedias through the Renaissance cosmographies, baroque globes and atlases, and Enlightenment museums. It opens with the aphorism "History is the Fourth Dimension of Geography—it gives it both time and meaning," and van Loon claims to be writing a geography that "connects" rather than "a jumble of badly digested recollections, like a museum too full of pictures." Thus, he will place "man in the centre of the stage," treating the physical earth only insofar as it is a fitting home for human life. But van Loon's is no hubristic celebration of the human conquest of the earth; it adopts the sympathetic but detached perspective of Erasmus or Pieter Bruegel. The opening image of a box measuring a half-mile on each side in which the entire human population might be packed and pushed over the edge of the Grand Canyon in Arizona suggests both his style and his view of humanity's relation with the earth:

> There would be a moment of crunching and ripping as the wooden planks loosened stones and shrubs and trees on their downward path, and then a low and even softer bumpity-bumpity-bump and a sudden splash when the outer edges struck the banks of the Colorado River.
> Then silence and oblivion!
> The human sardines in their mortuary chest would soon be forgotten.
> The canyon would go on battling wind and air and sun and rain as it has done since it was created.
> The world would continue to run its even course through the uncharted heavens.
> The astronomers on distant and nearby planets would have noticed nothing out of the ordinary.
> A century from now, a little mound, densely covered in vegetable matter, would perhaps indicate where humanity lay buried.
> And that would be all.[28]

This is Seneca's insight, put in simple and memorable terms—the insignificance of humanity against the vastness of the whole terrestrial globe, moving in interstellar space.

There follows a cosmographic outline of the planetary motions, calendrical and climatic patterns, the geophysical forces that have shaped the global surface, and methods of making globes and maps. The distribution of continents and oceans is followed by sketches of European countries, from West to East. Other parts of the world are treated at continental scale. The style is accessible, humorous, and anecdotal, tolerant of difference but unreflexively ethnocentric in celebrating "civilization" and "humanity" as European achievements. In its gentle irony the work is often more subtle than many of the essays in *Compass of the World* in recognizing the limits of Western humanism, and its final chapter, "A New World," clearly anticipates a post-colonial globe. Acknowledging the atrocities that accompanied Western exploration and expansion, van Loon remarks that as "the Great Era of Exploitation has definitely come to an end . . . the unwillingness of the victims to play that *rôle* any longer is causing uneasiness in many high places."[29] The heroes of van Loon's postimperial world are not airmen but medical scientists. He calls attention to Ronald Ross and Walter Reed, working to control malaria and yellow fever in tropical regions, men who "neither 'took' nor 'gave'—they 'co-operated.'" They demonstrate "planetary thinking," generated by the vision of a single earth and its variety, which van Loon's book seeks to inculcate. It is a consciousness that pulls local and global into a single frame: "When you come to think of it, is there really such a very great difference between the world at large and your own town or village? If there is any difference it is one of quantity rather than quality. And that is all!"[30]

Written more than two decades before large-scale decolonization, van Loon's sentiments are prescient. They represent the liberal humanism that characterized many of his contemporaries, such as Albert Schweitzer, and underpinned the global institutions of postwar reconstruction. They hark back to earlier Flemish humanists such as Abraham Ortelius and Gerardus Mercator, writing of the civilizing benefits of traveling the globe through the medium of the atlas. As in the case of their works, the most striking feature of *Van Loon's Geography* is its graphic illustration. Van Loon's own drawings capture the quirky humor of the text; like Harrison's maps, they eschew mathematical cartography and blur the boundaries between conventional mapping, the aerial view, and the striking advertising image.[31] Colored drawings expose the suboceanic forms of the Atlantic and Pacific basins with the continental land surfaces looming over them (Fig. 9.3). Axiometric designs offer perspectives on global features, raising the viewer high over the Atlas Mountains or the poles, swooping low between the edges of the Grand Canyon, and often incorporating the wing of a plane or the image of an airship to hint at the airman's vision. Van Loon's images bring the globe into view through the techniques of popular-culture media—newspaper journalism and the comic book.

Van Loon's illustrations contrast with those employed by *National Geographic,* his most obvious rival in shaping a popular geographical imagination and knowledge in twentieth-century America. *National Geographic* had an estimated 50 million regular readers in the mid-twentieth century.[32] From its origin in 1888, the magazine's goal has been to represent for Americans the physical and ethnographic variety of the globe in the most dramatically

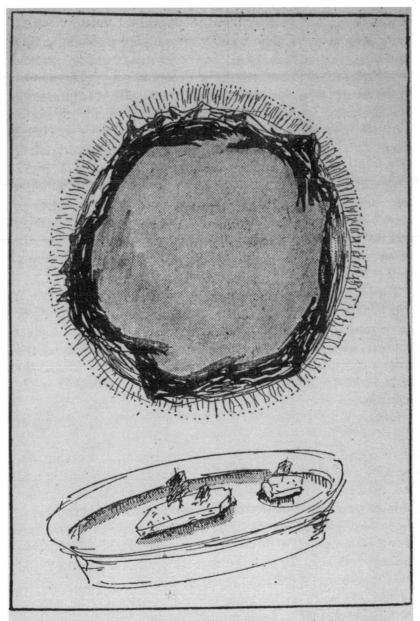

ARE OUR PROUD CONTINENTS PERHAPS ISLANDS OF SOME LIGHTER
MATERIAL WHICH FLOAT UPON THE HEAVIER SUBSTANCE OF THE
EARTH'S INTERIOR AS PIECES OF CORK WILL FLOAT ON WATER IN
A BASIN?

9.3. "Are our proud continents perhaps islands. . . ?" Hendrick van Loon takes a sideways look at global geography. From Hendrik Willem van Loon, *The Home of Mankind* (London: George G. Harrap, 1933), originally published as *Van Loon's Geography.*

visual way. The National Geographic Society was intimately involved in the connected technologies of flight and photography in the search for new images of the earth. Its 1928 series, "Geography from the Air," which included reports on Balbo's *Argonauti,* celebrated the new perspective aerial photography had brought to the geographical imagination. The society financed Byrd's 1920s Antarctic surveys and Albert Stevens's stratospheric balloon flights in the 1930s, which first provided photographic images of the earth's curvature. *National Geographic* also paid careful attention to the maps with which it represented terrestrial space to Americans. From 1922 the magazine employed the Van der Grinten projection for its world and continental maps, a projection that greatly exaggerates the temperate regions while diminishing the tropical and equatorial latitudes of the globe. Its maps were used almost exclusively in American schools and colleges and, derivatively, in newspapers and on television, and their radical distortion was heavily criticized by Richard Harrison in *Compass of the World.* The projection, however, remained unchanged until 1988.[33]

National Geographic's picture of a global humanity has been constructed through stunning ethnographic images. The magazine has relied strongly upon the eyewitness authority and the aesthetic appeal of photographs of places and peoples from "the ends of the earth" in order to bring the globe's physical and human diversity into American domestic space. The implicit cultural imperialism of *National Geographic*'s photography has been closely examined: the topographic images are glossily exotic; non-Western peoples are rendered "natural," classless, and somehow "outside" history because partially clothed, passive to the camera, and apparently unselfconsciously engaging in timeless patterns of conduct and ritual. The selection of photographic subjects consistently emphasizes bodily physique, translocating an aesthetic, sexualized ideal of "Greek" male beauty, with the effect of bringing exotic subjects into a unidimensional "Family of Man." Photographic subjects tend to adopt stances and locations that reinforce longstanding Western ideas of peoples situated on the edges of an ethnocentric world and thus backward in time and inferior in culture with respect to the viewer.[34] Like Schedel's *Chronicle* and Ortelius's *Theatrum, National Geographic* meets a voyeuristic curiosity about the globe's unity in diversity and satisfies the urge to see through vicarious travel.

Photography's mimetic claims were established early in its development, and they have been seriously challenged only in recent decades. They rest in part on the camera's capacity to record an apparently undistorted image of human presence, an objective (because mechanical) testimony whose pictorial authority is rooted in the ocularism that has long characterized Western culture.[35] The mechanical operation of shutter, lens, and light-sensitive plate that produces the photograph authorizes its claim to escape subjectivity. From its invention photography has been closely connected to European exploration and the commerce of images between the domestic and familiar and the exotic and different. This is perfectly captured in Antoine F. J. Chaudet's 1851 stereoscopic daguerrotype *The Geography Lesson,* a Victorian interpretation of Longhi's theme, although the gender and sightline patterns are very different. In a carefully framed and comfortably furnished domestic interior a tight pyramid of girls and young women pose around instruments of geographical education under the tutelage of a male teacher who "heads" the group, his hand resting on the globe. A textbook is open below the globe, while held in the hands of the two girls nearest the camera is the French work *Excursions Daguerriennes,* "a lesson in geography according to a Paris-centred, intellectually informed and class-structured concept of the world."[36]

Like steel engraving and lithography before it, photography was rapidly put into service as eyewitness to the Western geographical imagination. This capacity of the photograph to record an objective image of distant places was promoted by Halford Mackinder in a project for popular imperial education.[37] In 1907 he secured funding for the photographer Hugh Fisher to undertake a series of journeys through the British Empire making photographic images for use as lantern slides to illustrate geographical lectures to British schoolchildren. Visualizing the territorial, environmental, and ethnographic diversity of a global empire within the domestic spaces of a classroom complemented the Colonial Office Visual Instruction Committee's (COVIC) distribution of photographic images of Britain throughout the schools of its empire.

Like the photography for *National Geographic* or the marvels, monsters, and natural wonders that framed Renaissance and baroque world maps, Fisher's photographic images range the margins of Mackinder's imaginative world map. Fisher's explicit remit was to capture "the native characteristics of the country and its people and the super-added characteristics due to British rule."[38] According to Mackinder's geopolitical thesis it was the "crescent" of peninsulas and islands on the margins of the world island that constituted Britain's imperial bulwark against the heartland's thrust for global empire. Like the contents of the sixteenth-century cosmography, the photographs work in complex ways as global representations. Their naturalizing claims to eyewitness truth and their realistic reproduction lend support and authenticity to visions of global space through difference by pushing the strange, the exotic, and the marvelous to the margins of Eurocentric space, while the local, fragmentary, and arbitrary character of individual images serves to undermine the unitary vision they are intended to illustrate.[39] Messter's overlapping cinematic composite, photographing the terrestrial surface from above, and futurist aeropaintings anticipate photography's most powerful contribution to a late-twentieth-century global imagination more effectively than does Mackinder's collage of place impressions and iconic features scattered across an empire.

Realizing Apollo's View: Civil Airlines and the Lunar Missions

World War II was indeed global, signified by the fifty-inch globes presented in 1942 by General George C. Marshall to Roosevelt and Churchill "that the great leaders of this crusade may better follow the road to victory [and] chart the progress of the global struggle . . . to free the world of terror and bondage."[40] The war gave the United States global dominance in flight. The transfer of German rocketry expertise to the United States and its subsequent contribution to the U.S. space program is a familiar story. Less often noted are the implications of the war for the development of air passenger transportation. The two-thousand-mile range for a powered aircraft was broken immediately before the war, but only for planes with tiny payloads. In 1939 only six aircraft in the world were capable of carrying significant weight over such a distance. Global flight was a wartime achievement. The logistical demands of fighting a world war from a U.S. provisioning base and the practical decision to divide Allied aircraft production between American bombers and transport planes and British fighters gave the U.S. Air Transport Command a reservoir of machines and human skills that ensured its dominance in the postwar civilian airline competition. The sole difficulty for the United States had to do with airfields and landing rights. Americans looked with envy at Britain's imperial scatter of territories and islands upon which the landing strips and airports essential to construct the air globe could be built. The American response was to promote and expand the international accords over air space and landing rights that had been agreed on at the 1928 International Air Traffic Agreement (IATA) conference in Paris.[41]

Not tied to an imperial network as state-operated European national carriers were, the free-enterprise American international airlines adopted a consciously global market strategy, reflected in their names and logos—Trans World Airways, Pan American Airlines. Combining American consumers' disposable incomes with America's dominance in airliner manufacturing, these airlines came to signify America's postwar global reach. With the introduction of commercial jet aircraft such as the DC-8 and the Boeing 707 in the late 1950s, and especially the heavy-payload, long-distance "jumbo" jets from the early 1970s, the airman's vision became increasingly available to substantial numbers of Westerners able to travel in speed and comfort between continents and to view the earth's surface from a window more than thirty-five thousand feet above it. Struggling to promote the commercial market, postwar American airlines sought quite explicitly to encourage what they called "high-altitude thinking." The Consolidated Vultee Aircraft Corporation printed 350,000 pamphlets illustrated by Harrison's maps, while TWA developed a program of "flying seminars" in the 1950s for American students majoring in education. Future teachers were flown to various European cities, where they visited the iconic sites of Western high culture and attended lectures delivered in front of a large illuminated globe carrying the company logo. The lectures' theme of "air-mindedness" invoked a vision of global citizenship, reducing isolationism and provincialism while promoting students' recognition of "the consequences of the daily shrinking process of time and space on our globe. . . . Of course, you should remain good patriots of your own nation . . . but at the same time you must try to think globally." Global thinking was explicitly connected to air travel, and while TWA's commercial motives are obvious, the rhetoric of universal brotherhood in a shrinking world—"Our globe, formerly the image of mysticism and unknown remoteness, shrinks before our eyes to a tiny apple"—echoed MacLeish and van Loon and the effects of universal exhibitions.[42]

Globe and graticule were adopted by postwar airlines as marketing emblems, capable of evoking their complex register of associations with adventure and discovery, universal communication, human communication, and progress, which had passed into popular culture over more than a century of use through world exhibitions, newspapers, and missionary bodies. The globe was an obvious advertising device for international airlines, allowing them both to map their network across its surface and to promote a benign mission of human freedom, unity, and equality through universal communication. Thus Pan American commissioned a series of globes to decorate its offices, the largest ten feet in diameter and at a scale of 1:4 million.

American civil airlines represented the commercial side of a global reach sustained strategically by American military bases scattered across the world and the ICBMs that underpinned the Mutual Assured Destruction (MAD) global balance of the cold war, a forty-five-year struggle between compet-

ing versions of a common cultural inheritance. Both the United States and
the Soviet Union deployed the rhetoric of universal freedom and common
human rights, and each displayed a missionary conviction in seeking to extend these to every corner of the globe. John F. Kennedy's inaugural address
as president eloquently restated the American globalism initiated by Harry
Truman: "We shall bear any burden, pay any price to ensure the cause of
freedom in the world." The human price was actually paid in large measure
by those whom the competing superpowers sought to incorporate into their
respective ideological empires.

A highly symbolic expression of competitive cold-war globalism was its
projection into cosmic space.[43] The 1957 Soviet success in launching a satellite that would orbit Earth was an immense propaganda coup at the opening of the International Geophysical Year. It was immediately framed within
an imperial narrative by the future president Lyndon B. Johnson. Speaking
in the Senate debate on the establishment of the National Aeronautics and
Space Administration (NASA) in 1958, Johnson said: "The Roman Empire
controlled the world because it could build roads . . . the British Empire
was dominant because it had ships. In the air age we were powerful because
we had airplanes. Now the Communists have established a foothold in outer
space."[44] The so-called space race thus inaugurated was given a specific goal
by John F. Kennedy's 1961 commitment "to achieving the goal, before this
decade is out, of landing a man on the moon and returning him safely to
the earth." The appropriately titled Project Apollo reworked many of the
themes so long associated with Western global images and meanings, converging with particular clarity around a tiny selection of photographs.

The story of Project Apollo is well known.[45] It lasted a mere decade, ending in December 1972 with the Pacific splashdown of *Apollo 17*. The tale
had its epic qualities: the deaths of the *Apollo 7* astronauts in a launch-pad
conflagration, the near loss in deep space of *Apollo 13,* and above all the
Apollo 11 lunar landing in July 1969. The entire project, although by no means
uncontested at the time, was a highly choreographed affair in which the
rhetorics of scientific exploration and human endeavor coincided as neatly
with its role as spectacular public entertainment as they had in the earlier
expeditions of Cook, Livingstone, Stanley, and Peary. Live television trans-
missions, globalized through new satellite technology, connected the astronauts directly to a mass audience; night launches increased the drama of a
burning inferno from which the elegant rocket escaped into pure, ethereal
space; and the dates of missions were linked to public holidays, when mass
audiences were better guaranteed. A conscious effort was made to place the

project and its personalities within a progressive narrative, traced back to Columbus, of Western exploration to the frontiers of known space. For this reason, and despite engineers' oft-repeated objections to manned spacecraft, named astronauts were employed on each mission. Those chosen were uniformly young, married, white males, fighter pilots and heirs to the heroic mantle of World War II's aviation heroes.[46] On their shoulders rested both the honor and the destiny of the American nation. The Apollo program fulfilled exploration myths to the edges of space and time traceable to Homeric epic.

The Apollo Photographs of Earth

The Apollo space project was justified in terms of lunar exploration. Its most enduring cultural impact has not been knowledge of the Moon, but an altered image of the earth. A mere four years separate *Apollo 8*'s first escape from Earth's orbit in 1968 and the last lunar landing in 1972. Those two flights, however, left a pair of photographs of the globe that have subsequently achieved iconic status as images of the earth, partly through deliberate promotion, less consciously through their capacity to incorporate and frame the Western inheritance of global meanings, from the Ciceronian *somnium* and Senecan moral reflection, through Christian discourse of mission and redemption, to ideals of unity and harmony.[47] The images are *Earthrise,* a view of the half-shadowed earth rising over a lunar landscape, taken by *Apollo 8* astronauts in late December 1968, and an unnamed *Apollo 17* image of the whole, unshadowed globe floating in the blackness of space and given NASA number AS17–22727 (Figs. 9.4 and 9.5). These two views of the terrestrial globe have been more widely reproduced than any others from the program or indeed any other single space images of the earth. They remain in wide circulation today, used for an array of purposes from commercial advertising, book illustration, emblems, and symbols of "global" educational, humanitarian, and ecological issues. They have become the image of the globe, simultaneously "true" representations and virtual spaces.

Earthrise comes from *Apollo 8,* launched on 21 December 1968 and splashing into the Pacific on 27 December. The first space mission to escape Earth's orbit, coast to the Moon, orbit, and return safely, its public impact was enhanced by spanning the Christmas period. A huge television audience saw the image on screen, and it harnessed to the Apollo mission the complex meanings attached to this key Christian festival—of peace and good will among all peoples, of domesticity and harmony, of rebirth and renewal. The

9.4. Realizing the
global vision from
space: *Earthrise*, 1968.
Courtesy NASA.

astronauts themselves reinforced these connections by reciting from lunar
orbit the Genesis cosmogonic narrative, the same lines conventionally illustrated in the opening pages of medieval chronicles and Renaissance cosmographies. Comparing the lunar surface, apparent in the landscape foreground
of *Earthrise,* to gray, lifeless sand, Frank Borman contrasted it to his view of
the earth—the "home planet"—witnessed for the first time over the lunar
horizon, in a swirl of blues, browns, and whites, "a grand oasis in the big vastness of space." His response echoed exactly, if rather inelegantly, Seneca's
words: "When you're finally up at the moon looking back at the earth, all
those differences and nationalistic traits are pretty well going to blend and
you're going to get a concept that maybe this is really one world and why
the hell can't we learn to live together like decent people."[48] Achieving the
Apollonian perspective, so long anticipated in imagination, produced an unconscious but perhaps predictable set of responses—marvel at a vast yet tiny
earth, reflection on the insignificance of self, and yearning for human unity.

Borman's responses were reinforced in a *New York Times* editorial of 25
December 1968, widely reproduced by the press across America and the
English-speaking world. "Riders on the Earth" was a short essay by Archibald
MacLeish, who had associated the airman's vision and world peace a quar-
ter-century earlier. He drew upon exactly the themes he had elaborated at

that time, reworking them in response to *Earthrise:* "For the first time in all
of time men have actually *seen* the earth: seen it not as continents or oceans
from the little distance of a hundred miles or two or three, but seen it from
the depths of space; seen it whole and round and beautiful and small." Such
a vision, claimed MacLeish, would remake mankind's self-image. Humans
were henceforth neither grand actors at creation's center stage nor helpless
creatures at its margins. The lunar view of the earth allowed a *true* gauging
of human proportion and relations with the planet: "To see the earth as it
truly is, small and blue and beautiful in that eternal silence in which it floats,
is to see ourselves as riders on the earth together, brothers in that bright love-
liness in the eternal cold—brothers who know now that they are truly brothers."[49] Resting his argument on the combined authority of the eyewitness
and the objectivity of the camera, MacLeish located *Earthrise* firmly within
the Western imaginative tradition and the Apollonian perspective. The lack
of evident human presence in the image frees its imperial inclusiveness from
all contingency. The small disk hanging delicately in velvet space becomes

9.5. Whole earth:
NASA's photo AS17-
22727, 1972. Courtesy
NASA.

visual confirmation of American democracy's redemptive world-historical mission, namely, to realize the universal brotherhood of a common humanity. When the American magazine *Time* heralded the *Apollo 8* crew as "Men of the Year," pictured against the *Earthrise,* and gave the image the caption "Dawn," the citation was "not merely for the dazzling technology of their achievement, but for the larger view of our planet and the fundamental unity of mankind."[50]

If MacLeish's reading of *Earthrise* represents the universalist, implicitly imperial associations of global images, there were other interpretations with similar pedigree. During the *Apollo 11* Moon landing of July 1969 the astronaut Michael Collins remained alone in the command module, circling the Moon while his companions walked its surface. Recording his feelings as his tiny craft crossed into lunar shadow and out of radio contact with the earth, Collins wrote: "I am alone now, truly alone, and absolutely isolated from any known life . . . I am it. If a count were taken, the score would be three billion plus two on the other side of the Moon, and one plus God only knows what on this side." Then, as the earth rose over the lunar horizon, it seemed "so small I could blot it out of the universe simply by holding up my thumb. It suddenly struck me that that tiny pea, pretty and blue, was the earth. . . . I didn't feel like a giant. I felt very, very small."[51] Collins's words echo those of Cicero's hero at Carthage, Scipio Africanus, recognizing the limits to Rome's empire.

Photo AS17–22727, taken during the final Apollo flight in 1972, locates a perfectly circular earth within a square frame, mimicking the *mappa mundi* or the hemispheric planisphere. The edges of the floating globe seem to dissolve into the surrounding black, an impression produced by the earth's atmosphere. Within the circular frame, the continents, oceans, and atmospheric circulation compose downward-curving arcs and swirls to produce an image of almost tearful intensity. Dominantly blue, brown, and white, its colors clearly define the landmasses of Africa and the Arabian peninsula, the South Atlantic and Indian Oceans, and the island continent of Antarctica. The desert and savanna lands are free of cloud, while the Gulf of Aden, the Red Sea, and the Horn of Africa are etched against the intense blue of the tropical ocean. Wreaths of white cloud circle the southern middle latitudes, matching the frosting of the polar continent.

The image's geographical, compositional, and tonal qualities give it unusually strong imaginative appeal, aesthetic balance, and formal harmony. In the photographic frame the globe is oriented to the cardinal points, its balance of lands and oceans roughly proportional to that of the sphere as a whole. The warm browns of the Sahara contrast with the cold whites of Antarctica, and both are mediated by the moist greens of equatorial Africa. An appearance of rotational movement is provided by the sickle-shaped weather systems, leaving a central pool of blue punctuated by the subcontinental island of Madagascar. But while the photo can readily be matched to the mapped image of the earth, it upsets conventional Western cartographic conventions in significant ways. Primarily, it strips away the graticule, principal signifier of Western knowledge and control, radically challenging a global image dominant over four hundred years. Thus liberated, and with no signs of naming, boundary marking, or possession, Earth appears to float free as a *sui generis* organism. The diameter of the photographic circle follows the line of the tropic of Capricorn rather than the equator, while the central meridian of the image runs more or less along the 45°E latitude line. The result is to centralize Africa and Antarctica on the global surface, revealing a restricted part of Asia and the tiniest slivers of Australia and South America. The North Atlantic world centered on the Greenwich meridian and the conventional West disappears, while antiquity's ecumene around the Mediterranean Sea is a marginal arc in the upper margins. In terms of the long-evolving Western global image, the world is radically decentered.

Decentering seems appropriate for an image whose burgeoning popularity since 1972 has coincided with a broader political and cultural thrust to imagine and articulate a globe without a privileged center and subordinated periphery, in which all voices across a decolonized globe, regardless of location, claim equal rights to announce their unique place, memory, and vision. Photo 22727's apparent absence of cultural signifiers has made it a favored icon for environmental and human-rights campaigners and those challenging Western humanism's long-held assumption of superiority in a hierarchy of life.[52] These arguments have been explicitly associated with photo 22727, dramatically in the series of photomontages by the English artist Peter Kennard blending the earth's image with images of living trees, nuclear missiles, and the human foetus (Fig. 9.6) and harnessing it to various "global" issues.

The frequency with which photo 22727 is reproduced in reverse or inverted suggests that its status is iconic rather than cartographic. While it is instantly recognized as an image of the earth, few register its precise geographical contents. Most respond primarily to its cosmographic and elemental qualities. Both *Earthrise* and photo 22727 have recharged the emblematic status that global images have historically held in the West. As MacLeish recognized, in both images the globe floats isolated in a lifeless

9.6. Photomontage of
NASA photo 22727 and
foetus by Peter Kennard,
1990. Courtesy of the artist.

void. The comfortably enclosing spheres of the pre-Copernican cosmos are
absent, as is evidence of the regular mechanical motion of a Newtonian cosmos or of the relative space-time of Einsteinian physics. The globe itself has
become an island. Even the random dusting of stars that have populated cosmic dreams from Cicero to Kircher and beyond is absent from these images;
only the dead dust of lunar landscape appears, underlining Earth's cosmic
isolation. But this same void serves to intensify the circle of intense color
where the elements of water, air, and earth dissolve into one another, recalling earlier creation images from Genesis. We should not be surprised that
the Apollo images have been linked to both foetal origins and eschatological fears of ecological doom.

Associations of the Virtual Globe

Since the 1970s two related discourses have framed the Apollo space images,
each reworking the genealogy of global meanings I have traced. A "whole-
earth" discourse stresses the globe's organic unity and matters of life, dwell-

ing, and rootedness. It emphasizes the fragility and vulnerability of a corpo-real earth and responsibility for its care. It can generate apocalyptic anxiety about the end of life on this planet or warm sentiments of association, com-munity, and attachment. Such a discourse has to confront the globe's island-ness in the oxymoron of global localism. A "one-world" discourse, by con-trast, concentrates on the global surface, on circulation, connectivity, and communication. It is a universalist, progressive, and mobile discourse in which the image of the globe signifies the potential, if not actual, equality of all locations networked across frictionless space. Consistently associated with technological advance, it yields an implicitly imperial spatiality, con-necting the ends of the earth to privileged hubs and centers of control.

The Apollo images have served as graphic signifiers for both discourses. In the 1970s, photo 22727 quickly became the Earth Day logo in the United States, while the environmental lobby group Friends of the Earth used it effectively to convey a message of global dwelling, care, and fragility. It has subsequently illustrated antinuclear, environmental, and animal-rights cam-paigns. This image illustrated James Lovelock's 1978 book *Gaia,* perhaps the most eloquent and influential statement of whole-earth globalism, whose hypothesis represents the planet as a self-sustaining and self-regulating or-ganism characterized by biochemical homeostasis.[53] Despite Lovelock's de-nials, many find in his thesis distinct echoes of animist and vitalist theories reaching back to Neoplatonic to Pythagorean philosophies. Pointing to the contradictions implicit in the whole-earth embrace of both attachment and universalism, Timothy Ingold has pointed out that all "global" thinking sig-nals a replacement of cosmology by technology: "the relation between peo-ple and world is turned inside out."[54] He reasons that because envisioning the earth as a globe is possible only by an act of disengagement—Apollo's eye—globalism is always the opposite of dwelling and attachment. By con-trast, envisioning the earth as a sphere set within other spheres, as in pre-Copernican cosmology, for example, does signify a dwelling perspective. But to oppose Apollonian and Dionysian perspectives so starkly neglects their in-terdependence. In pre-Copernican cosmology the ascent of the soul through the spheres dissolved self into the harmony of a single creation precisely to escape the material rootedness of the dwelling body, fixed in place. And the imperial gaze *ad termini orbis terrarum* is always implicitly localized through its authorizing centrality.

Just as effectively, the Apollo images or, increasingly, their derivatives pic-ture a connected world. Anticipated by Richard Harrison's creative images reproduced by postwar air companies to map their routes and by the giant globes that signaled the universal, futurist themes of world exhibitions and trade fairs, the Apollo images quickly replaced the graticule or conventional map to appear regularly in the promotional copy of airlines, telecommunications companies, finance houses, and computer corporations. The networks they display have extended beyond the physical patterns of cables or air routes to become a free-floating sign. The network of interconnecting lines signifies communication between points that increasingly convey no material objects; they are virtual and purely informational, operating through satellites arrayed above the global surface and unconstrained by its physical barriers to flow. Their most dramatic expression is the Internet, and, predictably, Internet providers and companies are among the most enthusiastic users of global images for their publicity material. And in this material the global image itself increasingly dematerializes. Commercial "image libraries" from which advertisers routinely select publicity materials devote entire catalogue sections to images of the globe and space. In using these "cartographic" images for corporate promotional purposes, the *idea* of "being global" is far more important than the actual ways in which a company operates across the world.[55] Conventional cartographic images of the globe, and indeed the Apollo images themselves, are increasingly deemed too specific and insufficiently arresting to the eye to succeed in expressing the abstract values required in corporate advertising at the millennium. Today, global images do not have to be either cartographically or photographically accurate or even realistic. They are increasingly fragmented or distorted in various inventive ways. Represented through impossible angles, views, colors, resolutions, and situations, "the globe and the map have become so successful as symbolic images that their 'shadows' can be presumed in images which contain no map form at all" and rely for meaning on preformed global imagery to trigger messages and associations.[56] Virtual images of the globe perform for a visually literate audience purely as gestures toward its material spaces. This is equally true for architectural realizations of the globe. Far removed from Boullée's monument to Newtonian reason, the design by the architect Daniel Libeskind for the Imperial War Museum North in Manchester, England, takes "shards" of the globe, placing these shell fragments randomly on the site to connect the form of the earth's curvature and the idea of its twenty-four-hour revolution to signify the globalization of conflict in the twentieth century.

The freedom to manipulate, re-present, even fragment a global image that has long stood for unity and harmony is distinctly new, although it was anticipated by previous mathematical manipulations such as the cordiform projection. It is increasingly possible and accessible through advances in information technology. Information technology itself, the relentless digitizing and spatial referencing of information in geographical information systems (GIS), and the Internet all employ the globe, as word and as image, to signify their universalizing goals and planetary embrace.[57] Overwhelmingly Western in both origin and access, these systems have been readily incorporated into the American globalist discourse initiated in the 1940s and elaborated in the rhetoric of the space race. In the 1990s, American Vice President Al Gore regularly represented information technologies as the foundation for a global participatory democracy: "The Global Information Infrastructure (GII) will be an assemblage of local, national and regional networks. . . . This GII will circle the globe with information superhighways on which all people can travel. . . . The GII will be a metaphor for democracy itself."[58] Elsewhere, Gore has referred to the idea of a "digital earth," envisaging a child visiting a local museum, donning a headset to see the earth from space, and using virtual-reality controls to circle the planet, zooming in and out at will, able to call upon limitless data sets of global information from cyberspace and to generate his or her own digitized globe. Similarly, Bill Gates, chairman of the most successful software provider of the 1990s, has celebrated "a frictionless capitalism. . . . In the great planetary marketplace, we as social creatures will sell, negotiate, invest, bargain, choose, discuss, stroll, meet."[59] Such proclamations stand squarely in a tradition of "universal association," which from the time of Alexander the Great has followed the dream of imperium.[60]

In their different but complementary ways both one-world and whole-earth discourses inherit the most persistent and contradictory feature of the Western global imagination, its sense of global mission. From the joint inheritance of Aristotelian humanism, Alexandrian and Augustan imperialism, and Judeo-Christian monotheism the West has forged a flexible but deeply resistant sense of missionary *telos*: to redeem the world *ad termini orbis terrarum*. Redemption has been variously articulated, in both spiritual and secular terms. The world to be redeemed may be "humanity," that abstraction which has gradually expanded from property-owning, Greek-speaking males to both sexes and all language groups, cultures, and individuals, and today an ever more inclusive "life on earth," in which all boundaries between life forms dissolve into the organic amalgam of photo 22727. In projecting ideas and beliefs forged in one locale across global space, the liberal mission of universal redemption is inescapably ethnocentric and imperial, able to admit "other" voices only if they speak and are spoken by the language of

IN TUTTO AMARE E SERVIRE

9.7. "In tutto amare e servire": Jesuit globalism in a photomontage of NASA photo 22727. 1991 poster.

the (self-denying) center. Desire for a perfect, universal language has been a persistent companion of Western globalism.[61] Claims that cyberspace, the Internet, new hybridities, quasi objects, or cyborgs will offer genuine poly-vocality and equivalence must therefore be treated with some skepticism in the light of the genealogy of Apollonian vision.

A final image serves to register the continuities, disruptions, and conceits that come with the West's Apollonian inheritance. It is a photomontage produced to celebrate the quincentenary of the birth in 1491 of St. Ignatius, the founder of the Jesuit order (Fig. 9.7). The saint's eyes turn heavenward, while his finger points to the globe, pictured by the Apollo image, carto-graphically inverted. The caption, "In tutto amare e servire" (To love and to serve in all things), links the individual believer to the world in a missionary vision of love and service. The figure of St. Ignatius is of the statue erected above his tomb in Rome, a baroque confection in gold and silver, initially extracted from the edges of Spain's New World empire. The poster celebrates the Christian imperative of global redemption, pursued by the Vatican with no less vigor today than in 1540. During the pontificate of Gregory XVI (1830–46) the Jesuit order was allowed to reestablish its global missionary activity, repressed in the years of Enlightenment. That activity has subsequently advanced hand in hand with both Western imperial politics and modern technology, as a missionary statement of 1919 made clear: "We see the forces of evil fall before the European conquests and the development of missions. The railways and the telegraph wires cross deserts, steppes, forests, and plateaus previously unknown by the white man, and from one ocean to another, the Christian traveller can . . . pray before the altar of the true God."[62]

In the 1920s Marconi, pioneer of radio, established Vatican radio as the first international broadcasting network, to be organized and run by the Jesuit order. And it was the American Jesuit Edmund Walsh whose theories of threatened global secularism and fierce brand of "spiritual geopolitics" influenced American anticommunists such as Joe McCarthy at mid-century. The millennial year 2000 was the greatest *Giubileo* in the history of the Roman church, with pilgrims converging on St. Peter's from across the globe, an event of global association promoted and organized through the most advanced technologies of video and Internet.[63] Vatican globalism still deploys the most sophisticated technical means to pursue the dream of global redemption.

But this poster is also an emblem. The globe is presented as an object of individual vision and piety, echoing the words addressed to Scipio Africanus in Cicero's *somnium:* "Man was given life that he might inhabit that sphere called Earth, which you see in the centre of this temple; and he has been given a soul out of those eternal fires which you call stars and planets, which being round and globular bodies animated by divine intelligences, circle about in their fixed orbits with marvellous speed."[64] In his *Commentary* on these lines the Neoplatonist Macrobius wrote: "In the above passage we are also informed . . . that such a divinity is present in the human race, that we are all of us ennobled by our kinship with the heavenly mind."[65] Apollo's eye is located neither beyond the globe nor rooted in its earth. In the words of the Florentine humanist and Neoplatonist Giovanni Pico della Mirandola: "Medium te mundi posui, ut circumspiceres inde comodius quicquid est in mundo."[66]

NOTES

Chapter One. Imperial and Poetic Globe

> *Epigraph:*
> Who can understand? Here the circle speaks,
> The circle without beginning or end, lacking wings
> And separating its own space from the world
> To make a perfect day.
>
> Here one does not know
> Falsehood. An unknown concept
> Clear in its placings within the perfect circle.
>
> Who traced it? the bird, the geometer?
> A man chases after his own self. Athlete
> With no rest, with no other goal than to be,
> To connect across his own oppositions.
>
> It is the eyeball. It is sun and moon.
> Its territory traced by the stars.
> None but a god may efface it, and voice the word of End.
>
> We celebrate, we venerate, the circle.

Robert Sabatier, from *Icare* (Paris: Albin Michel, 1976), reprinted in Jean-Pierre Luminet, *Les poètes et l'universe* (Paris: Le Cherche Midi, 1996), 203, author's translation.

1. Apollo's sexuality is ambiguous, in patriarchal Greek culture his association with the intellect connecting him to maleness and allowing a contrast with Dionysus, whose association with the passions connected this androgynous figure with his female train.

2. "West" and "Europe" are themselves the historically constituted outcome of a "continental" geography connected with the globalism discussed here. See Martin Lewis and Karen Wigen, *The Myth of Continents* (Berkeley and Los Angeles: University of California Press, 1997).

3. The *locus classicus* is Macrobius, *Commentary on the Dream of Scipio,* trans. William Harris Stahl (New York: Columbia University Press, 1952). Carl Jung's work

on dreams touches frequently on flying and Apollonianism. See C. G. Jung, *The Archetypes and the Collective Unconscious* (London: Routledge & Kegan Paul, 1959).

4. Leo Spitzer, *Classical and Christian Ideas of World Harmony: Prolegomena to an Interpretation of the Word "Stimmung"* (Baltimore: Johns Hopkins Press, 1963); Fernand Hallyn, *The Poetic Structure of the World: Copernicus and Kepler* (New York: Zone, 1993).

5. Plato, *Phaedo,* quoted in Brian Harley and David Woodward, *The History of Cartography,* 2 vols. (Chicago: University of Chicago Press, 1987–94), vol. 1, *Cartography in Prehistoric, Ancient, and Mediaeval Europe and the Mediterranean,* 137–38.

6. The changing meanings of cosmography are discussed in chapters 4 and 5.

7. Lisa Jardine, *Worldly Goods: A New History of the Renaissance* (London: Macmillan, 1996); Jerry Brotton, *Trading Territories* (London: Reaktion, 1997).

8. Edward Grant, *Planets, Stars, and Orbs: The Medieval Cosmos, 1200–1687* (Cambridge: Cambridge University Press, 1994); Hallyn, *Poetic Structure of the World.*

9. Franco Farinelli, *I segni del mondo: Immagine cartografica e discorso geografico in età moderna* (Florence: La Nova Italia, 1992); Svetlana Alpers, *The Art of Describing: Dutch Art in the Seventeenth Century* (London: John Murray, 1983), 119–68.

10. Recent histories of geography in English have concentrated on the modern period, e.g., David N. Livingstone, *The Geographical Tradition: Episodes in the History of a Contested Enterprise* (Oxford: Blackwell, 1992). On classical and medieval geography see E. H. Bunbury's classic two-volume text of 1879, *A History of Ancient Geography among the Greeks and Romans from the Earliest Ages till the Fall of the Roman Empire* (reprint, New York: Dover, 1959); J. Oliver Thompson, *A History of Ancient Geography* (Cambridge: Cambridge University Press, 1948); George H. T. Kimble, *Geography in the Middle Ages* (London: Methuen, 1938); and Clarence J. Glacken, *Traces on the Rhodian Shore: Nature and Culture in Western Thought from Ancient Times to the End of the Eighteenth Century* (Berkeley and Los Angeles: University of California Press, 1967). French writers have given this period more attention, e.g., Jean-François Staszak, *La géographie d'avant la géographie: Le climat chez Aristote et Hippocrate* (Paris: L'Harmattan, 1995). See also Harley and Woodward, *The History of Cartography;* and Denis Cosgrove, ed., *Mappings* (London: Reaktion, 1999).

11. Lewis and Wigen, *Myth of Continents.*

12. Matteo Florini, *Iconologia di Cesare Ripa Perugina,* 2 vols. (Siena, 1613), 1:202.

13. Ibid., 2:62. Ripa offers an alternative, older image for *mundus:* the satyr figure of Pan.

14. Richard Helgerson, "The Folly of Maps and Modernity" (paper delivered at the conference "Paper Landscapes: Maps, Texts, and the Construction of Space, 1500–1700," Queen Mary and Westfield College, London, 1997).

15. James Lovelock, *Gaia: A New Look at Life on Earth* (Oxford: Oxford University Press, 1979); Donna Haraway, *Simians, Cyborgs, and Women: The Reinvention of Nature* (San Francisco: Free Association Books, 1991); Timothy Ingold, "Globes and Spheres: The Topology of Environmentalism," in *Environmentalism: The View from Anthropology,* ed. K. Milton, ASA Monographs 32 (London: Routledge, 1993), 31–42; John Pickles, *Ground Truth: The Social Implications of Geographical Information Systems* (New York: Guilford, 1995).

16. Spitzer, *Classical and Christian Ideas of World Harmony;* Hallyn, *Poetic Structure of the World.*

17. On early modern maps Fortune appears personified with a wind-filled sail on the ocean or, sometimes, standing on a shell to signify the uncertainties of navigation.

18. Catherine Hofman, Danielle Lecoq, Eve Netchine, and Monique Pelletier, *Le globe et son image* (Paris: Bibliothèque Nationale de France, 1995).

19. Brotton, *Trading Territories,* 125–26, 147–48.

20. John Dee's 1570 "Mathematical Praeface" to the English translation of Euclid's *Elements* makes this distinction, a commonplace by the late sixteenth century.

21. Lewis and Wigen, *Myth of Continents.*

22. The proposal to relocate the prime meridian was made by the German historian Arno Peters, better known for his equal-area world map.

23. John Hale, *The Civilization of Europe in the Renaissance* (London: Harper Collins, 1993), 7–11; Michael J. Heffernan, *The Meaning of Europe: Geography and Geopolitics* (London: Arnold, 1995), 9–48.

24. Francis Bacon, quoted in James Romm, "A New Forerunner for Continental Drift," *Nature,* no. 367 (1994): 407–8.

25. Victor N. Sholpo, "The Harmony of Global Space," *Geografity* 1 (1993): 6–15.

26. See chapters 4 and 5. On the association of the geographically unknown and the psychologically repressed or feared "Other" see Mary Helms, *Ulysses' Sail: An Ethnographic Odyssey of Power, Knowledge, and Geographical Distance* (Princeton: Princeton University Press, 1988).

27. Staszak, *La géographie d'avant la géographie,* 34–36.

28. Christian Jacob, *L'empire des cartes: Approche théorique de la cartographie à travers l'histoire* (Paris: Albin Michel, 1992), 229–31; see also below, chapter 7.

29. Michael Bury, *Giulio Sanuto: A Venetian Engraver of the Sixteenth Century* (Edinburgh: National Gallery of Scotland, 1990).

30. The word *cartography* is a nineteenth-century invention, as Harley and Woodward point out in their introduction to *History of Cartography,* vol. 1.

31. *Ecumene,* a form of *oikoumene,* is used hereafter to refer to the habitable earth.

32. For historical perspectives on globalization see Immanuel Wallerstein, *The Modern World System,* 3 vols. (London: Academic Press, 1974–88); and J. M. Blaut, *The Coloniser's Model of the World: Geographical Diffusionism and Eurocentric History* (New York: Guilford, 1993). For more sociological perspectives see Michael Featherstone, ed., *Global Culture: Nationalism, Globalization, and Modernity* (London: Sage, 1990); and M. Albrow and E. King, eds., *Globalization, Knowledge, and Society* (London: Sage, 1990). On the geopolitical economy of globalization see J. Agnew and S. Corbridge, *Mastering Space: Hegemony, Territory, and International Political Economy* (London: Routledge, 1995). On the cultural geography of globalization see Doreen Massey, "A Global Sense of Place," *Marxism Today,* June 1991, 24–29; and David Harvey, *The Condition of Postmodernity: An Inquiry into the Origins of Cultural Change* (Oxford: Blackwell, 1989).

33. Blaut, *Coloniser's Model of the World.*

34. Peter Hulme, *Colonial Encounters: Europe and the Native Caribbean, 1492–1797* (London: Routledge, 1986); Anthony Pagden, *European Encounters with the New World: From Renaissance to Romanticism* (New Haven: Yale University Press, 1993).

35. Derek Gregory, *Geographical Imaginations* (Oxford: Blackwell, 1994); David Harvey, *Justice, Nature, and the Geography of Difference* (Oxford: Blackwell, 1996).

36. Ingold, "Globes and Spheres."

37. Martin Kemp, *The Science of Art: Optical Themes in Western Art from Brunelleschi to Seurat* (New Haven: Yale University Press, 1992); W. J. T. Mitchell, ed., *Landscape and Power* (Chicago: University of Chicago Press, 1994).

38. Recent scholarship properly emphasizes the hybrid construction of European culture (like all other cultures) as a consequence of external contact, but it is difficult to ignore the discursive significance of Genesis and Greek natural, moral, and political philosophy.

39. The ideas of "Renaissance," "modern man," and a modern "self" derive from Jules Michelet and Jacob Burckhardt's nineteenth-century historical writings, which radically underplayed external contributions, especially from Byzantium and Islam. See Peter Burke, "Representations of the Self from Petrarch to Descartes," in *Rewriting the Self: Histories from the Renaissance to the Present,* ed. Roy Porter (London: Routledge, 1997); Jardine, *Worldly Goods;* and Brotton, *Trading Territories.*

40. The claim translates literally as "global empire, to which all peoples, monarchs, nations . . . consent." The words are taken from Augustus, *Res gestae* (see ch. 2), and quoted in Claude Nicolet, *Space, Geography, and Politics in the Early Roman Empire* (Ann Arbor: University of Michigan Press, 1991), 31.

41. The words "to the city and to the world" are still intoned in the papal Easter blessing.

42. Robert David Sack, *Human Territoriality: Its Theory and History* (Cambridge: Cambridge University Press, 1986); Henri Lefebvre, *The Production of Space* (Oxford: Blackwell, 1991). On the mutual construction of identity and otherness see J. Fabian, *Time and the Other: How Anthropology Makes Its Object* (New York: Columbia University Press, 1983).

43. Claudio Magri, *Danube* (New York: Farrar, Straus & Giroux, 1989), 97–98; Simon Schama, *Landscape and Memory* (New York: Knopf, 1995), esp. 75–99. Edward Gibbon's *Decline and Fall of the Roman Empire* (1776) opens with a discussion of setting limits to Rome's expansion.

44. Edward Said's term *Orientalism,* denoting Europe's imaginative construction of an Islamic East as its Other, has stimulated a large literature. See Said, *Orientalism* (London: Verso, 1978).

45. The Egyptian and Sumerian origins of Greek astronomical and geometrical science provided the foundation for the "Black Athena" thesis, which claims African foundations for Western culture. See Martin Bernal, *Black Athena: The Afroasiatic Roots of Classical Civilization,* vol. 1, *The Fabrication of Early Greece* (London: Free Association Books, 1997).

46. Nicolet, *Space, Geography, and Politics,* 15.

47. Ibid., 33.

48. Christian Jacob, "Mapping in the Mind: The Earth from Ancient Alexandria," in Cosgrove, *Mappings,* 24–49.

49. Mircea Eliade, *The Sacred and the Profane: The Nature of Religion,* trans. W. S. Trask (London: Routledge & Kegan Paul, 1959); Paul Wheatley, *The Pivot of the Four Quarters: A Preliminary Enquiry into the Origins and Character of the Ancient Chinese City* (Edinburgh: Edinburgh University Press, 1971); Yi-Fu Tuan, *Cosmos and Hearth: A Cosmopolite's View* (Minneapolis: University of Minnesota Press, 1996).

50. Felix Driver and David Gilbert, "Heart of Empire? Landscape, Space, and Performance in Imperial London," *Environment and Planning D: Society and Space* 16 (1998): 11–28.

51. Aeneas's rejection of Dido is a complex issue, raising questions of gender, movement and encounter, domestic and exotic space.

52. Trading colonies, such as Venice and Genoa in the Levant or European treaty ports in nineteenth-century China, do not fit this model; the merchant is a cosmopolite, unbound by localism.

53. Historical examples include disputes between the Spanish court and settlers of New Spain over the treatment of native peoples recorded by Bartolomé de las Casas, between William Penn and the frontier Scotch-Irish of western Pennsylvania over armed response to Indian attacks, and more recently between white settlers in British East and South Africa and London over African "readiness" for self-rule.

54. James S. Romm, *The Edges of the Earth in Ancient Thought: Geography, Exploration, and Fiction* (Princeton: Princeton University Press, 1992), 37.

55. Anthony Pagden, *The Fall of Natural Man: The American Indian and the Origins of Comparative Ethnography* (Cambridge: Cambridge University Press, 1982); idem, *European Encounters with the New World;* Haraway, *Simians, Cyborgs, and Women;* Bruno Latour, *We Have Never Been Modern* (London: Harvester Wheatsheaf, 1993).

56. John B. Friedman, *The Monstrous Races in Medieval Art and Thought* (Cambridge: Harvard University Press, 1981).

57. This hierarchy is generalized from Aristotle, Vitruvius, Alberti, and Serlio. Later stadial theories and hierarchies—hunter-gatherers, pastoralists, settled agriculturists, urban traders and manufacturers—did not fundamentally transform it.

58. Richard Sennett, *Flesh and Stone* (London: Wiley, 1996), 31–86.

59. Pagden, *European Encounters with the New World,* 141–81.

60. The apocryphal question "Angles or angels?" supposedly posed about "barbarians" brought to Rome from Anglia by St. Augustine, refers to their physical appearance and hair color but gestures to a theological positioning of humans between angels and animals.

61. Frank Lestringant, *Mapping the Renaissance World: The Geographical Imagination in the Age of Discovery* (Cambridge: Polity, 1993), 123.

62. According to Herodotus, geometry originated in annual remeasurement of property lines in the Nile delta by land surveyors with ropes; the gnomon, an upright pole whose shadow allowed time and latitude to be fixed, also came to Greece from Egyptian and Babylonian astronomers. M. R. Wright, *Cosmology in Antiquity* (London: Routledge, 1995).

63. S. K. Heninger Jr., *The Cosmographical Glass: Renaissance Diagrams of the Universe* (San Marino, Calif.: Huntington Library, 1977).

64. Kemp, *Science of Art;* see also the discussion of Roger Bacon in David Woodward, "Roger Bacon's Terrestrial Coordinate System," *Annals of the Association of American Geographers* 80 (1990): 109–22.

65. Spitzer, *Classical and Christian Ideas of World Harmony.* On *anima mundi* see P. More, "Anima Mundi, or the Bull at the Centre of the World," *Spring* 48 (1987): 116–31.

66. Relating Apollo to the figurative arts and Dionysus to music, Ernst Cassirer suggested that the Dionysian is the overwhelming element, the most authentical in revealing the nature of things, because it consists of the eternal essence of a primordial unity with nature, whereas the Apollonian is the eternity of appearance, the pure illusion of nature. Local and universal spatialities are provocatively discussed in Tuan, *Cosmos and Hearth.*

67. John Donne, "Good Friday 1613 Riding Westward," from *John Donne: A Selection of His Poetry,* ed. John Hayward (Harmondsworth: Penguin, 1950), 174. On Camões, see chapters 3 and 4.

68. Hallyn, *Poetic Structure of the World;* Robert Lawlor, *Sacred Geometry* (London: Thames & Hudson, 1982).

69. Philippe Oulmont, ed., *Le terre et la ciel* (Paris: Bibliothèque Nationale de France, 1998).

70. Giorgio Mangani, "Abraham Ortelius and the Hermetic Meaning of the Cordiform Projection," *Imago Mundi* 50 (1998): 59–83.

Chapter Two. Classical Globe

1. Seneca, from *Medea,* choral 3, cited in Romm, *Edges of the Earth,* 170 (see ch. 1, n. 54).

2. Romm, *Edges of the Earth,* 31.

3. Christian Jacob, "Quand les cartes réfléchissent," *Espaces/Temps* 62–63 (1996): 37–49.

4. Ken Dowden, *The Uses of Greek Mythology* (London: Routledge, 1992).

5. Despite Pythagorean belief in a spherical earth, earlier notions of a cylindrical or disk-shaped planet remained at least until Aristotle's time.

6. The Farnese Atlas, at the Museo Archeologico Nazionale in Naples, is a late-second-century C.E. work, taken to be a copy of the first recorded globe, by Eudoxus (c. 408–355 B.C.E.). "We must conclude that the celestial globe of Eudoxus . . . was an authentic instrument. It probably helped to conventionalise the figures of the constellations . . . and it also gave the Greeks a taste for the mechanical description of the universe." W. Dilke, "Foundations of Classical Cartography in Archaic and Classical Greece," in Harley and Woodward, *History of Cartography,* 1:142 (see ch. 1, n. 5); see also Vladimiro Valerio, "Historiographic and Numerical Notes on the Atlante Farnese and Its Celestial Sphere," *Der Globusfreund: Wissenschaftliche Zeitschrift für Globen- und Instrumentenkunde* 35–37 (1987): 97–124. A brass celestial globe in the

Römisch-Germanisches Zentralmuseum at Mainz dates from about the same period as the Farnese. See Ernst Künzl, "The Globe in the 'Römisch-Germanisches Zentralmuseum Mainz' the Only Complete Celestial Globe to Date from Classical Greco-Roman Antiquity," ibid. 45–46 (1988): 81–153.

7. In the Renaissance Heracles/Hercules was often represented with armillary sphere and compasses.

8. Homer and Hesiod represent the earliest Greek cosmographies; Virgil and Ovid represent Roman imperial autocelebration.

9. Not all ancient writers accepted such a theistic view. The first-century B.C.E. Epicurean Lucretius, in *De rerum naturae,* adopted a strictly materialist view: "Multitudinous atoms, swept along in multitudinous courses through infinite time by mutual clashes and their own weight, have come together in every possible way and realized everything that could be formed by their combinations. So it comes about that a voyage of immense duration, in which they have experienced every variety of movement and conjunction, has at length brought those whose sudden encounter normally forms the starting point of substantial fabrics—earth and sea and sky and the races of living creatures." Lucretius, *The Nature of the Universe,* trans. R. E. Latham (Harmondsworth: Penguin, 1958), 183–84.

10. Hesiod, *Theogony,* in *Theogony, Works and Days, Shield,* trans. A. N. Athanassakis (Baltimore: Johns Hopkins University Press, 1983), lines 116–18, 123–28. On the relations between models and myths of cosmic order, see Wright, *Cosmology in Antiquity,* 37–55 (see ch. 1, n. 62).

11. Hesiod, *Theogony,* lines 517–19.

12. Hesiod, *Works and Days,* in *Theogony, Works and Days, Shield,* lines 168–71.

13. The text is from Alexander Pope's translation of *The Iliad* (London, 1715–20). For a more recent translation, see A. T. Murray's (London: Heinemann, 1976), bk. 18, lines 468–608, quotations from 483–88, 606–8. In the *Aeneid,* Virgil has Vulcan (the Roman Hephaestus) make a similar shield and armor for the book's eponymous hero.

14. Ovid, *Metamorphoses,* trans. Mary M. Innes (Harmondsworth: Penguin, 1955), bk. 1, p. 29.

15. Ibid., 30. Pliny's *Natural History* draws upon the Aristotelian scheme of climates while incorporating the poetic harmony of the world.

16. Ibid., 31. Ulysses' *Odyssey* and Aeneas's voyages are structured by the unpredictable and uncontrollable winds.

17. Ted Hughes's acclaimed translation of Ovid's *Metamorphoses* (London: Faber & Faber, 1997) indicates the work's continued cultural significance.

18. Virgil, *Aeneid,* prose trans. David West (Harmondsworth: Penguin, 1990), bk. 8, lines 720–28; the full description occupies lines 626–728. David Quint, *Epic and Empire: Politics and Generic Form from Virgil to Milton* (Princeton: Princeton University Press, 1993), finds in Virgil's shield story an "orientalist" distinction between a dynamic West and an exotic, feminized East.

19. E. Baldwin Smith, *The Dome: A Study in the History of Ideas* (Princeton: Princeton University Press, 1950); John Gillies, *Shakespeare and the Geography of Difference* (Cambridge: Cambridge University Press, 1994).

20. Dowden, *Uses of Greek Mythology,* 124. Spatial gendering derives from Dionysus's power to evoke a wildness associated with women. See ibid., 123–33.

21. Gregory Vlastos, *Plato's Universe* (Oxford: Clarendon, 1975), 10. Vlastos points out the division between those *physiologoi,* including Plato, who believed that an intelligence ordered the world, and atomists, "whose rigorous materialism totally excludes the possibility of an ordering cosmic mind" (22). The root meaning of *kosmos* is "ornament," echoed in the modern *cosmetic.* See Wright, *Cosmology in Antiquity,* 3–8.

22. See, e.g., Plato's comments in *The Republic,* trans. H. D. P. Lee (Harmondsworth: Penguin, 1962), 10.595–602, where art is placed at a third remove from "reality" and Homer's "falsehood" is adversely compared with Pythagoras's "truth." On Plato's use of myth in *Timaeus,* his principal cosmological text, see Wright, *Cosmology in Antiquity,* 25–26.

23. Vlastos, *Plato's Universe,* 31.

24. Jacob, "Quand les cartes réfléchissent," states that maps circulating this period consisted of a compass-drawn circle with equator and two hemispheres.

25. Romm, *Edges of the Earth,* 60–67; Hesiod, *Works and Days,* 168–74, describes the Hyperboreans thus: "Yet others of them father Zeus, son of Kronos, settled at earth's ends, apart from men, and gave them shelter and food. They lived there with hearts unburdened by cares in the islands of the blessed, near stormy Okeanos, these blissful heroes for whom three times a year the barley-giving land brings forth grain as sweet as honey."

26. Jacob, "Quand les cartes réfléchissent," 42.

27. Charles H. Cotter, *The Astronomical and Mathematical Foundations of Geography* (London: Hollis & Carter, 1966).

28. Virgil, *Georgics,* 1.276–85, in *The Eclogues and Georgics of Virgil,* trans. T. F. Royds (London: Dent, 1907). Like Ovid, Virgil describes the sphere with the polar axis running left to right, which is how it was commonly described into the seventeenth century. Eratosthenes used the conceit of Hermes' flight around the earth to describe the zones, giving each a characteristic color: "Five encircling zones were girt around it: two of them darker than greyish-blue enamel, another one sandy and red, as if from fire. . . . Two others there were, standing opposite one another, between the heat and the showers of ice; both were temperate regions, growing with grain, the fruit of the Eleusian Demeter; in them dwelt men antipodal to one another" (quoted in Romm, *Edges of the Earth,* 128).

29. Staszak, *La géographie d'avant la géographie,* 25, 30–45 (see ch. 1, n. 10).

30. For Aristotle, meteorology included all phenomena located between Earth and the Moon. In addition to wind, temperature, and precipitation, these included earthquakes, volcanoes, and other phenomena that demonstrated the interaction of the elements and humors of the body of the earth.

31. Staszak, *La géographie d'avant la géographie,* 56–57.

32. On climatic and environmental determinism see Glacken, *Traces on the Rhodian Shore* (see ch. 1, n. 10). Staszak, in *La géographie d'avant la géographie,* distinguishes between Aristotle's theoretical work and Hippocrates' "geographical" empiricism.

33. Staszak, *La géographie d'avant la géographie,* 33. The absence of map references in Aristotle forms the central problematic of Staszak's text.

34. H. C. Baldry, *The Unity of Mankind in Greek Thought* (Cambridge: Cambridge University Press, 1965), 11. The emphasis on bread has to do with the symbolic spatial order beyond the polis: those beyond culture supposedly dwelt in caves (not houses) and ate *a*corns (i.e., "not corn").

35. Dowden, *Uses of Greek Mythology,* 81.

36. Baldry, *Unity of Mankind,* 16. Of course, Elizabethans were not particularly effective explorers.

37. Democritus, quoted in ibid., 52.

38. Bramante's mapped area extends the coast of Africa beyond Cape Verde, reflecting Portuguese navigation.

39. For a contrasting interpretation of Alexander's imperialism see A. B. Bosworth, *Alexander and the East: The Tragedy of Triumph* (Oxford: Clarendon, 1996); and John Maxwell O'Brien, *Alexander the Great: The Invisible Enemy* (London: Routledge, 1992).

40. Romm, *Edges of the Earth,* 108–9.

41. O'Brien, *Alexander the Great,* 21.

42. Helms, *Ulysses' Sail* (see ch. 1, n. 26).

43. O'Brien, *Alexander the Great,* 165–66.

44. Romm, *Edges of the Earth,* 37.

45. In the late Victorian era "natives" were contrasted to civilized but decadent Europeans as both barbarous and childlike yet physically and morally strong, for example, in Leni Riefenstahl's 1934 film *The Last of the Nubia,* discussed by Susan Sontag in "Fascinating Fascism," in *Under the Sign of Saturn,* ed. Susan Sontag (New York: Farrar, Straus & Giroux, 1980), 73–105.

46. See Bosworth, *Alexander and the East.*

47. Baldry, *Unity of Mankind,* 99.

48. Ibid., 114.

49. See Schama's discussion of Tacitus's *Germania* in *Landscape and Memory,* 76–86 (see ch. 1, n. 43).

50. "The lands of other nations have fixed boundaries, the circuit of Rome is the circuit of the world," quoted in Nicolet, *Space, Geography, and Politics,* 44 (see ch. 1, n. 40).

51. Mariella Cagnetta, "'Mare nostrum' un mito geopolitico da Pompeo a Mussolini," *Limes* 2 (1994): 251–57.

52. Sennett, *Flesh and Stone,* 92–117 (see ch. 1, n. 58).

53. Barbara A. Kellum, "The Construction of Landscape in Augustan Rome: The Garden Room at the Villa *ad gallinas,*" *Art Bulletin* 76 (1994): 211–24.

54. "Statement of the works of the divine Augustus, wherein the whole earth is subject to the empire of the Roman people," quoted in Nicolet, 17.

55. Nicolet, *Space, Geography, and Politics,* 9.

56. Äsa Boholm, "Reinvented Histories: Medieval Rome as Memorial Landscape," *Ecumene: Environment, Culture, Meaning* 4 (1997): 247–72.

57. Sarah Whatmore and Lorraine Thorne, "Wild(er)ness: Reconfiguring the Geographies of Wildlife," *Transactions of the Institute of British Geographers,* n.s., 23 (1998): 435–54.

58. Nicolet, *Space, Geography, and Politics,* 73, 83 n. 52.

59. Pliny the Elder, quoted in Jacob Isager, "Man and Nature: A Moral Approach," in *City and Nature: Changing Relations in Space and Time,* ed. Thomas Moller Kristensen et al. (Odense, Denmark: Odense University Press, 1994), 198.

60. Kellum, "Construction of Landscape in Augustan Rome," 221.

61. On the symbolism of the Pantheon see Sennett, *Flesh and Stone,* 102–6. For a parallel attitude toward rulers' building works in Sri Lanka see James S. Duncan, *The City as Text: The Politics of Landscape Interpretation in the Kandyan Kingdom* (Cambridge: Cambridge University Press, 1990).

62. Baldry, *Unity of Mankind,* 167.

63. Jacob, "Quand les cartes réfléchissent," 46.

64. Romm, *Edges of the Earth,* 29–31.

65. Ibid., 32.

66. The English word *survey* derives from the French *sur voir,* "to view from above."

67. Nicolet, *Space, Geography, and Politics,* 67.

68. Cicero, *De Republica,* trans. C. W. Keyes, Loeb Classical Library (London: Heinemann, 1928), 267.

69. "Iam ipsa terra ita mihi parva vista est, ut me imperii nostri, quo quasi punctum eius attingimus paeniteret" (ibid., 269).

70. Romm, *Edges of the Earth,* 135.

71. Seneca, *Medea,* lines 375–79, quoted in ibid., 220. This prophecy is taken up by the Portuguese Renaissance poet Camões in *The Lusiads* (see ch. 5).

72. See the discussion in Romm, *Edges of the Earth,* 167–70.

73. Lucian, *The Works of Lucian of Samosata,* vol. 1, trans. H. W. Fowler and F. G. Fowler (Oxford: Clarendon, 1905), 171.

74. Jacob, "Quand les cartes réfléchissent," 47.

75. On macrocosm and microcosm in classical thought, see Wright, *Cosmology in Antiquity,* 56–73. On Vitruvius's significance for the architectural language of empire see John Onians, *Bearers of Meaning: The Classical Orders in Antiquity, the Middle Ages, and the Renaissance* (Cambridge: Cambridge University Press, 1988).

76. It is a twentieth-century commonplace (e.g., for Martin Heidegger, Jacques Derrida, and Richard Rorty) that the relationship between the sense of sight and reasoned knowledge is fundamental to Western philosophy to the extent that even Aristotle scarcely subjected this aspect of *theoria* to critical reflection.

Chapter Three. Christian Globe

1. Virgil, *Eclogue* 4.8–14, 57–59, in *The Eclogues and Georgics of Virgil* (see ch. 2, n. 28).

2. The death and rebirth of a god, man, or king as a metonym for the season cycle

is a common mythic trope and fundamental to the Christian narrative. It was the central theme of Sir James Frazer's *The Golden Bough: A Study in Religion and Magic*, 3rd ed., 12 vols. (London: Macmillan, 1911–15).

3. The date chosen for Easter was the first Sunday after the first full moon after the vernal equinox, that is, as close as possible to the first day of the year having twenty-four hours of light—twelve of sunlight, twelve of moonlight. Evelyn Edson, "World Maps and Easter Tables: Medieval Maps in Context," *Imago Mundi* 48 (1996): 25–42.

4. Henceforth I use the modernized word *ecumene*, partly for ease of understanding, partly as reference to the Christian claim of *ecumenism*, wherein the physical habitability of the earth cedes place to its spiritual unity as a home for a redeemed humanity under the authority of Rome. On *ecumenical* and *catholic* as descriptors for the Latin Church see Pierre Chaunu, "Rome catholique . . . et oecumenique," *Géopolitique* 58 (1997): 54–58.

5. Peter Brown, *The Rise of Western Christianity: Triumph and Diversity, AD 200–1000* (Oxford: Blackwell, 1996).

6. Discussing the historical evolution of papal claims, Pierre Béhar emphasizes the historical significance of geopolitical divisions in the classical world between Rome and Carthage and between Rome and Germania. On the subsequent evolution of papal power see Béhar, "Pour une géopolitique de la paupeté," *Géopolitique* 58 (1997): 8–20.

7. The term *Christendom* was first used to define Europe by Pius III in 1548. Hulme, *Colonial Encounters*, 84 (see ch. 1, n. 34); Hale, *Civilization of Europe in the Renaissance*, 3–15 (see ch. 1, n. 23).

8. Béhar, "Pour une géopolitique de la paupeté," 12.

9. The relation between Son and Holy Spirit was a key point of theological division between Western and Eastern Christianity, subject of the so-called *filioque* debate and the Photian schism of 867 C.E.

10. On the spatiality Christianity inherited from Judaism, which disconnects the believer from local attachment—"My kingdom is not of this earth"—see Sennett, *Flesh and Stone*, 129–32 (see ch. 1, n. 58).

11. The Ebstorf map dates from about 1234; the original was destroyed in 1943. See the summary monograph and sources at *www.henry-davis.com/MAPS/EMwebpages/224mono.html*.

12. See the illustration of God with the compasses in Osterreichische National-bibliotek, Vienna, Latin MSS 2554, fol. 1r, reproduced in Grant, *Planets, Stars, and Orbs*, 86 (see ch. 1, n. 8). Plato's *Timaeus* was the only text of classical cosmography continuously available in the Latin West after the fall of Rome.

13. Luis Revenga, ed., *Los Beatos*, catalogue for exhibition "Europalia 85 España" (Madrid: Grafican, 1985); John Williams, "Isidore, Orosius, and the Beatus Map," *Imago Mundi* 49 (1997): 7–32.

14. Staszak, *La géographie d'avant la géographie*, 23, 38–40 (see ch. 1, n. 10).

15. Christian Jacob, *L'empire des cartes*, 150–54 (see ch. 1, n. 28); Thomas Raff, "Die

Ikonographie der Mittelalterlichen Windpersonifikationen," *Aachener Kunstblätter* 48 (1978–79): 71–218.

16. Rev. 6:12–15.

17. Rev. 7:1. Matt. 24:31 anticipates this: "And he shall send his angels with a great sound of a trumpet, and they shall gather together the elect from the four winds, from one end of heaven to the other."

18. David Woodward, "Medieval *Mappaemundi,*" in Harley and Woodward, *History of Cartography,* 1:307, 319, 328 (see ch. 1, n. 5).

19. Ahmet T. Karamustafa, "Cosmographic Diagrams," and Joseph E. Schwartzberg, "Cosmographical Mapping," in Harley and Woodward, *History of Cartography,* vol. 2, *Cartography in the Traditional East and South-east Asian Societies,* 71–89 and 332–87, respectively (see ch. 1, n. 5).

20. Boholm, "Reinvented Histories," 255 (see ch. 2, n. 56).

21. Pomponius Mela, *Cosmographi geographia* (Venice: Erhart Ratdolt, 1482); Macrobius, *Commentary on the Dream of Scipio,* 214–16 (see ch. 1, n. 3); Woodward, "Medieval *Mappaemundi,*" 300–312; Leslie B. Cormack, "Flat Earth or Round Sphere: Misconceptions of the Shape of the Earth and the Fifteenth-Century Transformation of the World," *Ecumene: Environment, Culture, Meaning* 1 (1994): 363–85.

22. See Williams, "Isidore, Orosius, and the Beatus Map," 23–25.

23. Friedman, *Monstrous Races in Medieval Art and Thought* (see ch. 1, n. 56); idem, "Cultural Conflicts in Medieval World Maps," in *Implicit Understandings: Observing, Reporting, and Reflecting on the Encounters between Europeans and Other Peoples in the Early Modern Era,* ed. S. B. Schwartz (Cambridge: Cambridge University Press, 1994), 64–95.

24. Pliny the Elder, *Natural History,* trans. M. Rackham (London: Heinemann, 1963). I owe my interpretation in part to Jacob Isager, *Pliny on Art and Society* (London: Routledge, 1991); and idem, "Man and Nature," 191–200 (see ch. 2, n. 59).

25. Friedman, *Monstrous Races in Medieval Art and Thought;* Woodward, "Medieval *Mappaemundi,*" 332; Stephen Greenblatt, *Marvellous Possessions: The Wonder of the New World* (Oxford: Clarendon, 1991).

26. Umberto Eco, *The Search for a Perfect Language* (London: Fontana, 1997), 7–19, discusses Christian interpretations of the Babel story—"And the whole earth was one language, and of one speech," but the people had the fear of being "scattered abroad on the face of the earth" (Gen. 11:1, 4)—in relation to attempts to reunify humanity by rediscovering the initial language of Adam.

27. Jacob, *L'empire des cartes,* 181.

28. Alessandro Scafi, "Mapping Eden: Cartographies of the Earthly Paradise," in Cosgrove, *Mappings,* 50–70 (see ch. 1, n. 10).

29. The *isolario* is discussed in chapter 4.

30. According to St. Adamnan, abbot of Iona, a high column in the center of Jerusalem "fails to cast a shadow at midday during the summer solstice, when the sun reaches the center of the heavens," proving that Jerusalem was the navel of the world. Quoted in Woodward, "Medieval *Mappaemundi,*" 340. Egeria's early-fifth-century

text, discovered in 1887, is discussed in Mary B. Campbell, *The Witness and the Other World: Exotic European Travel Writing, 400–1600* (Ithaca: Cornell University Press, 1988).

31. Samuel B. Edgerton Jr., "From Mental Matrix to *Mappamundi* to Christian Empire: The Heritage of Christian Cartography in the Renaissance," in *Art and Cartography: Six Essays,* ed. David Woodward (Chicago: University of Chicago Press, 1987), sees Atlantic navigation as a response to the Ottoman capture of Jerusalem, the Holy Land, and Byzantium itself. See also Loren Baritz, "The Idea of the West," *American Historical Review* 66 (1960–61): 618–40.

32. Campbell, *Witness and the Other World,* 6, 8. Woodward, "Medieval *Mappaemundi,*" 298, indicates that the number of earth images in manuscripts increased from fewer than 12 in the eighth century to nearly 50 in the ninth century and from 150 in the twelfth century to 330 in the fifteenth. Other representations are found in decorative sculpture, mosaics, and painting.

33. Paul of Venice, quoted in Jacob, *L'empire des cartes,* 181.

34. Macrobius, *Commentary on the Dream of Scipio.* On fifteenth-century connection of Neoplatonism and global mapping see chapter 5.

35. Johannes Eschuid, *Summa astrologiae judicialis* (Venice: Johannes Lucilius Santritter for Franciscus Bolanus, 1489). The text contains a woodcut Macrobian map transposing the ecumene, with Europe to the east and India to the west. The text is not reversed, suggesting a disjuncture between text and graphic image even at this stage in the evolution of European global cosmography.

36. Hofman et al., *Le globe et son image,* 8 (see ch. 1, n. 18).

37. A *monstrance* is a proof or demonstration dependent upon placing an object in front of the eyes. This usage remains only for a gold or silver holder in whose transparent center the sanctified host may be raised for public view. The frame of the Hereford map has a similar shape to such a monstrance. The word shares the same root as *monster,* a strange and fearful sight. *Monstrous races* were denoted by their visible features.

38. Woodward, "Medieval *Mappaemundi,*" 324 and n. 189.

39. Campbell, *Witness and the Other World,* 86, 89.

40. Ibid., 44.

41. Karamustafa, "Cosmographic Diagrams," 73.

42. Karamustafa points out that "the Qur'àn . . . does not contain a systematic cosmology" and that "materials of cosmological import that appear in the Qur'àn are as a rule devoid of descriptive detail." Ibid., 71.

43. Ibid., 75, 87.

44. Ibid., 76.

45. Grant, *Planets, Stars, and Orbs,* 29, emphasizes the critically central role of Aristotle's natural philosophy into the seventeenth century. Eighteen treatises on *De caelo* date from the thirteenth century, 21 from the fourteenth, 32 from the fifteenth, and 108 from the sixteenth.

46. Sacrobosco, *De sphaera,* quoted in ibid., 115.

47. Aristotle's *De caelo et mundo* was translated by Gerard of Cremona in Toledo, and Euclid's *Elements* was translated by Adelard of Bath, both in the twelfth century.

See Woodward, "Medieval *Mappaemundi*," table 18.3. My discussion of Roger Bacon is drawn from Woodward, "Roger Bacon's Terrestrial Coordinate System" (see ch. 1, n. 64). See also David Woodward and Herbert M. Howe, "Roger Bacon on Geography and Cartography," in *Roger Bacon and the Sciences: Commemorative Essays,* ed. Jeremiah Hackett (Leiden: Brill, 1997), 199–222.

48. The lines are from Mandeville's *Travels,* written in 1356, the most popular prose work of the later Middle Ages. Quoted in Campbell, *Witness and the Other World,* 161; see also Grant, *Planets, Stars, and Orbs,* 629.

49. In Erhard Ratdolt's 1482 edition of Sacrobosco's *Sphaera mundi,* published at Venice, this argument is illustrated by a hand drawing. A woodcut clearly derived from the drawing appears in the Venice edition by Johannes Santritter and Hieronymous de Sanctis of 1488.

50. Grant, *Planets, Stars, and Orbs,* 626–27.

51. Cormack, "Flat Earth or Round Sphere," 363–85; Woodward, "Medieval *Mappaemundi,*" 319–21; Grant, *Planets, Stars, and Orbs,* 623–24, 630–37.

52. Grant, *Planets, Stars, and Orbs,* 117.

53. Ibid., 681–741; see esp. questions 383, 386–88, and 396–400.

54. Ibid., 59.

55. From Pierre d'Ailly, *Ymago mundi,* translated and quoted in ibid., 621.

56. Ibid., 49.

57. Jardine, *Worldly Goods,* 35–90 (see ch. 1, n. 7); Brotton, *Trading Territories,* 17–26 (see ch. 1, n. 7); Patricia Fortini Brown, *Venice and Antiquity: The Venetian Sense of the Past* (New Haven: Yale University Press, 1996), 55–74.

Chapter Four. Oceanic Globe

1. Luis Vaz de Camões, *The Lusiads* (1572), trans. Sir Richard Fanshawe, ed. Geoffrey Bullough (London: Centaur Press, 1963), canto 3, verses 6, 20.

2. Jardine, *Worldly Goods,* 53–56, 296 (see ch. 1, n. 7), details the costs involved in the overland spice trade.

3. Hulme, *Colonial Encounters,* 96–101 (see ch. 1, n. 34).

4. The impacts of the Portuguese entry into the Indian Ocean on preexisting trading patterns have been exaggerated by European writers according to K. N. Chaudhuri, *Trade and Civilization in the Indian Ocean: An Economic History from the Rise of Islam to 1750* (Cambridge: Cambridge University Press, 1985).

5. Brotton, *Trading Territories,* 87–103 (see ch. 1, n. 7); Jardine, *Worldly Goods,* 249. Jacopo Bellini worked in Istanbul on Ottoman commissions that may have included a map of the city. Ian Manners, "Constructing the Image of a City: The Representation of Constantinople in Cristoforo Buondelmonti's *Liber Insulario Archipelagi,*" *Annals of the Association of American Geographers* 87 (1997): 72–102. See also Hale, *The Civilization of Europe in the Renaissance,* 38–43 (see ch. 1, n. 23).

6. It was in the course of undertaking such a *volta* that Pedro Álvares Cabral touched upon the coast of Brazil in 1500. See Alfred William Crosby, *Ecological Impe-*

rialism: The Biological Expansion of Europe, 900–1900 (Cambridge: Cambridge University Press, 1986), 112–15.

7. Islam had extended by land as far as what was then called the Gold Coast in West Africa; Arab trading colonies stretched the length of the continent's eastern coast, while the Congo, the Kalahari, and the South African coasts offered little incentive for trade or missionary activity. See ibid. on the pathogenic power of intertropical African environments to resist European colonization, unlike the Americas.

8. J. H. Parry, *The Age of Reconaissance, Discovery, Exploration, and Settlement, 1450–1650* (Berkeley and Los Angeles: University of California Press, 1981); Crosby, *Ecological Imperialism;* Livingstone, *Geographical Tradition,* 37–41 (see ch. 1, n. 10).

9. Hulme, *Colonial Encounters,* points out that these peoples were interpreted through the spatiality attributed to their islands, as floating detritus on the edge of Christendom's body politic. In the Canaries, the native population *(guanches)* adopted the guerrilla tactics of military resistance used by indigenous peoples throughout European imperial history: they surrendered the plains and open country to mounted invaders while using their knowledge of the more rugged terrain to ambush and exhaust the enemy (Crosby, *Ecological Imperialism,* 81–100). Given the elaborate European codes of warfare, such tactics were seen as immoral, unchristian, and evidence of subhuman nature.

10. The writer in question is Francesco Chiregato in a summary of Magellan's circumnavigation, quoted in Jardine, *Worldly Goods,* 367.

11. Martin Waldseemüller, *Cosmographiae introductio cum quibus dam geometriae ac astronomiae principiis ad eam rem necessariis* (St. Die, France, 1507; n.p.: Readex Microprint, 1966), 70.

12. Eviatar Zerubavel, *Terra Cognita: The Mental Discovery of America* (New Brunswick, N.J.: Rutgers University Press, 1992), pls. 5, 13, 20, 24.

13. Zerubavel, *Terra Cognita;* see also Lewis and Wigen, *Myth of Continents,* 25–27 (see ch. 1, n. 2).

14. William D. Phillips Jr. and Carla Rahn Phillips, *The Worlds of Christopher Columbus* (Cambridge: Cambridge University Press, 1992), 187–88.

15. See Hulme, *Colonial Encounters,* 45–87, on the association of the South American mainland and cannibals in Columbus's anthropological distinction between gentle Arawaks figured in Virgilian terms as golden-age peoples and the aggressive, man-eating Caribs (cannibals), their monstrous opposites.

16. Brotton, *Trading Territories,* 119–50. Jardine, *Worldly Goods,* 271–74, discusses Diego Ribero's 1529 map.

17. Called *rutters* or *portolanos* in English, these maps were also produced by Islamic chartmakers. See the catalogue *XIV–XVIII yüzyil portolan ve deniz haritalari* (Istanbul: Topkapi Sarayi Müzesi ve Venedik Correr Müzesi Koleksiyonlarindan, 1994).

18. Tony Campbell, "Portolan Charts from the Late Thirteenth Century to 1500," in Harley and Woodward, *History of Cartography,* 1:371–463 (see ch. 1, n. 5). On the Catalan Atlas, see Woodward, "Medieval *Mappaemundi,*" 315 (see ch. 3, n. 18). No direct lineage has been established between *portolani* and *periploi.* O. A. W. Dilke, *Greek and Roman Maps* (1985; reprint, Baltimore: Johns Hopkins University Press, 1998), 143.

19. Theodore de Bry's 13-part *Historia americae* was published beginning in 1590 and continuing until 1634; Galle's *Triumph of Magellan* appears in *Pars Quarta* (Frankfurt, 1594), pl. 15. *Apotheosis of Columbus* appears in Galle's *Speculum diversarum imaginum speculativarum* (Antwerp, 1638).

20. The Miller Atlas map of Brazil is discussed in Jacob, *L'empire des cartes,* 194–202, fig. 17 (see ch. 1, n. 28).

21. Projecting in that the image that appears on the crusading flag borne by Columbus in the Galle image also appears, for example, on the c. 1550 portolan chart by Bartolomeo Olivo, reproduced in Sandra Sider, ed., *Maps, Charts, Globes: Five Centuries of Exploration (a New Edition of E. L. Stevenson's "Portolan Charts" and Catalogue of the 1992 Exhibition)* (New York: Hispanic Society of America, 1992), fig. 12; and on the anonymous late-sixteenth-century chart reproduced as figs. 27 and 28 in Susana Biadene, ed., *Carte da navigar: Portolani e carte nautiche del Museo Correr, 1318–1732* (Venice: Marsilio, 1990).

22. Campbell, "Portolan Charts," 387–88.

23. David Turnbull, "Geography and Science in Early Modern Europe: Mapping and the Construction of Knowledge Spaces," *Imago Mundi* 48 (1996): 5–24.

24. Alongside the coast of Brazil are inserted the words "hesta terra describo fernando de magalhaes," indicating knowledge of the circumnavigation completed in the year of its production. The Straits of Magellan and Tierra del Fuego are not indicated. The map, housed in the Topkapi Museum at Istanbul, is reproduced in *XIV–XVIII yüzyil portolan,* pl. 12.

25. Hildegard Binder Johnson, *Carta Marina: World Geography in Strasbourg, 1525* (Minneapolis: University of Minnesota Press, 1963).

26. "Isolario et Portolano de tuto el Mare Mediterraneo di Antonio Millo nel qual si ragiona di tutte le isole dil ditto Mare con sui porti cita sorzitori sache scholgi distancie da l'una a l'altra et qual vento et quanto circhondano longeze et largeze con il portolano qual chomincia dal streto gi gibiltara per tuta la costa di tuta la europa fino ala citta di constantinopoli et poi la costa di l'asia fino al fiume nilo et la costa di Africa fino a cuetta in stretto / con il portolano dil mare oceano principiando dal stretto di gibiltara fino tuta la costa di fiandra con lochi porti seche distanzie da locho a locho le sonde dil fondo li sengali con tute le corente di aque de flussi et reflussi con la luna diligientemente." MS Correr 904, in Biadene, *Carte da navigar,* pl. 42.

27. Francesco Prontera, "Géographie et mythes dans l'*isolario* des Grecs," in *Géographie du monde au moyen age et à la renaissance,* ed. Monique Pelletier (Paris: Editions du CTHS, 1989), 169–200; Lestringant, *Mapping the Renaissance World,* 109 (see ch. 1, n. 61).

28. Cristoforo Buondelmonti, *Librum insularum archipelagi* (1420), ed. G. R. L. De Sinner (n.p.: Lipsiae et Berolini, 1824); Eco, *Search for a Perfect Language,* 145–46 (see ch. 3, n. 26); Fortini Brown, *Venice and Antiquity,* 77–81 (see ch. 3, n. 57). Buondelmonti stimulated the Western fascination with Egyptian hieroglyphics as a source of original wisdom by bringing Horapollo's *Hieryoglyphica* from Andros to Florence in 1419. His classicism led him to see these Aegean islands with a poetic eye for *locus amoenus* (beautiful nature). Benedetto Bordone's *isolario* has been directly connected

to the famous archaeological romance *Hypnerotomachia Polifili,* published by Aldus Manutius in 1499. See Lilian Armstrong, "Benedetto Bordone, *Miniator,* and Cartography in Early Sixteenth Century Venice," *Imago Mundi* 48 (1996): 65–92.

29. Stephen Greenblatt, in *Marvellous Possessions* (see ch. 3, n. 25) discusses Mandeville but makes no particular link between the marvelous and the island as a representational space, for example, in his key quotation from Columbus: "And there I found very many islands filled with people innumerable, and of all of them I have taken possession" (168–69).

30. Bartolommeo dalli Sonetti (Zamberto), *Isolario* (Venice: Guilelmus Anima Mia, [before 1485]); Fortini Brown, *Venice and Antiquity,* 160–61.

31. In my discussion on islands I draw on the following works by Frank Lestringant: *Cartes et figures de la terre* (Paris: Centre Georges Pompidou, 1980), 470–75; *Mapping the Renaissance World,* 106–8; and "Isles," in Pelletier, *Géographie du monde,* 165–67. See also W. H. Babcock, *Legendary Islands of the Atlantic: A Study in Medieval Geography* (New York: American Geographical Society, 1922).

32. Albrecht Altdorfer's *Battle of Actium* and Peter Bruegel the Elder's *Fall of Icarus* show Cyprus and Crete, respectively. See chapter 5.

33. Even after longitude became fixed, imaginary islands continued to appear on maps and charts. Distortions of vision as sea by mist, mirage, and heat haze, not to mention human errors of observation caused by tiredness, alcohol, or simple mendacity, ensured that British Admiralty charts registered nonexistent islands in the Pacific Ocean into the era of satellite observation.

34. Prontera, "Géographie et mythes dans *l'isolario* des Grecs."

35. Lestringant, *Cartes et figures de la terre,* 203–4. Allegorical mapping is discussed in chapter 5.

36. Greenblatt, *Marvellous Possessions,* distinguishes between a medieval attitude toward exotic marvels that did not entail possession of the marvelous object and a Renaissance one, typified by Columbus, that did.

37. See Leonardo Olschki, *Storia letteraria delle scoperte geografiche* (Florence: Leo S. Olschki, 1937), 31–42, on the literary sources for Bordone's entries.

38. Philip W. Porter and Fred E. Lukermann, "The Geography of Utopia," in *Geographies of the Mind,* ed. D. Lowenthal and M. Bowden (London: Oxford University Press, 1975), 197–224.

39. Carlo Ginzburg, *The Cheese and the Worms: The Cosmos of a Sixteenth-Century Miller* (London: Routledge & Kegan Paul, 1980), 41–51, gives an example of peasant utopias based on the island.

40. Lestringant, *Mapping the Renaissance World,* 123, makes this point.

41. Quoted in Armstrong, "Benedetto Bordone," 72–73.

42. Hulme, *Colonial Encounters,* 89–136. The link between Ulysses' odyssey and Atlantic discovery appealed to humanists more than to navigators themselves. Waldseemüller's poetic prologue to Amerigo Vespucci's voyage reports explicitly compares him to the Homeric hero, and Abraham Ortelius's map of the central Mediterranean and Aegean revealed "Ulysses' errors" in his *Parergon* (c. 1609–24), companion volume to the *Theatrum orbis terrarum* (London, 1606).

43. Francesco Chierigato, quoted in Jardine, *Worldly Goods,* 367.

44. Lestringant, *Mapping the Renaissance World,* 12. On cosmography see also Heninger, *Cosmographical Glass* (see ch. 1, n. 63); and Marica Milanesi, *Tolomeo sostituito: Studi di storia delle conoscenze geografiche nel XVI secolo* (Milan: Unicopli, 1984).

45. On Regiomontanus see Jardine, *Worldly Goods,* 201–2, 350–55.

46. Waldseemüller, *Cosmographiae introductio.*

47. On cosmography in university teaching see Leslie B. Cormack, *Charting an Empire: Geography at the English Universities, 1580–1620* (Chicago: University of Chicago Press, 1997).

48. Lestringant, *Mapping the Renaissance World,* 32–36. The image of the whole earth as a marvelous jewel recurs in descriptions of the *Apollo* space photographs, where it is compared to a Christmas tree bauble (see ch. 9). See Jardine, *Worldly Goods,* on jewels as a stimulus to Renaissance navigation.

49. See the discussion of Dati's work in Anthony Grafton, *New Worlds, Ancient Texts: The Power of Tradition and the Shock of Discovery* (Cambridge: Harvard University Press, Belknap Press, 1992), 61–68.

50. Dati probably learned his trade at the *abacco,* or merchant's school, whose curriculum derived from the Schools but emphasized practical skills such as gauging, measuring, etc. See Kemp, *Science of Art* (see ch. 1, n. 37).

51. Grafton, *New Worlds, Ancient Texts,* 68.

52. Waldseemüller, *Cosmographiae introductio.* The lines from Ovid and Virgil, in chapters 4, 5, 7, and 8, describe the sphere, the celestial zones, the climates, and the winds, respectively.

53. Ibid., 70.

54. Woodward, "Medieval *Mappaemundi,*" points out that in Russia similar maps continued to be produced into the eighteenth century and to be reproduced into the nineteenth. Images of the world are controlled as much by traditions of graphic representation as by empirical knowledge and scientific theory.

55. Ibid., 314; see also Brotton, *Trading Territories,* 30–31.

56. On Thevet and his cosmography see Lestringant, *Mapping the Renaissance World;* the quotation is from p. xvii.

57. Ibid., 106.

58. Ibid., 107.

59. Waldseemüller, *Cosmographiae introductio,* 32.

60. Kemp, *Science of Art;* Martin Jay, "Scopic Regimes of Modernity," in *Modernity and Identity,* ed. S. Lash and J. Friedman (Oxford: Blackwell, 1992), 178–95.

Chapter Five. Visionary Globe

Epigraph: Camões, *The Lusiads,* canto 10, verse 77 (see ch. 4, n. 1).

1. Brotton, *Trading Territories,* 93, 98–99 (see ch. 1, n. 7).

2. On the complex history of the text in Arab translation see ibid., 98–103. On Mehmed II's multiple copies of the *Geography* see Jardine, *Worldly Goods,* 137–39 (see ch. 1, n. 7).

3. Elizabeth Eisenstein, *The Printing Press as an Agent of Change: Communications and Cultural Transformations in Early Modern Europe,* 2 vols. (Cambridge: Cambridge University Press, 1979). Forty-three Latin manuscripts and at least eight printed editions appeared before 1500. Leo Bagrow, *The History of Cartography,* rev. R. Skelton (Cambridge: Harvard University Press, 1964), 78, 91–93.

4. *Tabulae* were maps derived from Ptolemy's coordinates. Seven *tabulae novae* were attached to the codex prepared for Federigo; Waldseemüller included twenty new maps in his edition of 1513; and the largest number included in one place was forty-eight, in Sebastian Münster's 1540 revision of the Strasbourg edition. On the substitution of new geographical knowledge for the Ptolemaic world image see Milanesi, *Tolomeo sostituito* (see ch. 4, n. 44); and Eisenstein, *Printing Press as an Agent of Change,* 1:192–93, 514–18.

5. See, e.g., the c. 1453 portrait from the vellum transcription of the *Geography* at Venice, reproduced in Brotton, *Trading Territories,* fig. 7, or the frontispiece to Ruscelli's Italian translation of 1558. Jardine, in *Worldly Goods,* 355, claims that Ptolemy is the middle figure in Giorgione's *Three Philosophers* (c. 1506), positioned between Regiomontanus and Aristotle. See also Stephen M. Buhler, "Marsilio Ficino's *De stella magorum* and Renaissance Views of the Magi," *Renaissance Quarterly* 43 (1990): 348–71.

6. Eco, *Search for a Perfect Language,* 145–58 (see ch. 3, n. 26).

7. Samuel B. Edgerton Jr., *The Renaissance Rediscovery of Linear Perspective* (New York: Basic Books, 1975), 105.

8. Ptolemy, *The Geography,* trans. Edward Luther Stevenson (1930; reprint, New York: Hispanic Society of New York, 1991), bk. 1, p. 25.

9. Woodward, "Roger Bacon's Terrestrial Coordinate System." (see ch. 1, n. 64).

10. Edgerton, *Renaissance Rediscovery of Linear Perspective;* Joan Gadol, *Leon Battista Alberti: Universal Man of the Early Renaissance* (Chicago: University of Chicago Press, 1969); Michael Kubovy, *The Psychology of Perspective in Renaissance Art* (Cambridge: Cambridge University Press, 1986); Kemp, *Science of Art* (see ch. 1, n. 37); Alpers, *Art of Describing* (see ch. 1, n. 9).

11. Christopher S. Wood, *Albrecht Altdorfer and the Origins of Landscape* (London: Reaktion, 1993), discusses the fifteenth-century creation of framed pictorial space in the Danube cities of Ulm, Nuremberg, and Regensberg, where the *Geography* had particular impact.

12. Dana Bennet Durand, in *The Vienna-Klosterneuburg Map Corpus of the Fifteenth Century: A Study in the Transition from Medieval to Modern Science* (Leiden: Brill, 1952), 7, claims that "the persistence of [the classical ecumene] is truly astonishing."

13. Trepuzuntios, quoted in Brotton, *Trading Territories,* 92.

14. Jardine, *Worldly Goods,* 137–40. Montefeltro's library and the manuscript entered the Vatican under Pope Alexander VII in 1658.

15. Lorenzetti's cosmographic image was part of his 1338–40 iconographic scheme for Siena's new Palazzo Pubblico, which included the fresco of city and countryside known as the *Allegory of Good Government,* illustrating Siena's territorial authority.

16. *Dizionario biografico degli Italiani* (Rome: Istituto della Enciclopedia Italiana,

1960), 9:121–24; Roberto Almagià, "Osservazioni sull'opera geografica di Francesco Berlinghieri," in *Scritti geografici (1905–1957),* ed. Roberto Almagià (Rome: Edizioni Cremonese, 1961), 497–526; R. A. Skelton, ed., *Francesco Berlinghieri: Geographia* (Amsterdam: Theatrum Orbis Terrarum, 1966); Brotton, *Trading Territories,* 87–103.

17. A painting of Democritus and Heraclitus similar to Bramante's at Milan may have hung in Ficino's study at Careggi, the meeting place of the Platonic Academy. Arthur Field, *The Origins of the Platonic Academy of Florence* (Princeton: Princeton University Press, 1988), 189. See also Francis Ames-Lewis, "Cosimo de'Medici and the Study of Geography" (London: Birkbeck College, Department of Art History, 1996, mimeo); and S. Gentile, S. Niccoli, and P. Viti, eds., *Marsilio Ficino e il ritorno di Platone* (Florence: Biblioteca Medicea Laurenziana, Istituto Nazionale di Studi sul Rinascimento, Le Lettere, 1984).

18. G. Fowden, *The Egyptian Hermes: A Historical Approach to the Latin Pagan Mind* (Princeton: Princeton University Press, 1993); Frances Yates, *Giordano Bruno and the Hermetic Tradition* (London: Routledge & Kegan Paul, 1964).

19. Francesco Berlinghieri, *Septe giornate della geografia di Francesco Berlinghieri Fiorentino* (Florence, 1482). Brotton, in *Trading Territories,* 90, points out that the printed version was originally dedicated to Mehmed II, who died in 1482, before the work could be presented. The choice of Federigo was also ill starred as he died later the same year.

20. Almagia, "Osservazioni sull'opera geografica di Francesco Berlinghieri," 508.

21. Eco, *Search for a Perfect Language,* 147–48.

22. Ma perche in tanti exempli ne ritardo
    di chi la magna copia na fatica
    Che pie[ne] reggio il mo[n]do qua[n]do il guardo
    Ne sol la militare arte nutrica
    ma la philosophia et la scriptura
    historica et poetica lo dica
    In dolce vita della agricoltura
    la medicina et l'arte quale ha in seno
    Delli animanti in prima la natura
    In somma la notitia del terreno
    Sichome questa ogni altra facultata
    Non ha bisogno veramente meno
Berlinghieri, *Septe giornate.*

23. Da ingengno humano anzi mente divina
    si chome stia naturalmente il cielo
    monstrar si puo convera disciplina
    Voltando intorno anoi sanxa alchun velo
    informa tal che raghuardar possiano
    per imago laterra nota adpelo
    Laquale & vera & maxima diciano
    ne tutta o parte noi circundi puote

perergrinarsi dacoloro pelpiano
Da quali il cielo & lestellante rote
Ibid.

24. Charles H. Lohr, "Metaphysics," in *The Cambridge History of Renaissance Philosophy,* ed. Charles B. Schmitt (Cambridge: Cambridge University Press, 1988), 537–78; Erwin Panofsky, "The Neoplatonic Movement in Florence and North Italy," in *Studies in Iconology: Humanistic Themes in the Art of the Renaissance* (1939; reprint, New York: Harper & Row, Icon Editions, 1972), 129–69.

25. Bagrow, *History of Cartography;* Rodney W. Shirley, *The Mapping of the World: Early Printed Maps, 1472–1700* (London: Holland, 1984).

26. Nicolaus Laurentii (Germanus) was a German Benedictine working in Italy who "passed off as his own innovations maps made earlier in Klosterneuburg," the monastery near Vienna. There, in the 1430s, the earliest true projections were developed, originating the maps for all the early printed editions of Ptolemy. Durand, *Vienna-Klosterneuburg Map Corpus of the Fifteenth Century,* 83–85.

27. Hildegard Binder Johnson, in *Carta Marina,* 26, points out that humanist scholars, among them Albrecht Dürer, ridiculed the decoration of the Grüninger German edition of Ptolemy, which was "pretty and quite small in size for the laymen, so that he could carry it in his vest pocket." Carnival scenes in the margins were intended to popularize it. See also Miriam Usher Chrisman, *Learned Culture, Lay Culture: Books and Social Change in Strasbourg, 1480–1509* (New Haven: Yale University Press, 1982).

28. Wood, *Albrecht Altdorfer and the Origins of Landscape,* 13, 128.

29. Kirsten Seaver, "Norumbega and Harmonia Mundi in Sixteenth-Century Cartography," *Imago Mundi* 50 (1998): 34–58.

30. Apian's *Cosmographicus liber* first appeared in Latin in 1524, to be translated and republished throughout the sixteenth century. On Nuremberg's intellectual significance see Jardine, *Worldly Goods,* 333, 347–55.

31. See, e.g., Petri Apiani, *Cosmographia, per gemmam phrisium, apud lovaniensis medicum ac mathematicum insignem, restituta* (Antwerp: Arnoldo Berckmano, 1539).

32. Jardine, in *Worldly Goods,* 366, comments that many Lutherans saw the Ottoman expansion as a divine judgment on a corrupt Christendom. At the same time Lutheranism encouraged a more quietist, familist idea of concord sympathetic to the contemplative aspects of Neoplatonic cosmography.

33. Giorgio Mangani, "Abraham Ortelius and the Hermetic Meaning of the Cordiform Projection" (see ch. 1, n. 70). I deal with emblematic mapping in chapter 6.

34. M. Watelet, ed., *Gerardus Mercator cosmographe: Le temps et l'espace* (Antwerp: Fonds Mercator Paribas, 1994).

35. Durand, *Vienna-Klosterneuburg Map Corpus of the Fifteenth Century,* 165–68.

36. Jacques Paviot, "Ung mapmonde rond, en guise de pom[m]e," *Der Globusfreund: Wissenschaftliche zeitschrift für globen- und instrumentenkunde* 43–44 (1995): 19–29; Józef Babicz, "The Celestial and Terrestrial Globes in the Vatican Library, Dating from

1477, and Their Maker, Donnus Nicolaus Germanus (ca. 1420–ca. 1490)," ibid. 35–37 (1987): 155–65.

37. Brotton, *Trading Territories;* Jardine, *Worldly Goods,* 397–98, 425–36.

38. David Woodward, "Maps and the Rationalisation of Geographic Space," in *Circa 1492: Art in the Age of Exploration* (New Haven: Yale University Press; Washington, D.C.: National Gallery of Art, 1991), 83–87.

39. As already indicated, cosmography was an ill-defined Renaissance practice. Ptolemy's *Geography* was initially titled *Cosmography,* which could include geographical and chorographic mapping. See Lestringant, *Mapping the Renaissance World,* 12–36 (see ch. 1, n. 61).

40. On Dürer's self-reflection see Jonathan Sawday, "Self and Selfhood in the Seventeenth Century," in Porter, *Rewriting the Self,* 29–48, 41–43 (see ch. 1, n. 39). On the Renaissance self more generally see Burke, "Representations of the Self from Petrarch to Descartes" (see ch. 1, n. 39).

41. Hartman Schedel, *Liber chronicarum* (Nuremberg: Anton Koberger, 1493). The woodcut illustrations in some copies are hand colored. Schedel's humanism connects Regiomontanus's Nuremberg to Pierckheimer's and Dürer's. Jardine, *Worldly Goods,* 202.

42. Illustrations of the days of Creation connected encyclopedias to Bibles. Many printed Bibles contained such illustrations, e.g., the *Biblia Teutonica* of c. 1475, 1483, and 1494 and the *Biblia Rhenica* of 1478. See D. S. Berkowitz, *In Remembrance of Creation: Evolution of Art and Scholarship in the Medieval and Renaissance Bible* (Waltham, Mass.: Brandeis University Press, 1986); and Catherine Delano-Smith and Elizabeth Morley Ingram, *Maps in Bibles, 1500–1600: An Illustrated Catalogue* (Geneva: Droz, 1991).

43. Jardine, *Worldly Goods,* 49–50.

44. Durand, *Vienna-Klosterneuburg Map Corpus of the Fifteenth Century,* 252–64.

45. Eco, *Search for a Perfect Language,* 70.

46. Lohr, "Metaphysics," 548.

47. On the debate about images and substance, which later would be central to the debate on world systems, see Brian P. Copenhaver, "Astrology and Magic," in Schmitt, *Cambridge History of Renaissance Philosophy,* 281–94.

48. Copernicus's interest in the metaphysics of light was Pythagorean. See Waldemar Voisé, "The Great Renaissance Scholar," in *The Scientific World of Copernicus,* ed. B. Bienkowska (Dordrecht: D. Reidel, 1973), 84–94; and R. Hookykaas, "The Rise of Modern Science: When and Why?" *British Journal for the History of Science* 20 (1987): 453–73, esp. 463–66.

49. Cusanus's mapmaking is discussed in Durand, *Vienna-Klosterneuburg Map Corpus of the Fifteenth Century,* 39–44, 252–63. For his links with Byzantium see Jardine, *Worldly Goods,* 49–50; and for his links with Lullian combinatorial art see Eco, *Search for a Perfect Language,* 70–72.

50. Lohr, "Metaphysics," 552–53.

51. Paul Kristeller, *Renaissance Thought and the Arts* (Princeton: Princeton University Press, 1980), 97. See also Copenhaver, "Astrology and Magic," esp. 274–85.

52. Giovanni Pico della Mirandola, *On the Dignity of Man; On Being and the One; Heptaplus,* trans. G. G. Wallis (Indianapolis: Bobbs Merrill, 1965), 5.

53. Jardine, *Worldly Goods,* 324; the phrase "cosmographic ascension" is from Lestringant, *Mapping the Renaissance World,* 20.

54. Lestringant, *Mapping the Renaissance World,* 21.

55. Jill Kray, "Moral Philosophy," in Schmitt, *Cambridge History of Renaissance Philosophy,* 314. The Latin reads: "Homo nichil est ominium et a Natura extra omnia factus est creatus est: ut multividus fiat sitque omnium expressio et naturale speculum, abiunctum et seperatum ab universorum ordine, eminus et regione omnium collocatum, ut omnium centrum."

56. Heninger, *The Cosmographical Glass,* xv (see ch. 1, n. 63).

57. Copenhaver, "Astrology and Magic."

58. Jardine, *Worldly Goods,* 135–80.

59. Sebastian Münster, *Cosmographiae universalis* (Basel, 1552). See the discussion in Grafton, *New Worlds, Ancient Texts,* 97–111 (see ch. 4, n. 49).

60. Mariano Cuesta, *Alonso de Santa Cruz y su obra cosmografica,* 2 vols. (Madrid: Consejo Superior de Investigaciones Cientificos, Instituto "Gonzalo Fernandez de Oviedo," 1983).

61. Ursula Lamb, *A Navigator's Universe: The Libro de Cosmographia of 1538 by Pedro de Medina* (Chicago: Newberry Library, 1972); William Cuningham, *The cosmographical glasse, conteinyng the pleasant principles of cosmographie, geographie, hydrographie, or navigation* (Norwich: Joan Day, 1559). On Dee see N. H. Clulee, *John Dee's Natural Philosophy: Between Science and Religion* (London: Routledge, 1988).

62. Gastaldo is discussed in chapter 6; I return to Danti's work below.

63. Joaquim Barradas de Carvalho, *A la récherche de la specificité de la renaissance Portugaise* (Paris: Fondation Calouste Gulbenkian, Centre Culturel Portugais, 1983).

64. Tethys's island is an imaginative confection of Circe's island from Homer's *Odyssey* and Tenerife, whose peak was long regarded as the highest mountain on the globe, used by mapmakers as the prime meridian. Camões's Tethys is a hybrid of Circe and Dido, just as his da Gama is a hybrid of Ulysses and Aeneas.

65. Camões, *The Lusiads,* canto 10, verse 91.

66. William Atkinson comments in *The Lusiads* (Harmondsworth: Penguin, 1952), 12, that the work "heralded . . . the revolutionising of men's ideas of empire; the old basis of territorial conquest suffering eclipse with the development of trade and sea power."

67. A. Doroszlaï, J. Giudi, M.-F. Piéjus, and A. Rochon, *Espaces réeles et espaces imaginaires dans le Roland Furieux* (Paris: Centre Interuniversitaire des Récherches sur la Renaissance Italien, C.N.R.S., 1991). Girolamo Ruscelli's edition of *Orlando* is discussed in chapter 6.

68. On de Holanda see Jorge Segurado, *Francisco D'Ollanda: Da sua vida e obras, arquitecto da Renascença ao serviço de D. João III, pintor, desenhador, escritor, humanista, facsimile da carta a Miguel Angelo (1551) e dos seus tratados sobre Lisboa e desenho (1571)* (Lisbon: Ediçoes Excelsior, 1970); J. B. Bury, *Two Notes on Francisco de Holanda,* Warburg Institute Surveys, 7 (London: Warburg Institute, 1981).

69. De Holanda's illustrations were rediscovered in the Biblioteca Nacional at Madrid in the early 1950s. See Sylvie Deswarte, *Les "De Aetatibus Mundi Imagines" de Francisco de Holanda,* in Fondation Eugène Piot, *Monuments et mémoires* (Paris: Presses Universitaires de France, 1983); and J. B. Bury, "Francisco de Holanda and His Illustrations of Creation," *Portuguese Studies* 11 (1985–86): 15–48.

70. Bury, "Francisco de Holanda," 38. The similarity to William Blake's image of the patriarch has been noted by more than one observer.

71. Cuningham, *Cosmographical glasse,* fol. 9, emphasizes the *visibility* of creation: "The world is an apte frame, made of heaven, and earth and of things in them conteyned. This comprehendeth all things in itself, neither is there anything without the limits of the visible."

72. Thomas Blundeville, *Exercises* (London, 1594), quoted in Heninger, *Cosmographic Glass,* 7.

73. Ibid.

74. Milanesi, *Tolomeo sostituito;* Livingstone, *Geographical Tradition,* 63–101 (see ch. 1, n. 10).

75. Fame, the true goal of the Renaissance man, was often personified seated upon a globe. James Hall, *Hall's Dictionary of Subjects and Symbols in Art,* rev. ed. (London: John Murray, 1974), 119. The emblem of the Accademia della Fama at Venice, founded in 1557, showed Fame resting her left foot on a globe and the motto "Io volo al ciel per risposarmi in Dio" [I fly to the heavens to rest in God]. The academy is discussed more fully in chapter 7.

76. Mario Biagioli, "Galileo the Emblem Maker," *Isis* 81 (1990): 230–58.

77. It is possible that the obelisk/gnomon at St. Peter's originally stood on a globe (F. Fiorini, conversation with author, June 1993). On Danti see *Dizionario biografico degli Italiani,* 32:659–62; Kemp, *Science of Art;* Florio Banfi, "The Cosmographic Loggia of the Vatican Palace," *Imago Mundi* 9 (1952): 23–34; and R. Thomassy, "Les papes géographes et la cartographie du Vatican," *Acta Cartographica* 7 (1970): 433–62.

78. Giorgio Vasari, *Ritratti dei piu famosi pittori. . . ,* vol. 5 (1550), quoted in Edward Luther Stevenson, *Terrestrial and Celestial Globes,* 2 vols. (1921; reprint, New Haven: Yale University Press for Hispanic Society of America, 1971), 1:161.

79. The true nature of the original scheme is somewhat obscured by repainting, in 1595 by Santucci and in the nineteenth century, using different tables from those employed by Danti. See Francesca Fiorani, "Post-Tridentine 'Geografia sacra': The Galleria delle carte geografiche in the Vatican Palace," *Imago Mundi* 48 (1996): 124–48.

80. Ibid., 138.

81. Walter S. Gibson, *"Mirror of the Earth": The World Landscape of the Sixteenth Century* (Princeton: Princeton University Press, 1989). "World landscape" is a direct English rendering of the German *Weltlandschaften;* an alternative is "cosmic landscapes." See also Wood, *Albrecht Altdorfer and the Origins of Landscape.*

82. Juan Vives, quoted in Gibson, *"Mirror of the Earth,"* 49–50.

83. Gibson, *"Mirror of the Earth,"* xx.

84. Erasmus, quoted in ibid., 57. Although their roots are different, the linguistic

similarities between *cosmos* and *cosmetic* were sometimes remarked upon to figure the earth as a jewel.

85. Gibson points out that Altdorfer almost certainly used a world map designed by Stabius and issued by Albrecht Dürer in 1515 as the source for his image. Ibid., 53; see also Wood, *Albrecht Altdorfer and the Origins of Landscape,* 19–23 ff.

86. Jardine, *Worldly Goods,* 392–94.

87. W. H. Auden's lines on the painting in "Musée des Beaux Arts" make the point:

> In Bruegel's *Icarus,* for instance: how everything turns away
> Quite leisurely from the disaster; the ploughman may
> Have heard the splash, the forsaken cry,
> But for him it was not an important failure; the sun shone
> As it had to on the white legs disappearing into the green
> Water; and the expensive delicate ship that must have seen
> Something amazing, a boy falling out of the sky,
> Had somewhere to go and sailed calmly on.

W. H. Auden, *Collected Poems,* ed. Edward Mendelson (London: Faber & Faber, 1976).

88. The incorporation of the classical cycle of social development into the painting is not surprising. Bruegel took his narrative from Ovid's account of the fall of Icarus in *Metamorphoses,* where these figures are mentioned. The four elements are also incorporated, used as part of the painting's moral emblematics: plowman, shepherd, fisherman, and mariners make appropriate use of the elements, while Icarus misuses the air to escape from the earth and suffers the consequences. R. Baldwin, "Peasant Imagery and Bruegel's *Fall of Icarus,*" *Kunsthistorisk Tidskrift* 55 (1986): 101–14.

89. Robert Karrow, *Theatrum: Map Makers of the Sixteenth Century and Their Maps: Bio-bibliographies of the Cartographers of Abraham Ortelius* (Chicago: University of Chicago Press, 1993).

90. On Ferrando (Fernando, Ferrante) Bertelli and his family of Venetian engravers, cartographers, and book merchants see *Dizionario biografico degli Italiani,* 9:490–94. Bertelli's bound map collections were not without a certain logic, beginning with at least one world map, the others organized according to proximity. But works of different provenances and styles were bound together. An example is the collection that once belonged to John Locke and is now held at the Harry Ransom Humanities Research Center at Austin, Texas (HRC F912 B461a 1553). It contains 83 engraved maps, including 2 of the world, on ninety-four sheets. Twenty-five are signed by Giacomo Gastaldi, 27 by F. Bertelli, 17 by Camoccio, and 14 by Forlani.

91. Gerardus Mercator states in his *Historia mundi or Mercator's atlas containing his cosmographical description of the fabricke and figure of the world,* trans. W. S. Generosus, ed. Jodocus Hondius (London: Sparke & Cartwright, 1635), unpaginated: "Seeing personall travels in these tempestuous times, cannot be attempted with any safety, here you may in the quiet shade of your studies travell at home." Given the date, midway through the Thirty Years' War, this was not bad advice.

92. Abraham Ortelius, *Theatrum orbis terrarum* (Antwerp, 1570); published in Ger-

man as *Theatrum oder schawbuch des erdtrem* (Antwerp, 1580) with 117 colored maps. Mekerchus's poem reads in Latin:

Ortelius, quem quadrijugo super aëra curru
Phoebus Apollo vehi secum dedit, unde iacentes
Lustraret terres circumfusunmque profundum.
Hine olli Phoebum, qui conspicit omnia, prorsus
Ignotas, alio penitens penitusque sub Orbis
Axe sitas monstrasse plagas, solisque retectas
Indigenis, Orbemque novum gentesque, virosque
Detexisse ferunt, mundisque arcana remoti.

93. The use of an arch to signify entrance to a distinct knowledge space, initiated by Sebastiano Serlio in 1537, was common by this time. See Jonathan J. G. Alexander, *The Printed Page: Italian Renaissance Book Illumination, 1450–1550* (London: Prestel for the Royal Academy, 1994); M. Corbett and R. Lightbrown, *The Comely Frontispiece: The Emblematic Title Page in England, 1550–1660* (London: Routledge & Kegan Paul, 1979).

94. The figures of the continents are partially explained by Ortelius himself. Their iconography was fixed in 1593 by Cesare Ripa's *Iconologia*. Ripa himself was closely associated with Egnazio Danti.

95. Montaigne's "Of the caniballes" appeared in his first volume of essays, published in 1580. *The Essays of Montaigne,* trans. John Florio, vol. 1 (London: Everyman Library, Dent, 1910), 215–29.

96. "Quid ei potest videri magnum in rebus humanis cui aeternitas omnis totiusque mundi nota sit magnitudo."

97. "Equus vehendi causa, arandibos, venandi et custodiendi canis, homo autem ortus ad mundum contemplandum"; "Hoc est punctum quod inter tot gentes ferro et igni dividitur, o quam ridiculi sunt mortalium termini." These juxtaposed quotations from Seneca refer indirectly to the story of Heraclitus and Democritus, regularly referred to throughout the sixteenth century. The other medallions on the revised world map are also from Seneca: "If only the entire face of the world would appear in this form: then the whole of philosophy would confront us" [Utinam quem ad modum universa mundi facies in conspectum venit. Ita philosophia tota nobis posset occurrere] and "In conformity with this law were created men, who would witness that globe at the center which you inhabit, in this temple called earth" [Homines hac lege sunt generati, qui tuerrentur illum globum quem in hoc templo medium vives quae terra digitur].

98. J. Rabasa, "Allegories of the *Atlas,*" in *Europe and Its Others,* ed. Francis Barker et al., vol. 2 (Essex Sociology of Literature Conference, University of Essex, Colchester, 1985), 1–16. The full title of Mercator's work is *Atlas sive cosmographicae meditationes de fabrica mundi et fabricati figura.* The quotation is from the 1636 English edition, quoted in Rabasa, "Allegories of the *Atlas,*" 7.

99. Mangani, "Abraham Ortelius and the Hermetic Meaning of the Cordiform Projection."

100. Georg Braun and Franz Hogenberg, eds., *Civitates orbis terrarum,* 6 vols. (Cologne, 1572–1618). Like Ortelius's *Theatrum,* this was a vast undertaking, drawing upon chorographic maps from across Europe and published in the major European languages. The frontispieces of the six volumes offer a narrative of human social evolution from the crudest shelter, through the beginnings of cultivation, the attempt to construct Babel, the fruits of civilization, and the discovery of the continents, to the conflicts of peace and war. Lucia Nuti, "Alle origini del *grand tour:* Immagini e cultura della città italiana negli atlanti e nelle cosmograpfie del secolo XVI," in *Cultura del viaggio: Ricostruzione storico-geografico del territorio,* ed. Giorgio Botta (Milan: Unicopli, 1989), 209–52.

101. Lucia Nuti, "The Mapped Views by Georg Hoefnagel: The Merchant's Eye, the Humanist's Eye," *Word and Image* 4 (1988): 545–79.

102. *De locis et mirabilis mundi* first appeared in Petrus de Turre's 1490 Rome edition of the *Geography.*

103. Chapter 2 describes Eden; chapter 3, Asia; chapter 17, the Fortunate Isles: "propter soli fecunditatem easdem est paradisum putaverunt." See Pagden, *European Encounters with the New World* (see ch. 1, n. 34); Campbell, *Witness and the Other World* (see ch. 3, n. 30); and Hulme, *Colonial Encounters* (see ch. 1, n. 34).

104. Helms, *Ulysses' Sail* (see ch. 1, n. 26); Loren Bairitz, "The Idea of the West," *American Historical Review* 66 (1961): 618–40.

105. Jonathan Sawday, in *Body Emblazoned: Dissection and the Human Body in Renaissance Culture* (London: Routledge, 1995), 16–32, discusses the Renaissance counterposition of a sacred soul to a corrupt body and the metaphorical and actual links between mapping the body and mapping territory.

106. Pagden, *European Encounters with the New World,* 186.

107. On Columbus's literary use of medieval Romance and texts such as *De locis et mirabilis mundi* see Campbell, *Witness and the Other World,* 166–204: "Columbus' epistles from the New World acted as a prism through which an old cosmography was refracted into the colors of romance" (204). See also Hulme, *Colonial Encounters,* ch. 1; and Phillips and Phillips, *Worlds of Christopher Columbus* (see ch. 4, n. 14).

108. See the discussion in Pagden, *European Encounters with the New World,* 19–20. In a letter to Juana de la Torre, written during his third voyage, Columbus said: "Of the new heaven and the new earth, which Our Lord made, as St John writes in the Apocalypse, after he had spoken of it by the mouth of Isaiah, He made me the messenger and He showed me where to go" (quoted in Campbell, *Witness and the Other World,* 189–90).

109. Las Casas was the first bishop to the Indies. His *Brevíssima relación de la destrucción de las Indias occidentales,* of 1523, recounted in graphic and statistical detail the cruelties and injustices done by Spaniards in the New World. His writings became the basis for the "Black Legend," used as Protestant propaganda against the papacy and by English imperialists to cast themselves as redeemers of a fallen New World.

110. For accessible selections in English see Michael Alexander, ed., *Discovering the New World* (London: London Editions, 1976). The original volumes have been reprinted by the Hakluyt Society.

111. Early humanists such as Celtis in Germany and chorographers in Protestant nations such as John Speed and William Saxton in England idealized pre-Roman inhabitants as a counter to Italian cultural hegemony.

112. Frank Lestringant, "Une cartographie iconoclastique: 'La mappe-monde nouvelle papistique' et Jean Baptiste Trento (1566–1567)," in Pelletier, *Géographie du monde,* 99–120 (see ch. 4, n. 27).

113. Francis Bacon, "New Atlantis," in *The English Works of Francis Bacon,* vol. 1 (London: Methuen, 1905), 169.

Chapter Six. Emblematic Globe and the Poetics of the World

1. Henry Peacham, *Minerva Britannia, or a Garden of Heroic Devises, Furnished, and Adorned with Emblemes and Impressas of Sundry Natures, Newly Devised, Moralised, and Published by Henry Peacham, Mr of Artes* (London: Dight, 1612), emblem 26.

2. Examples of unresolved geographical questions included California's island status, the existence of the Strait of Anian between America and Asia, and the size and shape of Magellanica, the southern continent.

3. The longitude problem is discussed in chapter 7.

4. Stephen Toulmin, *Cosmopolis: The Hidden Agenda of Modernity* (Chicago: University of Chicago Press, 1990).

5. Kemp, *Science of Art* (see ch. 1, n. 37). The idea that a scientific revolution pioneered by Rene Descartes, Francis Bacon, Johannes Kepler, and Isaac Newton, based on systematic experimental method and empirical proof backed by mathematical logic, rapidly replaced speculation has been significantly revised among historians of science. See B. Vickers, *Occult and Scientific Mentalities in the Renaissance* (Cambridge: Cambridge University Press, 1984); and the discussion in Livingstone, *Geographical Tradition,* 14–23 (see ch. 1, n. 10).

6. The Tetragrammaton is the Hebrew script of the Creator's name in which letters also have numerical values. Unutterable, the Tetragrammaton was believed to hold profound secrets, accessible by numerical speculation, and to possess miraculous qualities. It was thus a common feature of emblems. Eco, *Search for a Perfect Language,* 74–75, 80–85 (see ch. 3, n. 26).

7. Hallyn, *Poetic Structure of the World,* 183–202 (see ch. 1, n. 4).

8. Denis Cosgrove, *The Palladian Landscape* (Leicester: Leicester University Press, 1993), 53–54.

9. John Martin, "Salvation and Society in Sixteenth-Century Venice: Popular Evangelism in a Renaissance City," *Journal of Modern History* 60 (1988): 205–33.

10. Cosgrove, *Palladian Landscape,* 226–32.

11. Corbett and Lightbrown, *Comely Frontispiece,* 185–89 (see ch. 5, n. 93); Livingstone, *Geographical Tradition,* 32–62.

12. Each challenged Aristotelian order: the tides made water rise toward the air; the volcano brought fire out of the earth, mixing spheres clearly separated in nature; the Nile's flood brought water from torrid to temperate zones in the midst of the summer drought.

13. Girolamo Fracastoro to Gianbattista Ramusio, 25 January 1533, quoted in Stevenson, *Terrestrial and Celestial Globes,* 1:137 (see ch. 5, n. 78).

14. On Zeno's collection of globes and maps see Manfredo Tafuri, *Venice and the Renaissance* (Cambridge: MIT Press, 1990), 121 n. 97. Zeno's 1558 claim (published in Giovanni Battista Ramusio, *Navigationi et viaggi,* 3 vols. [Venezia: Giunti, 1563–1606], vol. 1) that his ancestors had made a fourteenth-century Venetian voyage to America, supported by an almost certainly fabricated map, was accepted by Ortelius, who published the map from Francesco Marcolini's *Dei commentati del viaggio in Persia di M. Caterino Zeno & delle guerre fatti nel Imperio Persiano* . . . (Venice, 1558), which charts Greenland with remarkable accuracy. It was used by the English explorer Martin Frobisher in 1576. See W. Hobbs, "Zeno and the Cartography of Greenland," *Imago Mundi* 6 (1950): 15–19.

15. The academy's "objective . . . was pursued through . . . an almost obsessive concern for the thoroughness and universality of materials, above all for their definitive ordering." Tafuri, *Venice and the Renaissance,* 117. Fracastoro proposed the renewal of Venice through the remodeling of its physical environment and turning the lagoon into a freshwater lake on the model of Tenochtitlan, its New World equivalent (152). The concept of renewal had broad cultural purchase in sixteenth-century Europe, applied in equal measure by Reformers and post-Tridentine Catholics to spiritual matters and by utopians such as Tommaso Campanella and Francis Bacon to social and intellectual concerns. See Hallyn, *Poetic Structure of the World,* 282; and Toulmin, *Cosmopolis,* 45–87.

16. The rooms of the academy were named after parts of the human body. On the membership and work of the Accademia della Fama see Michele Maylender, *Storia delle accademie d'Italia,* vol. 5 (Bologna: Licinio Capelli, 1930), 436–43; Tafuri, *Venice and the Renaissance,* 114–15; A. Olivieri, "L'intelletuale e le accademie fra '500 e '600: Verona e Venezia," *Archivio Veneto* 130 (1988): 31–56, esp. 37–41.

17. Eugenia Bevilacqua, "Geografi e cosmografi," in *Storia della Cultura Veneta* 3, pt. 2, *Dal Primo Quattrocento al Consilio di Trento,* 1986, 355–74.

18. Michael Bury, "Engraved Maps and Prints in Sixteenth Century Venice" (lecture delivered at the National Gallery of Scotland, Edinburgh, December 1990). Horatio F. Brown, *The Venetian Printing Press, 1469–1800: An Historical Study Based upon Documents for the Most Part Hitherto Unpublished* (Amsterdam: Van Heusden, 1969), 102–5, points out that the presence of Arab, Greek, and Armenian scholars in Venice broadened the scale and interconnections of its cartographic and geographical publishing. David Woodward, *Maps as Prints in the Italian Renaissance: Makers, Distributors, and Consumers,* 1995 Panzini Lectures (London: British Library, 1996).

19. Ramusio, *Navigationi et viaggi;* Milanesi, *Tolomeo sostituito* (see ch. 4, n. 44).

20. S. Grande, "Le relazioni geografici fra P. Bembo, G. Fracastoro, G. B. Ramusio, G. Gastaldi," *Società Geografica Italiana* 12 (1905): 93–197.

21. Giacomo Gastaldi, *Universale descrittione del mondo* (Venice: Francesco de Tomaso di Sato, 1561). Gastaldi based his calculations on a division of the globe into 24 meridians of 15 degrees each.

22. Enrico Peruzzi, "Note e ricerche sugli 'Homocentrica' di Girolamo Fracas-

toro," *Rinascimento* 25 (1985): 247–68. Fracastoro had studied at Padua in the same years as Copernicus, but no direct connection has been established between them, although a fragment from Fracastoro's writings before 1531 ("sed jam [nos] qui nec longe abest, pulcherrimus rerum, sol nos vocat: sol, mundi lumen ac vitae pater, atque omni universo veluti cor, in medio planetarum situs") is tantalizingly close to Copernicus's words in *De revolutionibus orbis:* "In medio vero omnium residat Sol. Quis enim in hoc pulcherrimo templo lampadem hunc in alio vel meliori loco poneret, quam unde totum simul possit illuminare?" (quoted in ibid., 251). Fracastoro's correspondents included Oviedo, Aretino, and members of the Farnesi, Medici, and Spanish Habsburg courts.

23. Girolamo Fracastoro, *Opera omnia* (Venice: Giunta, 1555); Wm. van Wyck, *The Sinister Shepherd: A Translation of Girolamo Fracastoro's Syphilidis sive de morbo gallico libri tres* (Los Angeles: Primavera, 1934), quotation on 24. The poem, on p. 29, adopts a globalizing theme similar to Camões's:

> Over the waves, sea conquerors have thrust
> Their prows to find immensities beyond
> The sea of forbears, hardly more than pond.
> When Poseidon's proud empire was thus ploughed,
> Cape Verde became an unimportant goal;
> De Gado's Cape and Fortunate Isles less proud,
> All swept by storms assailing from the Pole
> Zambesi's banks were found to be a rift,
> Kerman brought many an unexpected gift.
> Ganges and Indus soon were found to be
> No universal limits. Men were free
> To push beyond these barriers of the world,
> Toward Phoebus and his morning flag unfurled.

24. Pietro Bembo, *Gli asolani,* trans. Rudolf B. Gottfried (1954; reprint, Freeport, N.Y.: Books for Libraries, 1971), 113. On Bembo, see Giancarlo Mazzacurati, "Pietro Bembo," *Storia della Cultura Veneta* 3, pt. 2, *Dal Primo Quattrocento al Consilio di Trento,* 1985, 1–59. Bembo died in 1547.

25. Pietro Bembo, *De Aetna: Il testo di Pietro Bembo tradotto e presentato da Vittoria Enzo Alfieri* (Palermo: Sellerio, 1981).

26. Pietro Bembo, *Pietro Bembo cardinales historiae venetae libri XII* (Venice: Aldi Filos, 1591), 6:82–99.

27. Pietro Bembo, *Della historia Venitiana libri XII* (Venetia: Giordao Zileti, 1570), 73r.

28. The group was contemporary with Navagero's academy on Murano, which debated the natural history of American plants. The Venetian patrician Daniele Barbaro was responsible for establishing the botanical garden at the University of Padua about 1545 with New World specimens. Lucia Tongiorgio Tomasi, "Geometric Schemes for Plant Beds and Gardens: A Contribution to the History of the Garden in the Sixteenth and Seventeenth Centuries," in *World Art: Themes of Unity in Diver-*

sity, ed. Irving Lavin (University Park: Pennsylvania State University Press, 1989), 211–15. Globes and maps were among the most popular decorations of Venetian *studioli.* F. Ambrosini, "'Descrittioni del mondo' nelle case venete dei secoli XVI e XVII," *Archivio Veneto* 112 (1981): 67–79. Ambrosini notes that pairs of globes, world maps, and collections of maps of the four continents increased in significance among Venetian patrician households into the seventeenth century. See also Woodward, *Maps as Prints in the Italian Renaissance.*

29. Girolamo Fracastoro, "Sive de anima (ad Ioannem Baptistam Rhamnusium)," in *Opera omnia,* 149–62.

30. Grande, "Le relazioni geografici," 126–30; Giuliano Luccetta, "Viaggiatori e racconto di Viaggi nel cinquecento," *Storia della Cultura Veneta* 3, pt. 2, 1985, 487–88. On the intellectual ramifications of the Nile's mysteries at this time, see Schama, *Landscape and Memory,* 252–55 (see ch. 1, n. 43).

31. Fracastoro, *Opera omnia,* 75.

32. Ambrosini, "Descrittioni del mondo," 72, describes the method of representing the globe in a typical patrician Venetian house: "Generally, paintings were favored that, either singly or together, succeeded in placing before the observer's eye an image of the world in its totality: thus, planispheres or four images each representing an individual continent."

33. Giuseppe Rosaccio, *Teatro del cielo e della terra* (Venice, [c. 1598]); idem, *Le sei età del mondo di Gioseppe Rosaccio con Brevità Descrittione* (Venice, 1595).

34. Giuseppe Rosaccio, *Fabrica universale dell'huomo . . .* (1627). Among Rosaccio's more than forty published works were a pilgrim's guide, *Viaggio da Venetia a Constantinople* (1598), and a world map, *Universale descrittione del mondo,* first published in 1597 but reprinted in 1647 decorated with ethnographic illustrations taken from Theodore de Bry. On Rosaccio see Giuliano Luccetta, "Viaggiatori, geografi e racconti di viaggio dell'età barocca," *Storia della Cultura Veneta* 4, pt. 2, 1985, 201–2.

35. Ramusio, quoted in Luccetta, "Viaggiatori, geografi e racconti," 489.

36. On Ruscelli, see William Eamon and Françoise Paheau, "The Accademia Segreta of Girolamo Ruscelli: A Sixteenth-Century Italian Scientific Society," *Isis* 75 (1984): 327–42.

37. Girolamo Ruscelli, *La geografia di Claudia Ptolomeo Alessandrino nuovemente tradutta di greco in italiano da Girolamo Ruscelli* (Venice: Vincenzo Valgrisi, 1561): "Ove l'huomo per se stesse col solo conoscimento, o guidicio humano, non potrebbe mai have piena contezza di tutta la terra, l'arte della Geografia, con l'aiuto della medesima scienze matematiche, che la dimonstra, & fa comprendere, non già in se tutta, cioè la terra stessa nell'esser suo, ma per la figura, o sembianza, come si vede nei globi, ò nelle palle, & ne i mappamondi" (11).

38. Peter Wagner, *Reading Iconotexts* (London: Reaktion, 1995).

39. Girolamo Ruscelli, *Tavola universal nuova, con descrittione di tutto il mondo,* bound into Ruscelli, *La geografia.* The map was probably constructed by the Sanuto brothers, who also produced a set of globe gores.

40. Shirley, *Mapping of the World* (see ch. 5, n. 25).

41. Girolamo Ruscelli, *Annotationi et avvertimenti di Girolamo Ruscelli sopra i luoghi*

difficili et importanti del furioso (Venice:VincenzoValgrisi, 1558); idem, *Le imprese illustri con espositioni, et discorsi del S.or Ieronimo Ruscelli al serenissimo et sempre felicissimo re Catolico Filippo d'Austria* (Venice: Francesco de Franceschi Senese, 1566).

42. Ludovico Ariosto, *Orlando furioso,* trans. John Harrington (London, 1591). Ruscelli's original reads:"Nelle figure, avvertano ancor quei che non sanno le regole della pictura, ch'elle non son fatte tutte con molta ragione di perspettiva, & che da piede di tutto il quadro le figure di gli huomini de'cavalli, & dell'altri cosi sono fatte più grandi, & poi quanto più vanno verso l'alto, più si vengono diminuendo. E queste perchè quelle figure che nel foglio stannno cosi colcate si imaginano nella perspettiva che stiano in piedi, & chi tiene il libro in mano viene ad have li più basse per più vicine à lui, & cosi à dilungarseli di mano in mano."

43. Frank L. Borchardt,"The Magus as Renaissance Man," *Sixteenth Century Journal* 21 (1990): 57–86.

44. Peacham, *Minerva Britannia,* preface.

45. E. H. Gombrich, "*Icones Symbolicae:* Philosophies of Symbolism and Their Bearing on Art," in *Symbolic Images: Studies in the Art of the Renaissance* (London: Phaidon, 1975), 123–91; Carlo Ginzburg, *Myths, Emblems, Clues* (London: Hutchinson Radius, 1990).

46. Silvia Ferreti, *Cassirer, Panofsky, and Warburg: Symbol Art and History,* trans. Richard Pierce (New Haven:Yale University Press, 1989), 5.

47. In *Le imprese illustri* Ruscelli emphasizes the importance of maintaining a just proportion between material and spiritual matters, not so obscure as to require an oracle to interpret them nor so obvious that any "plebean" could interpret their meaning.

48. Carlo Ginzburg,"The High and the Low:The Theme of Forbidden Knowledge in the Sixteenth and Seventeenth Centuries," in Ginzburg, *Myths, Emblems, Clues,* 60–76.

49. Charles Moseley, *A Century of Emblems* (Aldershot: Scolar Press, 1989), 7–8.

50. There were ninety editions of Alciati alone between 1531 and 1600. Ibid., 15; Florini, *Iconologia di Cesare Ripa Perugina* (see ch. 1, n. 12). Ripa's work was written c. 1540, and an illustrated edition appeared in 1600.

51. Moseley, *Century of Emblems,* contains a useful anthology. See also Elizabeth Watson, *Achille Bocchi and the Emblem Book as Symbolic Form* (Cambridge: Cambridge University Press, 1993).

52. Florini, *Iconologia di Cesare Ripa Perugino.*

53. Cesare Ripa, *Iconologia* (Perugia, 1600).

54. Moseley, *Century of Emblems,* 26.

55. Biagioli,"Galileo the Emblem Maker" (see ch. 5, n. 76).The Medicean coat of arms had six spheres, allowing an analogy with Jupiter's satellites. Cosimo made much of the connections between his name and *cosmos,* and "what was new about Galileo's translating scientific concepts into the discourse of the court . . . was that he did so also as an attempt to legitimise scientific theories" (242). Allegorical meaning here precedes scientific meaning.

56. Ruscelli, *Le imprese illustri,* 15r (Phillip II), 23v (Ferdinand), 29r-v (Henry II).

Ruscelli uses Henry's emblem to introduce a summary discussion of the soul's ascent "di Cielo in Cielo, & di grado in grado fin a Dio" [from sphere to sphere and from step to step as far as God] and the downward flow of celestial influences onto the earthly globe, supporting his thesis by reference to both Plato and the story of Jacob's ladder.

57. Corbett and Lightbrown, *Comely Frontispiece,* 43. On Lady World see Richard Helgerson, "Folly of Maps and Modernity" (see ch. 1, n. 14).

58. Sir Walter Ralegh, *The History of the World in Five Books* (London, 1614). See also Buhler, "Marsilio Ficino's *De stella magorum* and Renaissance Views of the Magi," 364 (see ch. 5, n. 5). On the threshold image in Renaissance title pages see Norman K. Falmer Jr., "Renaissance English Title-Pages and Frontispieces: Visual Introductions to Verbal Texts," *Proceedings of the IXth Congress of the International Comparative Literature Association, Innsbruck, 1979* (Innsbruck: Innsbrücker Beiträge zur Kulturwissenschaft, 1981), 61–65.

59. *Twelfth Night* 3.2.79–80. The *Harvard Concordance* gives ten other references to globes in Shakespeare's work, ranging from direct references to the physical earth to references to various parts of the body (head, eye, breasts) and to a female body in a long sustained metaphor on the global distribution of lands and seas in *The Comedy of Errors* 3.2.114 ff. John Donne's literary use of the globe is found in "The Good Morrow" and the "Holy Sonnets." See Gillies, *Shakespeare and the Geography of Difference* (see ch. 2, n. 19); and Sawday, *Body Emblazoned,* 22–32 (see ch. 5, n. 105).

60. See, e.g., the two plates entitled *Spechio della vita humana,* an allegory on the ages of man with rhymed legends produced in Venice by Ferrando Bertelli in 1566, held in the Popular Imagery Collection of the Harry Ransom Humanities Research Center at the University of Texas at Austin, no. 29; and Giulio Sanuto's sea monster, which appears on both his globe and his engraving of Titian's *Perseus and Andromeda.*

61. The multiplicity of languages testified to the arbitrary relation between words and things. The confusion of tongues was traced to the time of Babel, replacing an Adamite language that truly expressed the nature of things. These beliefs explain the significance of Egyptian hieroglyphics, whose antiquity and fusion of word and image seemed to represent an original language. Eco, *Search for a Perfect Language.*

62. Moseley, *Century of Emblems,* 6.

63. Alpers, *Art of Describing,* esp. 229–33 (see ch. 1, n. 9).

64. Joseph Moxon studied under Joan Blaeu in Amsterdam. He was hydrographer to King Charles II and a fellow of the Royal Society. He traded in scientific instruments and maps at his shop, Sign of the Atlas. His text, *A Tutor to Astronomy and Geography, or, an easie and speedy Way to understand the Use of both the Globes, celestial and terrestrial,* went through a number of editions.

65. Mercator, *Atlas sive cosmographicae,* published in English as *Historia mundi or Mercators atlas. . . ,* preface p. 3 (see ch. 5, n. 91).

66. Mercator, *Historia mundi or Mercators atlas,* 1.

67. Ibid.

68. Alpers, *Art of Describing,* xxv. Alpers reproduces emblems from Jacob Cats, *Silenus Alcibiades* (Middelburg, 1618), among the most widely read Dutch emblem

books (232). See also Simon Schama, *The Embarrassment of Riches: An Interpretation of Dutch Culture in the Golden Age* (London: Collins, 1987).

69. Alpers, *Art of Describing,* 166–67.

70. Hallyn, *Poetic Structure of the World,* 231–52.

71. Delano-Smith and Ingram, *Maps in Bibles,* xxix (see ch. 5, n. 42).

72. Marijke de Vrij, *The World on Paper: A Descriptive Catalogue of Cartographical Material Published in Amsterdam during the Seventeenth Century* (Amsterdam: Totius Orbis Terrarum, 1967), 13. On Willem Blaeu see C. Koeman, "Life and Works of Willem Janzoon Blaeu: New Contributions to the Study of Blaeu Made during the Last Hundred Years," *Imago Mundi* 26 (1972): 9–16.

73. "Toutes les creatures visibles que Dieu a faites, sont comprises en ces deaux icy, l'Homme et le Monde. Cestuy-là a esté establi Seigneur de l'unvers: cestuy-ci est le siege de son Empire. Cestui-là est l'hoste et habitant du monde: cestuy-cy est la tres-magnifique et spatieuse maison d'un si grand hoste. Nous recognoissons en l'homme l'image de c'est excellent ouvrier qui l'a crée: & au onde celle de l'homme." Jan Jansson, *Nouvel atlas ou théâtre du monde,* 5 vols. (Amsterdam, 1646).

74. John Dryden, *The Works of John Dryden,* ed. Edward Niles Hooker and H. T. Swedenborg Jr., vol. 1 (Berkeley and Los Angeles: University of California Press, 1961), 59.

75. Joan Blaeu, *Atlas maior, sive cosmographia Blaviana, qua solum, salum, accuratissima describuntur* (Amsterdam, 1662). The French edition, *Le Grand Atlas; ov, Cosmographie Blaviane, en laqvelle est exactement descritte la terre, la mer, et le ciel* (Amsterdam: Blaeu, 1663), was dedicated to Louis XIV.

76. See Johannes Keuning, "Blaeu's Atlas," *Imago Mundi* 14 (1959): 74–89. The Dutch and German editions appeared in 9 volumes, the Latin, in 11 volumes; and the French, in 12.

77. Eleven copies still exist, including two in Tokyo. Minako Denbergh, "A Comparative Study of Two Dutch Maps, Preserved in the Tokyo National Museum: Joan Blaeu's Wall Map of the World in Two Hemispheres, 1648, and Its Revision *ca.* 1678 by N. Visscher," *Imago Mundi* 35 (1983): 20–36. The copy discussed here is of the first edition, at the University of Texas at Austin.

78. Toulmin, *Cosmopolis.* In 1553 Francesco Patrizi, a member of the Accademia della Fama, had written *La Citta Felice,* upon which Campanella's utopian *City of the Sun* was in part modeled.

79. Schama, *Landscape and Memory,* 303.

80. Emblem making was a typical Jesuit art of the seventeenth century, for example, *Imago primi seculi societatis Jesu a provincia Flandro-Belgica eiusdem societatis representata* (Antwerp: Malthasaris Moretti, 1640).

81. Jonathan D. Spence, *The Memory Palace of Matteo Ricci* (London: Faber & Faber, 1983).

82. The dome is an obvious, cross-cultural symbol of heaven. See K. Lehmann, "The Dome of Heaven," *Art Bulletin* 27 (1945): 1–24. An early example of its cosmographic decoration is the Genesis cupola of St. Mark's Basilica at Venice. Raphael initiated a tradition of illuminism in cupola decoration, especially in northern Italy

(Pozzo himself came from Trent), which peaked in Tiepolo's decorative ceilings, for example, at the Villa Maser and Würzburg. See Maria Dalai Emiliani, ed., *La prospettiva rinascimentale: Codificazioni e trasgressioni* (Florence: Centro Di Studi Rinascimentale, 1977).

83. "I came to bring light to the world." In his *Perspectiva pictorum et architectorum,* of 1702, Pozzo described the work thus: "Jesus illumines the heart of St. Ignatius with a ray of light, which is then transmitted by the Saint to the furthermost corners of the four quarters of the earth, which I have represented with their symbols in the four sections of the vault."

84. Valerie Shrimplin-Evangelides, "Sun Symbolism and Cosmology in Michelangelo's Last Judgement," *Sixteenth Century Journal* 21 (1990): 607–44.

85. Kircher had had a primary role in restoring St. Ignatius's companion Jesuit church in Rome, the Gesù, and as a professor of mathematics he was resident from 1638 at the college to which the Gesù is attached. He instructed Nicholas Poussin in perspective. Although Pozzo's work dates from half a decade after Kircher's death, Kircher's intellectual influence on Pozzo's decorative scheme is highly apparent. Many of Kircher's images are reproduced in Joscelyn Godwin, *Athanasius Kircher: A Renaissance Man and the Quest for Lost Knowledge* (London: Thames & Hudson, 1979).

86. See Alpers, *Art of Describing,* 1–25, on Christian Huygens and the telescope.

87. Paula Findlen, *Possessing Nature: Museums, Collecting, and Scientific Culture in Early Modern Europe* (Berkeley and Los Angeles: University of California Press, 1994), 81.

88. Ibid., 93.

89. Eco, *Search for a Perfect Language.* According to Plotinus, the hieroglyphics inscribed on obelisks expressed the "essence of things," as did the seventeenth-century emblem. Valeriano's *Hieroglyphica* (1556) comprised fifty-eight books of images and symbols widely used throughout Europe. Hieroglyphics continued to appeal as a mystical language until their deciphering in the 1820s.

90. The work was initially published in Rome and subsequently published under the title *Iter exstaticum* (Würzburg, 1660, 1671). Kircher based his maps of the Sun and the Moon (represented in the frontispiece to his *Ars magna lucis* as Apollo and Diana) and his eclipse diagrams on the widely reproduced images of his fellow Jesuit Christopher Scheiner, which were used, for example, in Carel Allard's *Planispherii coelestis hemisphaerum septrentionale* (Amsterdam, 1700).

91. Kepler's work was titled *Somnium seu de astronomia lunari.* It contains a view of Earth from the Moon: "The most beautiful of all the sights on Levania [the Moon] is the view of its Volva [Earth]. This they enjoy to make up for our moon." Quoted in Hallyn, *Poetic Structure of the World,* 272.

92. Athanasius Kircher, *Ars magna sciendi* (Amsterdam: Jannzon, 1669), 2. Hallyn, *Poetic Structure of the World,* 250–51, points out that unlike Kepler, Kircher did not take mathematics or music as logical principles underlying universal harmony but worked through analogy and symbol, appealing to imagination rather than intellect: "The title says it well: *ars.* All the knowledge, all the technique of the period are put in the

service of the production of a theatre of marvels directed principally to the imagination" (250). See also Schama, *Landscape and Memory*, 302.

93. Mangani, "Abraham Ortelius and the Hermetic Meaning of the Cordiform Projection," 71.

94. The dedicatory texts of the celestial globe are printed in Vincenzo Coronelli, *Atlante veneto, nel quale si contiene la descrittione geografica, storica, sacra, profana, e politica, degl'imperÿ, regni, provincie, e stati dell'universo, loro divisione, e confini, coll'aggiunta di tutti li paesi nuovamente scoperti, accresciuto di molte tavole geografiche, non più publicate*, vol. I (Venice, 1690).

95. Vincenzo Coronelli, quoted in Jacob, *L'empire des cartes*, 449 n. 112 (see ch. 1, n. 28).

96. François le Large, "Receuil des inscriptions, des remarques historiques et géographiques qui sont sur le globe terrestre de Marly exécuté par le P. Coronelli," 2 vols., Paris BN, MSS francais 13.365 and 13.366. I am grateful to Christian Jacob for a copy of a transcription by Gabrielle Duprat.

97. Petite Academie, quoted in Monique Pelletier, "Les globes du Louis XIV: Les sources françaises de l'oeuvre de Coronelli," *Imago Mundi* 34 (1982): 75.

98. Tommaso Campanella, *The City of the Sun*, trans. A. M. Elliott and R. Millner (London: Journeyman, 1981), 15. Each of the city's circles acts as the space for the taxonomic arrangement and display of natural phenomena—stones, flora, birds, land fauna—with the mechanical arts and lawmakers in the outer circles and mathematical figures in the innermost.

99. The link with Campanella's text is made by Pelletier, "Les globes du Louis XIV."

100. Connections between Neoplatonism, hermeticism, utopian thought, and contemporary scientific development in early modern Europe were the lifework of Frances Yates, from *Giordano Bruno and the Hermetic Tradition* (see ch. 5, n. 18) to her summary work, *The Occult Philosophy in the Elizabethan Age* (London: Routledge & Kegan Paul, 1979). A large critical literature has subsequently developed.

101. On relations between Le Nôtre and mapping see Thierry Mariage, *The World of André le Nôtre* (Philadelphia: University of Pennsylvania Press, 1999).

102. Schama, *Landscape and Memory*, 338–44, discusses the hydrogeography of Versailles.

103. Monique Pelletier, "Les globes de Marly, chefs-d'oeuvre de Coronelli," *Revue de la Bibliothèque Nationale* 47 (1992): 46–91.

104. Pelletier, "Les globes du Louis XIV," 88.

105. Stevenson, *Terrestrial and Celestial Globes*, 2:78.

106. Christian Jacob, "Lire les globes" (mimeo).

107. Le Large, "Receuil des inscriptions," 334–36.

108. Ibid., 119–24.

109. The impresa of the Argonauti was designed by Girolamo Antonio Parisotti of Castelfranco. See *Il P. Vincenzo Coronelli dei frati minori conventuali 1650–1718 nel III centenario della nascità* (Rome: Miscellanea Francescana, 1951) for detailed information on Coronelli.

Chapter Seven. Enlightened Globe

Epigraph: Immanuel Kant, *Principles of Lawful Politics: Immanuel Kant's Philosophic Draft "Toward Eternal Peace,"* trans. Wolfgang Schwarz (Aalen, Germany: Scientia, 1985).

1. In *The Music Exam,* now in the Gemäldegalerie in Berlin, the female subject also performs for male tutors, whose more than merely professional interest is indicated not only by the direction of their glances and gestures but quite explicitly by the nude, mythical scene hanging above the group.

2. The distinction between the gaze and the glance is discussed in Jay, "Scopic Regimes of Modernity."

3. The term *planetary consciousness* is taken from Mary Louise Pratt, *Imperial Eyes: Travel Writing and Transculturation* (London: Routledge, 1992), 29–30; see also Alpers, *Art of Describing,* 122 (see ch. 1, n. 9). Familiarity with globes and skill in their measurement and manipulation had become significant attributes for gentlemen in the later sixteenth century: "Knowledge of globes evidently became a fashionable ornament, and acquaintance with the sumptuous sets issued by Emery Molyneux a gentlemanly ornament." Stephen Johnston, "Mathematical Practitioners and Instruments in Elizabethan England," *Annals of Science* 48 (1991): 335. When such skill became regarded as appropriate for females also is not clear, but it was probably an eighteenth-century phenomenon.

4. Charles W. J. Withers and David N. Livingstone, "On Geography and Enlightenment," in *Geography and Enlightenment,* ed. Withers and Livingstone (Chicago: University of Chicago Press, 2000), 1-32; Dorinda Outram, *The Enlightenment* (Cambridge: Cambridge University Press, 1995).

5. David Fraser, "Fields of Radiance: The Scientific and Industrial Scenes of Joseph Wright," in *The Iconography of Landscape: Essays on the Symbolic Representation, Design, and Use of Past Environments,* ed. Denis Cosgrove and Stephen Daniels (Cambridge: Cambridge University Press, 1988), 119–41.

6. Sennett, *Flesh and Stone,* 293–94 (see ch. 1, n. 58).

7. Isaac Newton, "Queries," from *Optiks* (1717), quoted in Toulmin, *Cosmopolis,* 118 (see ch. 6, n. 4); Richard S. Westfall, *Never at Rest: A Biography of Isaac Newton* (Cambridge: Cambridge University Press, 1980); idem, "Newton and Alchemy," in Vickers, *Occult and Scientific Mentalities in the Renaissance,* 315–36 (see ch. 6, n. 5).

8. Westfall, "Newton and Alchemy," 317, 322.

9. Fraser, "Fields of Radiance"; Steven Shapin and Simon Schaffer, *Leviathan and the Air Pump: Hobbes, Boyle, and the Experimental Life* (Princeton: Princeton University Press, 1985). Dorinda Outram reproduces Wright's *Air pump* of 1768 as the cover illustration for her book *The Enlightenment.*

10. Barbara Stafford, "Desperately Seeking Connections: Four Scenes from Eighteenth-Century Laboratory Life," *Ecumene: Environment, Culture, Meaning* 2 (1995): 378–98; Keith Thomas, *Religion and the Decline of Magic* (London: Weidenfeld & Nicolson, 1971).

11. Outram, *Enlightenment,* 80–95, reproduces an image that is very different from

Longhi's, an image of a fashionably dressed mid-eighteenth-century woman holding the compass over a diagram of planetary geometry. It is a portrait of the marquise de Châtelet, the French translator of Newton's *Principia mathematica*.

12. Grant, *Planets, Stars, and Orbs*, 113–22 (see ch. 1, n. 8); David Woodward, "The Image of the Spherical Earth," *Perspecta 25: Yale Architectural Journal*, no. 25 (1989): 2–15.

13. Anne Marie Godlewska, *Geography Unbound: French Geographic Science from Cassini to Humboldt* (Chicago: University of Chicago Press, 1999), 53.

14. In astronomical mapping the same problem is solved conventionally by counting celestial longitude from the first point in Aries.

15. Brotton, *Trading Territories*, 119–50 (see ch. 1, n. 7).

16. Umberto Eco's "Longitudinum Optata Scientia," in *The Island of the Day Before* (London: Secker & Warburg, 1995), 181–97, offers an imaginative commentary on the significance of Tenerife's peak. The novel is based on a voyage to determine longitude at sea.

17. Matthew Edney, "Cartographic Culture and Nationalism in the Early United States: Benjamin Vaughan and the Choice for a Prime Meridian, 1811," *Journal of Historical Geography* 20 (1994): 384–95. More generally, see Dennis Wood, *The Power of Maps* (London: Routledge, 1993); and J. B. Harley, "Maps, Knowledge, and Power," in Cosgrove and Daniels, *Iconography of Landscape*, 277–312.

18. Louis XIII, quoted in Derek Howse, *Greenwich Time and the Discovery of the Longitude* (Oxford: Oxford University Press, 1980), 129.

19. See the discussion in P. D. A. Harvey, *Topographical Maps: Symbols, Pictures, and Surveys* (London: Thames & Hudson, 1980), esp. 162–64.

20. Le Large, "Receuil des inscriptions," 70–97 (see ch. 6, n. 96).

21. Derek Howse, "Navigation and Astronomy: The First Three Thousand Years," *Renaissance and Modern Studies* 30 (1986): 72.

22. William J. H. Andrews, ed., *The Quest for Longitude: The Proceedings of the Longitude Symposium, Harvard University, Cambridge, Massachusetts, 4–6, 1993* (Cambridge: Harvard Collection of Scientific Instruments, 1996). The story of John Harrison's perfecting of the chronometer has been popularized by Dava Sobel, *Longitude* (London: Fourth Estate, 1996).

23. Howse, "Navigation and Astronomy," 83.

24. Pierre Simon Laplace, quoted in Edney, "Cartographic Culture and Nationalism," 390.

25. Ibid., 388.

26. Benjamin Vaughan, quoted in ibid., 389.

27. Helen Proudfoot, "Town Plans and Their Impact on the Settlement Process in Australia, 1788–1849" (Ph.D. thesis, Graduate School of Environment, Macquarrie University, Sydney, N.S.W., 1996), 19–28.

28. Edmond Halley, *Atlas maritimus & commercialis or, A general view of the world so far as it relates to trade and navigation . . . with sailing directions and sea charts according to Mercator . . .* (London: printed for James and John Knapton, 1728).

29. Svetlana Alpers and Michael Baxandall, *Tiepolo and the Pictorial Intelligence* (New Haven:Yale University Press, 1995).

30. Bernard Smith, *European Vision and the South Pacific,* 2nd ed. (New Haven:Yale University Press, 1985).

31. Outram, *Enlightenment,* 63.

32. Paul Carter, "Darkness with an Excess of Bright: Mapping the Coastlines of Knowledge," in Cosgrove, *Mappings,* 125–47 (see ch. 1, n. 10).

33. Godlewska, *Geography Unbound,* 29–32, discusses the significance of Jesuit education for geographical knowledge in eighteenth-century France.

34. Anthony Pagden, *Lords of All the World: Ideologies of Empire in Spain, Britain, and France, c. 1500–1800* (New Haven:Yale University Press, 1995), 122.

35. Ibid., 119, 125.

36. Peter Hulme and Ludmilla Jordanova, *The Enlightenment and Its Shadows* (London: Routledge, 1990).

37. Pagden, *Lords of All the World,* 116.

38. Findlen, *Possessing Nature* (see ch. 6, n. 87); David Phillip Miler and Peter Hanns Reill, eds., *Visions of Empire:Voyages, Botany, and Representations of Nature* (Cambridge: Cambridge University Press, 1996).

39. Paul Carter, *The Road to Botany Bay:An Essay in Spatial History* (London: Faber & Faber, 1987).

40. Outram, *Enlightenment,* 26; Pratt, *Imperial Eyes,* 29.

41. Livingstone, *Geographical Tradition,* 129 (see ch. 1, n. 10).

42. My discussion is based on Carter, *Road to Botany Bay,* 175–201.

43. Luciana de Lima Martins, "Mapping Tropical Waters: British Views and Visions of Rio de Janeiro," in Cosgrove, *Mappings,* 148–68.

44. The phrase "explorer-artist-writers" is used by Barbara Maria Stafford in *Voyage into Substance:Art, Science, Nature, and the Illustrated Travel Account, 1760–1840* (Cambridge: MIT Press, 1984). The issue is summarized in Livingstone, *Geographical Tradition,* 125–33.

45. Carter, *Road to Botany Bay,* 179.

46. Martins, "Mapping Tropical Waters," explores the hybrid nature of British mappings of Rio de Janeiro through the sketchbooks of sailors whose works carry traces of places visited and sketched across truly global expanses in the course of their circulation through tropical waters.

47. Carter, *Road to Botany Bay,* 187.

48. For believers, the religious implications of Enlightenment discoveries were profound. The pre-Adamite controversy turned on reconciling Genesis accounts of the spread of peoples from original parents with the existence of peoples and languages separated by oceans, a problem initially raised by Beatus. See David Livingstone, "Science and Religion: Foreword to the Historical Geography of an Encounter," *Journal of Historical Geography* 20 (1994): 367–83.

49. Walter A. Goffart, "Breaking the Ortelian Pattern: Historical Atlases with a New Program, 1747–1830," in *Editing Early and Historical Atlases,* ed. Joan Winearls (Toronto: University of Toronto Press, 1995), 49–82.

50. Michael J. Heffernan, "On Geography and Progress: Turgot's *Plan d'un ouvrage sur la géographie politique* (1751) and the Origins of Modern Progressive Thought," *Political Geography* 13 (1994): 328–43.

51. J.-M. Degérando (1800), quoted in ibid., 338.

52. Edward Quin, *An historical Atlas in a Series of Maps of the World as known at different Periods* (London: R. B. Seeley & W. Burnside, 1830), quotations from the preface and p. 3. The maps were engraved by Sidney Hall, and the atlas went through numerous reprints and editions up to 1856.

53. Edmund Burke, quoted in Heffernan, "On Geography and Progress," 338.

54. Outram, *Enlightenment,* 53.

55. Stafford, "Desperately Seeking Connections."

56. Edney, "Cartographic Culture and Nationalism," 385.

57. Armand Mattelart, *The Invention of Communication,* trans. Susan Emanuel (Minneapolis: University of Minnesota Press, 1996), 9–12.

58. Anne Godlewska, "Jomard: The Geographic Imagination and the First Great Facsimile Atlas," in Winearls, *Editing Early and Historical Atlases,* 109–36.

59. Edme François Jomard, quoted in ibid., 117.

Chapter Eight. Modern Globe

Epigraph: Michel Serres, "Gnomon: Les debuts de la géometrie en Grèce," in *Elements d'histoire des sciences,* ed. Serres (Paris: Bords, 1989).

1. Timothy Mitchell, "The World as Exhibition," *Comparative Studies in Society and History* 31 (1989): 217–36, quotation on 227. See also Gregory, *Geographical Imaginations,* 37 (see ch. 1, n. 35).

2. Richard Gillespie, "Ballooning in France and Britain, 1783–1786: Aerostation and Adventurism," *Isis* 75 (1984): 249–86.

3. See the summary in Alan Pred, *Recognizing European Modernities: A Montage of the Present* (London: Routledge, 1995), 181–85.

4. J. B. Harley, "Deconstructing the Map," in *Writing Worlds: Discourse, Text, and Metaphor in the Representation of Landscape,* ed. T. Barnes and J. Duncan (London: Routledge, 1992), 231–47. Dennis Wood rehearses the argument in *The Power of Maps* (see ch. 7, n. 17).

5. Robert Lawson-Peebles, *Landscape and Written Expression in Revolutionary America* (Cambridge: Cambridge University Press, 1988), 196–230.

6. Martins, "Mapping Tropical Waters" (see ch. 7, n. 43); Mitchell, *Landscape and Power* (see ch. 1, n. 37); Matthew Edney, *Mapping an Empire: The Geographic Construction of British India* (Chicago: University of Chicago Press, 1997).

7. Kemp, *Science of Art,* 214–16 (see ch. 1, n. 37); Jonathan Crary, *Techniques of the Observer: On Vision and Modernity in the Nineteenth Century* (Cambridge: MIT Press, 1990).

8. Livingstone, *Geographical Tradition,* 131 (see ch. 1, n. 10).

9. W. Goetzmann, *New Lands, New Men: America and the Second Great Age of Discovery* (New York: Viking Penguin, 1986).

10. On Joseph Conrad's 1899 novel *Heart of Darkness* and the ways that geographical exploration was framed and represented see Felix Driver, "Geography's Empire: Histories of Geographical Knowledge," *Environment and Planning D: Society and Space* 10 (1992): 23–40; and idem, *Geography Militant* (forthcoming). On Werner Herzog, esp. *Fitzcarraldo* (1982), see Matthew Gandy, "Visions of Darkness: The Representation of Nature in the Films of Werner Herzog," *Ecumene: Environment, Culture, Meaning* 3 (1996): 1–22.

11. Senator James Harlan (1859), quoted in John Krygier, "Envisioning the American West: Maps, the Representational Barrage of 19th Century Expedition Reports, and the Production of Scientific Knowledge," *Cartography and Geographic Information Systems* 24 (1997): 31.

12. Krygier, in "Envisioning the American West," comments on the continued admiration among historians for "one of the most important detailed maps drawn before the Civil War" and for Egloffstein's "three-dimensional . . . rendering of the land's tormented topography" (27, 42).

13. Anne Godlewska, "Map, Text, and Image," *Transactions of the Institute of British Geographers,* n.s., 21 (1996): 5–28.

14. Krygier, "Envisioning the American West," 45; the quotation is from Edney, *Mapping an Empire,* 98. See also Godlewska, "Map, Text, and Image."

15. Global isotherms bear scant relationship to conventional zonal maps derived from Aristotelian theory. Thus "tropical" replaced "torrid zones" as a description of regions located between Cancer and Capricorn, denoting a geopolitical, racial, and moral evaluation rather than physical geography.

16. Anne Godlewska, "Humboldt's Visual Thinking: From Enlightenment Vision to Modern Science" (paper delivered at the conference "Geography and Enlightenment," Edinburgh, July 1996), 23; idem, "Map, Text, and Image."

17. Francis Bacon, quoted in Sholpo, "Harmony of Global Space," 8 (see ch. 1, n. 25).

18. Livingstone, *Geographical Tradition,* 308–10.

19. Alexander von Humboldt, *Cosmos: A Sketch of a Physical Description of the Universe,* trans. E. C. Otté, 2 vols. (London: Bohn, 1848), 1:36. Godlewska, *Geography Unbound,* ch. 3 (see ch. 7, n. 13), connects Humboldt's work to that of the French Jesuit cosmographer P. Jean François (1582–1668), a contemporary of Kircher whose *Géographie* was rooted in similar contemplative, mnemonic, and metaphysical principles to those of Kircher.

20. Humboldt, *Cosmos,* 1:36.

21. Ibid., 62–63.

22. Ibid., 63.

23. Gerhart Hoffmeister, "Goethe's *Faust* and the *Theatrum Mundi* Tradition in European Romanticism," in *Perspectives on Faust,* ed. M. Palencia-Roth (Cambridge: Alpha Academic, 1983), 42–55.

24. Schama, *Landscape and Memory,* 374–75 (see ch. 1, n. 43).

25. Michael J. Heffernan, "Bringing the Desert to Bloom: French Ambitions in the Sahara Desert during the Nineteenth Century—The Strange Case of 'la mer

intérieure,'" in *Water Engineering and Landscape: Water Control and Landscape Transformation in the Modern Period,* ed. Denis Cosgrove and Geoff Petts (London: Belhaven, 1990), 94–114, quotation on 99. See also Dorinda Outram, "On Being Perseus: The Explorer's Body and the Problem of Knowledge of Distant Places in the Enlightenment," in *Geography and Enlightenment,* ed. David Livingstone and Charles Withers (Chicago: University of Chicago Press, 2000), 281–94.

26. Armand Mattelart, *Invention of Communication,* 97 (see ch. 7, n. 57); Michael J. Heffernan, "The Science of Empire: The French Geographical Movement and the Forms of French Imperialism, 1870–1920," in *Geography and Empire,* ed. Anne Godlewska and Neil Smith (Oxford: Blackwell, 1994), 92–114, esp. 100–102.

27. Livingstone, *Geographical Tradition,* 196–212; idem, "Climate's Morality: Science, Race, and Place in Post-Darwinian British and American Geography," in Godlewska and Smith, *Geography and Empire,* 132–54.

28. Ronald E. Doel, "Expeditions and the CIW: Comments and Contentions" (1992, mimeo); Raphael Pumpelly, "Archaeological and Physico-Geographical Reconaissance in Turkestan," in *Explorations in Turkestan,* 2 vols., Carnegie Institute Publication No. 26 (Washington, D.C., 1905), 3–39.

29. William Davis's theory of landform evolution predated the expedition: "The Geographic Cycle," *Geographical Journal* 14 (1899): 481–504. His work on the expedition is published in idem, "A Journey across Turkestan," in *Explorations in Turkestan.* Ellsworth Huntington's works include *The Pulse of Asia* (Boston: Houghton Mifflin, 1907) and *Civilization and Climate* (New Haven: Yale University Press, 1924). Huntington published his expedition work on changing levels of the Caspian Sea in "The Historic Fluctuations of the Caspian Sea," *Bulletin of the American Geographical Society* 39 (1907).

30. Doel, "Expeditions and the CIW," 17.

31. Cosmographic ideas about the poles, including that of four rivers flowing from the North Pole, endured into the early nineteenth century.

32. See, for example, the map "Antarctic Regions" in *The Oxford Atlas* (Oxford: Oxford University Press, 1966), 10–11. On arctic mapping see Barry Lopez, *Arctic Dreams: Imagination and Desire in a Northern Landscape* (London: Bantam, 1987), 253.

33. The record of false geographical recording in Antarctica is long and complex. Some of it is summarized in Paul Simpson-Housley, *Antarctica: Exploration, Perception, and Metaphor* (London: Routledge, 1992), especially the case of Charles Wilkes, who was court-martialed for claiming to have discovered antarctic land (Wilkes Land) during his voyages of 1838–42 in regions subsequently shown to be ocean but was probably misled by superior mirages (61–68). See also Cindy Katz and Alex Kirby, "In the Nature of Things," *Transactions of the Institute of British Geographers,* n.s., 16 (1991): 259–71.

34. Patricia Fara, "Northern Possession: Laying Claim to the Aurora Borealis," *History Workshop Journal* 42 (1996): 37–57.

35. Lopez, *Arctic Dreams,* 15–17.

36. Fara, "Northern Possession," discusses the significance of the poles in the construction of Scandinavian nationalisms. On the place of the Tibetan Himalayas in

the imagination of nineteenth-century Europeans see Peter Bishop, *The Myth of the Shangri-La: Tibet, Travel Writing, and the Western Creation of Sacred Landscape* (London: Athlone, 1989). On parallels with mountaineering see Schama, *Landscape and Memory*, 463–506.

37. John McCannon, "To Storm the Arctic: Soviet Polar Exploration and Public Visions of Nature in the USSR, 1932–1939," *Ecumene: Environment, Culture, Meaning* 2 (1995): 15–31, quotation on 25.

38. The phrase is from Fara, "Northern Possession," 38.

39. Ernst Krenkel, quoted in McCannon, "To Storm the Arctic," 25. Peter Hoeg's novel *Miss Smilla's Feeling for Snow* (London: Harvill, 1993), based on the idea of a meteorite that has fallen into Greenland ice slowly releasing deadly extraterrestrial bacteria into human bodies, reworks many of these polar associations.

40. Klaus Dodds, "To Photograph the Antarctic: British Polar Exploration and the Falkland Islands Dependencies Aerial Survey Expedition (FIDASE)," *Ecumene: Environment, Culture, Meaning* 3 (1996): 63–89; idem, *Geopolitics in Antarctica: Views from the Southern Ocean Rim* (New York: Wiley, 1997).

41. The Antarctic Treaty was a direct outcome of the International Geophysical Year, 1957–58. There is similar territorial ambiguity at the northern polar regions, where the islands of Svalbard, although claimed as sovereign territory by Norway, can be exploited under certain conditions by any signatory to the Svalbard Treaty of 1925. Post-cold-war debates surround proposals to give the islands international wilderness status. M. I. Glassner, "The Frontiers of Earth—and of Political Geography: The Sea, Antarctica, and Outer Space," *Political Geography Quarterly* 10 (1991): 422–37.

42. Stephen Daniels, *Fields of Vision: Landscape Imagery and National Identity in England and the United States* (Cambridge: Polity, 1993), 146–73.

43. Steven Kern, *The Culture of Time and Space, 1880–1918* (Cambridge: Harvard University Press, 1983); Gerry Kearns, "Closed Space and Political Practice: Frederick Jackson Turner and Halford Mackinder," *Environment and Planning D: Society and Space* 2 (1984): 23–34.

44. Arno Peters has proposed revising the global graticule by shifting the zero meridian and the International Date Line to the same location, the current 190th meridian, and removing the present deviations as a postcolonial move toward representational justice.

45. Mattelart, *Invention of Communication*, 163–78.

46. Gerry Kearns, "*Fin de Siècle* Geopolitics: Mackinder, Hobson, and Theories of Global Closure," in *Political Geography of the Twentieth Century: A Global Analysis*, ed. Peter J. Taylor (London: Belhaven, 1993), 9–30.

47. *Consummation of Empire* is the title of Thomas Cole's third painting in the series "The Course of Empire"; it was preceded by *The Savage State* and *The Arcadian State* and followed by *Destruction* and *Desolation*.

48. Frederick Jackson Turner, "The Significance of the Frontier in American History" (1894), reprinted in *Frontier and Section: Selected Essays of Frederick Jackson Turner*, ed. Ray Allen Billington (Englewood Cliffs, N.J.: Prentice-Hall, 1961), 37–62; Kearns, "Closed Space and Political Practice."

49. See, e.g., Arnold J. Toynbee, *A Study in History,* 3 vols. (London: Oxford University Press, 1934); Ellsworth Huntington, *Mainsprings of Civilization* (London: Wiley, 1945); Griffith T. Taylor, "Racial Geography," in *Geography in the Twentieth Century,* ed. Taylor (New York: Philosophical Library, 1957); and the essays in pt. 3 of Godlewska and Smith, *Geography and Empire.*

50. Mark Bassin, "Imperialism and the Nation State in Friedrich Ratzel's Political Geography," *Progress in Human Geography* 11 (1987): 473–95.

51. "The Geographical Pivot of History" was the first of three statements by Mackinder on the theme of a global political order, made in 1902, 1919, and 1943, respectively. They are collected in Halford J. Mackinder, *Democratic Ideals and Reality* (New York: Norton, 1962).

52. Mackinder, "Geographical Pivot of History," 242.

53. Ibid., 262.

54. See, e.g., Mackinder's discussion of medieval *mappae mundi* and later cartographic images in *Democratic Ideals and Reality,* 89–95.

55. Lucio Gambi, "Geography and Imperialism in Italy: From the Unity of the Nation to the 'New' Roman Empire," in Godlewska and Smith, *Geography and Empire,* 90.

56. David Atkinson, "Arrows, Empires, and Ambitions in Africa: The Geopolitical Cartography of Fascist Italy," in *Maps and Africa,* ed. Jeffrey C. Stone (Aberdeen: Aberdeen African Studies Group, 1994), 43–65.

57. Harvey, *Condition of Postmodernity* (see ch. 1, n. 32).

58. Mattelart, *Invention of Communication,* 107.

59. L. W. Lyde, "The Teaching of Geography as a Subject of Commercial Instruction," *Geography Teacher* 4 (1907–8): 163–68, quotation on 165. See also Teresa Ploszajska, "Constructing the Subject: Geographical Models in English Schools, 1870–1944," *Journal of Historical Geography* 22 (1966): 388–99.

60. Alan Pred, in *Recognizing European Modernities,* provides a detailed discussion of the phenomenon in Stockholm and Sweden.

61. Illustrated in Fara, "Northern Possession," 46.

62. The claim is made by Garry Dunbar in "Elisée Reclus and the Great Globe," *Scottish Geographical Magazine* 90 (1974): 57–66, and repeated by Derek Gregory in *Geographical Imaginations,* 38 n. 55.

63. Irving Fisher and O. M. Miller, *World Maps and Globes* (New York: Essential Books, 1944).

64. Kay Anderson, "Culture and Nature at the Adelaide Zoo: New Frontiers in 'Human' Geography," *Transactions of the Institute of British Geographers,* n.s., 30 (1995): 275–95.

65. Mary Bronson Hartt, "The Play Side of the Fair" (1901), quoted in Barbara Rubin, "Aesthetic Ideology and Urban Design," *Annals of the Association of American Geographers* 69 (1979): 348.

66. Mitchell, "World as Exhibition." The semantic association between *monster* and *demonstration* (in Latin, *monstrare*) has been discussed above. *QED,* for *quod erat demonstrandum,* is used to signal the successful completion of a scientific experiment.

67. Ibid., 226.

68. This idea is discussed more fully in chapter 9.

69. Cyrus R. Teed, *The Cellular Cosmogony or the Earth as a Concave Sphere* (1905; reprint, Philadelphia: Porcupine, 1997), 11, 13.

70. The written description of the exhibition, signed simply "V.B.," is printed as appendix B to P. Abercrombie et al., *The Coal Crisis and the Future: A Study of Social Disorders and Their Treatment* (London: Williams & Norgate for the Sociological Society, 1926), v–xxviii. For a fuller discussion of this subject, see David Matless, "Regional Surveys and Local Knowledges: The Geographical Imagination in Britain, 1918–39," *Transactions of the Institute of British Geographers,* n.s., 17 (1992): 464–80.

71. V.B., appendix B in Abercrombie et al., *Coal Crisis,* x, emphasis added.

72. Ibid., xxii.

Chapter Nine. Virtual Globe

Epigraph: Umberto Eco, *The Island of the Day Before* (London: Secker & Warburg, 1995), 319–20.

1. "Nous voulons chanter l'homme qui tient le volent, dont la tige idéale traverse la terre, lancée elle-même sur le circuit de son orbite." The "Futurist Manifesto" appeared in *Le Figaro* on 20 February 1909.

2. Robert Wohl, *A Passion for Wings: Aviation and the Western Imagination, 1908–1918* (New Haven: Yale University Press, 1994), 256 ff.; Joseph J. Corn, *The Winged Gospel: America's Romance with Aviation, 1900–1950* (New York: Oxford University Press, 1983).

3. French pilot, quoted in Wohl, *Passion for Wings,* 257.

4. Ibid., 264.

5. Ibid., 286.

6. Claudio G. Segrè, *Italo Balbo: A Fascist Life* (Berkeley and Los Angeles: University of California Press, 1987). Creator of the Italian air force and its minister under Mussolini, Balbo used the resources of the Italian (Royal) Geographical Society to plan these flights. See *Bolletino del Reale Società Geografica Italiana* 7 (1940): 433–44.

7. Wohl, *Passion for Wings,* 281.

8. Couthino's *Santa Cruz* completes the display narrative of Portuguese discovery and empire in the Museo de Descubrimento at Lisbon.

9. Richard E. Byrd, "The Conquest of Antarctica by Air," *National Geographic* 58 (1930): 127–225; Dodds, "To Photograph the Antarctic" (see ch. 8, n. 40); idem, "Antarctica and the Modern Geographical Imagination (1918–1960)," *Polar Record* 33 (1997): 47–62.

10. David Atkinson and Denis Cosgrove, "Urban Rhetoric and Embodied Identities: City, Nation, and Empire at the Vittorio Emanuele II Monument in Rome, 1870–1945," *Annals of the Association of American Geographers* 88 (1998): 28–49; Felix Driver and David Gilbert, "Imperial Cities: Overlapping Territories, Intertwined Histories," in *Imperial Cities: Landscape, Display, and Identity,* ed. Driver and Gilbert (Manchester: Manchester University Press, 1999), 1–20.

11. Atkinson and Cosgrove, "Urban Rhetoric and Embodied Identities"; David

Atkinson, "Totalitarianism and the Street in Fascist Rome," in *Images of the Street,* ed. N. Fyfe (London: Routledge, forthcoming).

12. David Matless, "Visual Culture and Geographic Citizenship: England in the 1940s," *Journal of Historical Geography* 22 (1996): 424–39; Pyrs Gruffudd, "Reach for the Skies," *Landscape Research* 16, no. 2 (1991): 19–24.

13. The clearest articulation of ideas was in the works of the German biologist and geographer Friedrich Ratzel. Mark Bassin, "Race contra Space: The Conflict between German Geopolitik and National Socialism," *Political Geography Quarterly* 6 (1987): 115–34.

14. Eco, *Search for a Perfect Language,* 324–32 (see ch. 3, n. 26).

15. Catherine Nash, "Geocentric Education and Anti-Imperialism: Theosophy, Geography, Citizenship in the Writings of J. H. Cousins," *Journal of Historical Geography* 22 (1996): 399–411.

16. These journals included the *Zeitschrift für Geopolitik* in Germany and *Geopolitica* in Italy, for example. Geopolitical journals also existed in Spain, Bulgaria, and various Latin American countries. See David Atkinson, "Geopolitics and the Geographical Imagination in Fascist Italy" (Ph.D. thesis, Loughborough University, 1996).

17. Hans W. Weigert and Vilhjalmur Stefansson, eds., *Compass of the World: A Symposium on Political Geography* (London: George G. Harrap, 1944), x.

18. Isaiah Bowman, "Geography vs. Geopolitics," in Weigert and Stefansson, *Compass of the World,* 40–52; idem, *The New World: Problems in Political Geography* (Yonkers-on-Hudson: World Book Company, 1921). See also Neil Smith, "The Lost Geography of the American Century," *Scottish Geographical Journal* 115 (1999): 1–18.

19. Susan Schulten, "Richard Edes Harrison and the Challenge to American Cartography," *Imago Mundi* 50 (1998): 174–88, quotations on 174, 178, 177.

20. MacLeish's 1939 poem "America Was Promises" contains the following lines, universalizing "man's" destiny in seemingly vacant continental space:

> America was always promises.
> From the first voyage and the first ship there were promises
>
> America was promises—to whom?
> Jefferson knew:
> Declared it before God and before history:
> Declares it still in the remembering tomb.
> The promises were Man's: the land was his—
> Man endowed by his creator.

Archibald MacLeish, *New and Collected Poems, 1917–1976* (New York: Knopf, 1979).

21. Archibald MacLeish, "The Image of Victory," in Weigert and Stefansson, *Compass of the World,* 1–11, quotation on 3.

22. Ibid., 7.

23. Schulten, "Richard Edes Harrison and the Challenge to American Cartography," 180.

24. Eugene Staley, "The Myth of Continents," in Weigert and Stefansson, *Compass of the World*, 89–108. Schulten, "Richard Edes Harrison and the Challenge to American Cartography," 179–80, discusses the role of Harrison's 1940 *Atlas for the American Citizen* in making the same point graphically.

25. Editors of *Fortune*, "The Logic of the Air," in Weigert and Stefansson, *Compass of the World*, 121–36, quotation on 131.

26. Schulten, "Richard Edes Harrison and the Challenge to American Cartography," 180.

27. Van Loon wrote the introduction to a 1943 scholarly edition of Erasmus's *Moriae encomium*. He also wrote on Thomas Jefferson and Simon Bolivar. Van Loon's biography, *The Story of Hendrik Willem van Loon* (New York: Lippincott, 1972), was written by his son, G. W. van Loon. My thanks to Rex Walford for this reference. Van Loon was a personal friend of George G. Harrap, the British publisher of *Compass of the World*.

28. Hendrik Willem van Loon, *The Home of Mankind* (London: George G. Harrap, 1933), 17; originally published as *Van Loon's Geography*. Cf. Auden's lines on Bruegel's *Fall of Icarus*, written in the same years (ch. 5, n. 87).

29. Van Loon, *Home of Mankind*, 481.

30. Ibid., 485.

31. Van Loon's illustrations are in certain respects similar to those for the contemporary French aviator and writer Antoine de Saint-Exupéry's text *Le petit prince* (1943), which combines a celebration of modern technology with the Stoic moralism of the Apollonian perspective.

32. Tamar Y. Rothenberg, "Voyeurs of Imperialism: The *National Geographic Magazine* before World War II," in Godlewska and Smith, *Geography and Empire*, 155–72 (see ch. 8, n. 26); Howard S. Abramson, *National Geographic: Behind America's Lens of the World* (New York: Crown, 1987); Catherine A. Lutz and Jane L. Collins, *Reading National Geographic* (Chicago: University of Chicago Press, 1993).

33. The projection matched the cold-war ideology of the American Geographical Society since it exaggerated the size and thus the threat of the Soviet Union.

34. Lutz and Collins, *Reading National Geographic*. Photographs taken by the Fascist filmmaker and photographer Leni Riefenstahl (maker of *Triumph of the Will*) of "godlike Nuba, emblems of physical perfection, with large, well-shaped, partly shaven heads, expressive faces, and muscular bodies that are depilated and decorated with scars" form the subject of Susan Sontag's essay "Fascinating Fascism," quotation from 73 (see ch. 2, n. 45).

35. John Tagg, *The Burden of Representation: Essays on Photographies and Histories* (London: Macmillan, 1988), 8–11. See also Jean-Claude Lemagny and André Rouillé, *A History of Photography: Social and Cultural Perspectives* (Cambridge: Polity, 1987).

36. Joan Schwartz, "The Geography Lesson: Photographs and the Construction of Imaginary Geographies," *Journal of Historical Geography* 22 (1996): 16–45, quotation on 22.

37. James Ryan, "Visualising Imperial Geography: Halford Mackinder and the

Colonial Office Visual Instruction Committee, 1902–1911," *Ecumene: Environment, Culture, Meaning* 1 (1994): 157–76.

38. Ibid., 159.

39. Jay, "Scopic Regimes of Modernity," esp. 186 (see ch. 4, n. 60).

40. Arthur H. Robinson, "The President's Globe," *Imago Mundi* 49 (1997): 143–59.

41. Charles Hurd, "World Airways," in Weigert and Stefansson, *Compass of the World,* 140–41.

42. TWA publicity quoted in Denis Cosgrove, "Contested Global Visions: *One World, Whole Earth,* and the Apollo Space Photographs," *Annals of the Association of American Geographers* 84 (1994): 281.

43. Bruno Latour, *Science in Action: How to Follow Scientists and Engineers through Society* (Milton Keynes: Open University Press, 1987), 215–57, sees the NASA project as an example of his "centres of calculation" theory, reworking longstanding connections between scientific and geographical discovery in Western culture (247–49).

44. Lyndon B. Johnson, quoted in Cosgrove, "Contested Global Visions," 281.

45. William David Compton, *Where No Man Has Gone Before: A History of Apollo Lunar Space Exploration Missions* (Washington, D.C.: NASA, 1989). The movie *Apollo 13* narrates one of the project's most dramatic episodes, when key life-support systems on the command module failed.

46. Their piloting skills were of little practical value in what were essentially metal canisters passively propelled into space by Saturn rockets.

47. Earth photography was never formally incorporated into the Apollo program, whose objectives were consistently directed at a lunar landing. In the allocation of severely limited numbers of frames carried, Earth shots were placed in the lowest-priority category, "targets of opportunity." The images were very difficult to obtain, photographed in weightless conditions with a hand-held camera through the mist-covered windows of a turning space module. See Cosgrove, "Contested Global Visions."

48. Frank Borman, quoted in ibid., 282.

49. Archibald MacLeish, quoted in ibid., 283.

50. *Time* magazine, quoted in ibid., 284.

51. Michael Collins, quoted in ibid., 287.

52. Bruno Latour, for example, in *We Have Never Been Modern* (see ch. 1, n. 55), which challenges distinctions between nature and culture.

53. James Lovelock, *Gaia: A New Look at Life on Earth* (see ch. 1, n. 15). See also Jacques Arnould, "When Apollo Flirts with Gaia: Earth Observation from Space and the Ecological Crisis" (1996, mimeo).

54. Ingold, "Globes and Spheres."

55. *Tony Stone Images,* vol. 15 (London, 1996); Victoria Thompson, "New Globes for Old: Uses of the 'Oxford Globe'" (B.A. diss., Department of Geography, Royal Holloway University of London, 1999).

56. John S. Pickles, "Texts, Hermeneutics, and Propaganda Maps," in Barnes and Duncan, *Writing Worlds,* 193–230 (see ch. 8, n. 4).

57. Susan Roberts and Richard Schein, "Earth Shattering: Global Imagery and G.I.S." in Pickles, *Ground Truth,* 171–95 (see ch. 1, n. 15).

58. Al Gore, quoted in Armand Mattelart, "Mapping Modernity," in Cosgrove, *Mappings,* 188 (see ch. 1, n. 10).

59. Bill Gates, quoted in ibid., 191.

60. Mattelart, *Invention of Communication,* 104–7 (see ch. 7, n. 57); idem, *La mondialisation et la communication* (Paris: Presses Universitaires de France, 1996).

61. Eco, *Search for a Perfect Language.*

62. *Les missions catholiques,* 1919, quoted in Mattelart, *Invention of Communication,* 184.

63. The idea of a *Giubileo* was established by Boniface VIII in 1300. Various indulgences were granted to those who entered St. Peter's during the course of the jubilee year. Giovanni Morello, "Gli anni santi: Storia e immagini," *Tertium Millennium* 1 (1996): 77–82; Paolo Tripodi, "Rome Prepares for the Millennium," *Contemporary Review,* no. 270 (1997).

64. Cicero, *De Republica,* trans. C. W. Keyes, 269 (see ch. 2, n. 68).

65. Macrobius, *Commentary on the Dream of Scipio,* 142–43 (see ch. 1, n. 3).

66. "At the midpoint of earth I have placed you, that you may more easily see around you what is in the world." Giovanni Pico della Mirandola, *Pratio de hominis dignitate,* par. 5, line 21 (Pico Project, http://www.brown.edu/Departments/Italian_Studies/pico/oratio.html).

INDEX

Page numbers in italics refer to illustrations.

Denis Cosgrove was born in 1948 and raised in Liverpool, England. He received his B.A. in geography, with honors, at St. Catherine's College, Oxford University, his M.A. at the University of Toronto, and his Ph.D. at St. Catherine's College. He is currently a professor of geography at the University of California, Los Angeles.

In addition to being a joint founding editor of *Ecumene: Environment, Culture, Meaning*, he has published numerous articles in such professional journals as *Geographical Review, Canadian Geographer, Antipode, Progress in Human Geography, Transactions: Institute of British Geographers, Annals of the Association of American Geographers,* and *Urban Studies* and has contributed chapters to numerous books as well as articles in the *Encyclopaedia Britannica*. Professor Cosgrove has authored or edited ten books, among them *Social Formation and Symbolic Landscape* (1984), *The Iconography of Landscape: Essays on the Symbolic Representation, Design and Use of Past Environments* (1988), *Water, Engineering and Landscape: Water Control and Landscape Transformation in the Modern Period* (1990), *The Palladian Landscape: Geographical Change and Its Cultural Representations in Sixteenth-Century Italy* (1993), and *Mappings* (1999).